COLLECTIVE INTELLIGENCE
The Rise of Swarm Systems and their Impact on Society

Editors

Uwe Seebacher
Munich University of Applied Sciences
Munich, Germany

Christoph Legat
Technical University of Applied Sciences
Augsburg, Germany

CRC Press
Taylor & Francis Group
Boca Raton London New York

CRC Press is an imprint of the
Taylor & Francis Group, an **informa** business

A SCIENCE PUBLISHERS BOOK

First edition published 2025
by CRC Press
2385 NW Executive Center Drive, Suite 320, Boca Raton FL 33431

and by CRC Press
4 Park Square, Milton Park, Abingdon, Oxon, OX14 4RN

Library of Congress Cataloging-in-Publication Data (applied for)

ISBN: 978-1-032-69068-1 (hbk)
ISBN: 978-1-032-69070-4 (pbk)
ISBN: 978-1-032-69071-1 (ebk)

DOI: 10.1201/9781032690711

Typeset in Times New Roman
by Prime Publishing Services

Preface

As the esteemed CRC Press Francis Taylor Group approached me, Uwe Seebacher, to helm a seminal work on collective and swarm intelligence, I was reminded of the intricate dance between structure and method that has defined my career as a "methods and structural" scientist. With great enthusiasm, I embarked on the ambitious journey of editing this book, a venture that would crystallize the essence of this ever-evolving field.

In *Collective Intelligence: The Rise of Swarm Systems and Their Impact on Society*, we navigate the confluence of theory, application, and innovation. My journey in the academic and professional realm has consistently reinforced the belief that the fusion of methodical rigor and structural acumen is pivotal in harnessing the power of collective intelligence. It is this philosophy that has shaped the structure of this book, with each chapter meticulously crafted to reflect the multifaceted nature of swarm systems.

Engaging my colleague, Christoph Legat, with whom I have pursued thought-leading and award-winning ventures in predictive intelligence, has been instrumental in this endeavor. Our collaboration has been a testament to the synergy that drives innovation. Together, we have nurtured the growth of Predictores.ai, the 2023 leading student startup from the University of Applied Sciences in Munich and the Augsburg Technical University of Applied Sciences. The insights and breakthroughs from this vibrant incubator of ideas have been generously poured into the pages of this work.

The contents of this book are a distillation of the latest findings and advancements in the realm of collective and swarm intelligence. Our concerted efforts have been directed towards providing substantial added value and sustainable knowledge gains to you, the reader, and to the broader community. We present to you a tapestry woven with the threads of academic inquiry, practical application, and foresight.

It is our hope that this book will serve as a beacon for those navigating the complexities of collective intelligence, offering clarity and inspiration. May the insights contained within these pages empower innovators, practitioners, and scholars alike to push the boundaries of what is possible, forging a future that is smarter, more connected, and infinitely capable through the power of collective thought and action.

Uwe Seebacher
Christoph Legat

Acknowledgments

The fruition of this book would not have been possible without the initiation and unwavering support of Vijay Primalani and his dedicated team at CRC Press, Boca Raton, Florida, USA. Their enthusiasm and commitment to excellence have been the bedrock upon which this work was built. We are deeply grateful for their guidance, expertise, and the seamless collaboration that has made this publication a reality.

Heartfelt thanks are extended to the dynamic team of Predictores.ai across various esteemed institutions. Their diligence and innovative spirit have been instrumental in advancing the research and case studies that form the backbone of this book. The University of Applied Sciences Munich, the Augsburg Technical University of Applied Sciences, the University of Applied Sciences Vienna, and the Institute of Management Technology, led by the highly respected director Vishal Talwar, have provided an academic sanctuary for the cultivation of ideas and experimentation.

We are also indebted to our colleagues at the Indian Institute of Management Shillong, who have enriched this work with their insights and scholarly contributions. The collective wisdom of the Academy of Management (AoM), the Indian Academy of Management (INDAM), and the Federal German Association for Industrial Communication has been a beacon of inspiration and a source of rigorous academic discourse.

Each of these institutions and individuals have contributed to the rich tapestry of knowledge that this book represents. They have not only provided intellectual contributions but have also fostered an environment of learning and inquiry that is essential for the growth of collective intelligence as a field. For their trust, support, and collaboration, we are eternally thankful.

Introduction

This book discusses the advancements and challenges in the field of Collective Intelligence (CI), particularly Swarm Intelligence (SI), and its applications in various domains like industrial automation, business, management, and healthcare. It highlights the ability of swarm intelligence to improve (autonomous) decision-making and problem-solving through decentralized, self-organizing systems. The challenges addressed include scalability, adaptability to dynamic environments, and the need for robust communication protocols within swarm systems. The book also considers societal implications and ethical considerations in swarm intelligence applications highlighting the importance of artificial intelligence properties like transparency and fairness. The advancements are in the application of swarm intelligence principles to various industries. For a comprehensive analysis, please refer to the specific sections of the book.

The book is structured in the following contextual parts:

(1) Fundamentals of Collective Intelligence
(2) Industrial Applications of Collective Intelligence
(3) Applications of Collective Intelligence in Business and Management
(4) Applications of Collective Intelligence in Healthcare
(5) Perspectives on Societal Impact

As we embark on the journey through the compendium, *Collective Intelligence: The Rise of Swarm Systems and Their Impact on Society*, we delve into the intricate symphony of distributed artificial intelligence and its collective behavior providing profound implications on technology and society. As editors, it was our aim to curate a work that traverses the broad horizons of Swarm Intelligence and Collective Intelligence, shedding light on their fundamental principles and far-reaching applications.

The book begins by exploring the "Fundamentals of Collective Intelligence" and "Swarm Intelligence", providing readers with a comprehensive, in-depth understanding of these concepts' theoretical foundations. The narrative then takes an "Anthropological Turn", examining the impact of language and cognition on intelligence, both with and without the aid of communication. A "Social Turn" and "Technological Turn" signal shifts towards the societal and technological implications of CI, setting the stage for subsequent sections.

In "Swarm Intelligence: Applications and Implementations in Autonomous Systems", we are introduced to SI's transformative role in autonomous vehicles, drones, robotics, IoT ecosystems, and smart agriculture, culminating in a profound dive into real-world applications and innovative paradigms.

"Industrial Applications of Collective Intelligence" shift the focus towards practical implementations in the industrial domain, discussing federated learning in networked production and the vital concept of resilience in manufacturing and its future implications towards a resilient society fundamentally linked to the manufacturing and automation industry in the chapter "Resilience-X". Here, the intricate balance of robustness and adaptability in the face of adversity is dissected, offering a roadmap for navigating the tumultuous waters of industrial challenges.

The narrative continues with "Multi-agent Reinforcement-Learning for Solving Flexible Job Shop Scheduling Problems", where the synergy of machine learning and collective decision-making demonstrates its prowess in optimizing complex industrial processes.

A chapter on "Applications of Collective Intelligence in Business & Management" highlights the innovative use of Swarm Intelligence in portfolio optimization, stock price prediction, and risk management. The pioneering chapter "Towards Next Generation Data-Driven Management" underscores the utility of Predictive Swarm Intelligence, presenting a nuanced understanding of market dynamics by combining agent-based simulation with the predictive collective behavior for modeling economic systems.

"Applications of Collective Intelligence in Healthcare" offer a shift towards the medical domain, where the detection of Parkinson's Disease through advanced algorithms exemplifies the potential of Collective Intelligence in diagnostics and treatment.

The concluding chapters, such as "Emerald Evolution", discuss the societal impact of Collective Intelligence in general, mapping out the next phase of global sustainability driven by the collective wisdom encapsulated in this emergent intelligence.

Each chapter of this book not only presents the advancements and applications of Collective Intelligence and Swarm Intelligence but also critically examines the associated challenges, societal impact, and ethical considerations, fostering a discourse that is as multifaceted as the intelligence it aims to harness. This book serves as a testament to the power of unity in intelligence, heralding a future where the collective mind steers innovation and progress.

Contents

List of Contributors

Ahmad Waseem Ghauri
Ohio University, United States
Andreas W. Müller
Nuremberg Institute of Technology, Germany
Christoph Legat
Technical University of Applied Sciences Augsburg, Germany
Dominik Brunner
Munich University of Applied Sciences, Munich, Germany
Felix Schmalzel
Technical University of Applied Sciences Augsburg, Germany
Florian Kerber
Technical University of Applied Sciences Augsburg, Germany
Juergen Grotepass
Huawei Technologies Duesseldorf GmbH, Germany
Kirill Fridman
Huawei Technologies Duesseldorf GmbH, Germany
Luka Jovanovic
Singidunum University, Serbia
Marko Sarac
Singidunum University, Serbia
Martin Springer
Technical University of Applied Sciences Augsburg, Germany
Maximilian Schnitzler
Augsburg Technical University of Applied Sciences, Germany
Milos Antonijevic
Singidunum University, Serbia
Milos Dobrojevic
Singidunum University, Serbia
Miodrag Zivkovic
Singidunum University, Serbia
Muhammad Isfandyar Khan
Beaconhouse National University, Pakistan
Natasha Pankunni
University of Calicut, Kerala, India
Nebojsa Bacanin
Singidunum University, Serbia

Petar Bisevac
Singidunum University, Serbia

Praseetha Pankunni
Govt. Polytechnic College, Palakkad, Kerala, India

Saravanan Ravindran
Bharath Institute of Higher Education and Research, Chennai-600073, Tamil Nadu, India.

Sasidhar Bhimana
Bharath Institute of Higher Education and Research, Chennai-600073, Tamil Nadu, India.

Tamara Zivkovic
School of Electrical Engineering, Belgrade, Serbia

Tatjana Legler
Technologie-Initiative SmartFactory KL e. V., Germany

Thorsten Schöler
Technical University of Applied Sciences Augsburg, Germany

Uwe Seebacher
University of Applied Sciences Munich, Germany

Vinit Hegiste
Rheinland-Pfälzische Technische Universität Kaiserslautern-Landau (RPTU), Germany

1

Fundamentals of Swarm Intelligence

Ahmad Waseem Ghauri[1][*] and *Muhammad Isfandyar Khan*[2]

An Anthropological Turn

The parallels between swarm behaviors in nature and collective human behaviors are compelling. These behaviors, however, must be rooted in the dynamism of history to make sense of them. Historically speaking, it is the constant interaction of human nature with nature that has resulted in the formation of human nature and has also constituted its change. The question that sometimes arises is whether there ever was an inherent human nature. Much like animals in the kingdom manifest an obvious distinction between two categories of species, say for example a cat and a dog, physiologically and temperamentally, it can be said that an inherent nature must exist that led to the formation of human nature. While nature may be held responsible for many of the evolutionary changes that have occurred over centuries to reach the point of civilization in the 21st century, it must also be noted that an animal, other than humans, did not end up developing sophisticated technologies for survival and their species' advancement.

This may have likely evolved for finer manipulations because of an increased dependence on, and also an elaboration of, technology. This is sometimes challenged to be the actual case as most scientists contest that little or no change has occurred in the human hand over the years. The data is insufficient for understanding the evolution of hand proportions and for reconstructing the morphology of the last common ancestor (LCA) of humans and chimpanzees as Almécija, Smaers, &

[1] Ohio University, United States.
[2] Beaconhouse National University, Pakistan.
  Email : f2020-449@bnu.edu.pk
[*] Corresponding author: ag592521@ohio.edu

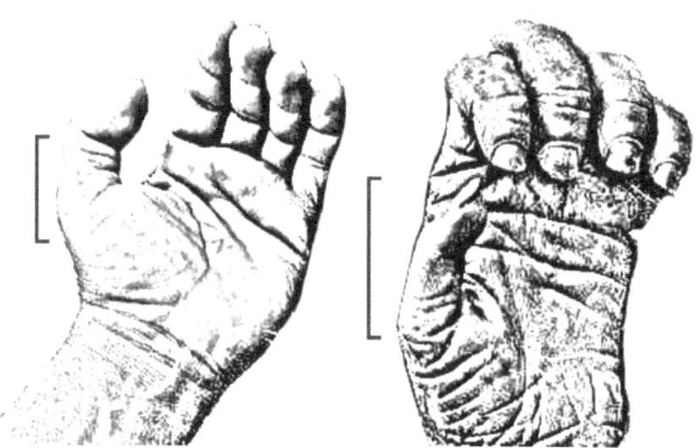

Fig. 1   The difference in human and chimpanzee hands.

Jungers (2015) evidenced (Figure 1). Young (2003) claims that throwing or clubbing to fight off adversaries gave the hominid lineage the advantage to form precision and power grips. This was perhaps necessary to create unforeseeable movements and to capitalize on their reproduction that may have led to how the human hands now function. Nowhere during the ontogenetic stage does the hand turn into a wing. A "unity of type" is postulated by the evolutionary biologist Darwin (1906) who claimed:

> *"WE have seen that the members of the same class, independently of their habits of life, resemble each other in the general plan of their organisation. This resemblance is often expressed by the term "unity of type"; or by saying that the several parts and organs in the different species of the class are homologous. The whole subject is included under the general term of Morphology. This is one of the most interesting departments of natural history, and may almost be said to be its very soul. What can be more curious than that the hand of a man, formed for grasping, that of a mole for digging, the leg of the horse, the paddle of the porpoise, and the wing of the bat, should all be constructed on the same pattern, and should include similar bones, in the same relative positions?".*

A semblance in the design of the human body with its predecessors can be warranted but the change in genetic material that only allows the hand to be produced and not a wing has remained constant for thousands of years. Nakamura et al., (2016) also advocate the close resemblance of the fin with the wing and the human hand in its developmental history. It still does not change the fact that humans living closer to aquatic regions do not develop fins in their embryonic stage. It was the development of digits that accelerated our ancestors to move out of water. Thus, a genetic makeup ensures a consistent output. As is seen, the manipulation of technology is not entirely being done through the use of limbs only. Material transformations require collaboration but these are not easily done without communication. Hence, a need for language was an inevitable outcome and one that continues to remain a crucial part of our collective collaborative tendencies.

Language and Cognition

The evolution of language, or 'languaging', is deeply intertwined with human development. Changes in hand structure due to tool-making activities may have concurrently transformed our vocal anatomy, leading to improved articulation and language production. Chomsky (1980) introduced the concept of an innate language faculty, suggesting that a genetic mutation might have enabled language production. This idea, countering the behaviorist approach of language as a stimulus-response mechanism championed by Skinner (1957), emphasized the structured, rule-based nature of language syntax.

Intriguingly, Lai et al., (2001) identified the FOXP2 gene in a London family afflicted with childhood apraxia of speech, a condition hindering coherent speech production despite normal cognitive functioning. While initially hailed as the "language gene", FOXP2 was later found in various species, including ancient hominins like Neanderthals. Although this discovery tempered claims of FOXP2 as the sole arbiter of language, its mutations have been linked to language disorders, highlighting its role in speech and language development. Human speech involves complex physical mechanisms, including the intricate control of the vocal apparatus, larynx, and movements of the mouth, tongue, and lips. Humans have greater control over these structures compared to chimpanzees and other great apes, underlining the genetic differences in speech capabilities (Enard et al., 2002). The FOXP2 gene offers insights into these evolutionary divergences, suggesting a genetic basis for humans' advanced spoken language abilities. This corresponds with Chomsky's claims of asocial origins of language that challenge the acquisition of such an elaborate and complex system via social mediation or the environment.

However, McBrearty & Brooks (2000) argue against the notion of a sudden 'revolution' in the Upper Paleolithic era leading to modern human behavior. They propose that the emergence of modern human traits was a gradual process, involving the accumulation of cognitive and cultural innovations over time. This view supports the idea of an evolutionary continuum rather than a discrete event marking the advent of modern human capabilities. Adding another dimension to this discussion, Hauser et al., (2002) differentiated between the faculty of language in the narrow sense (FLN) and the broad sense (FLB). FLN encompasses the uniquely human, narrowly linguistic components like recursion, enabling the embedding of clauses within clauses, a key feature of human language complexity. FLB includes broader cognitive and neurological mechanisms shared with other animals. The language faculty, according to this perspective, interacts with other cognitive systems, such as those involved in conceptual structuring and sensory-motor processing. This recent find was signed off by Chomsky as a possible hypothesis for language development. However, the role of the social in its formation remained contested.

This synthesis of ideas reflects the multidisciplinary nature of language evolution studies, involving linguistics, genetics, anthropology, and neurology. The field continues to evolve, driven by new discoveries and interpretations, illustrating the complexity of unraveling the origins and development of human language.

Nonetheless, human to human collaborations of all natures, be it centralized or decentralized, employ the use of language for more effective interactions with sophisticated goal-directedness.

Intelligence without Communication

Before the formation of formal language, it can be said that "swarm intelligence" could be at work since communicability was present without the use of words. Armstrong and Wilcox (2007) delved into evidence derived from sign languages to trace the roots of modern language. They propose that sign language, rather than verbal communication, was the primal form of human interaction. It is suggested that today's languages evolved from foundational actions and gestures, which gradually became recognized for their symbolic and communicative value. Essentially, the core of our linguistic capability is rooted in our knack for visual representation, like images or icons, as opposed to abstract linguistic symbols. The human fossil record further corroborates their theory, indicating that our ancestors had the anatomical capacity for gestures and signs before developing fluent speech capabilities. Though verbal communication emerged later and became the dominant mode of interaction, signs and gestures have never been fully overshadowed by speech. Johansson (2010) proposed a "chimp test" that checks if chimpanzees did something similar but did not evolve to generate formal language faculties. A chimpanzee may imitate an action correctly but rarely has a sense of why it was needed. The chimp test highlights a fundamental distinction between genuine understanding and sophisticated imitation. In the evolution of human language, understanding and context have been crucial. Language is not just about producing grammatically correct sentences but involves understanding context, culture, and subtle nuances. Hrdy (1999) claims that cooperative child care could have been a factor in evolution and Johansson (2005) agrees that that could have led to the development of language since childbirth was very hard without assistance from other members of the tribe. The cultural shift toward safely delivering through organizational efforts pushed for deeper understandings of the process with the help of language formation. Figure 2 shows the anatomical differences between a chimpanzee and a human. The upright posture, the design of the mouth cavity that facilitates better articulation of the tongue, more dexterous hands, and of course a larger brain—all of these features make us starkly different from our primate cousin.

However, in the natural world, Wohlleben (2016) believes communication without language is almost certainly done by trees (see Figure 3). In forests, trees have developed cooperative and mutualistic relationships, sustained through communication and a group intelligence akin to that of insect colonies. Even though Kingsland (2018) claims that many valid points about how ecological relationships operate in the forest are made, however, the use of the scientific literature to ascertain it often becomes a lift-off based on an imagined conclusion that does not coincide with scientific facts. It is, thereof, easier to underpin a strict utilitarian need for communication in the natural world in the absence of a formal

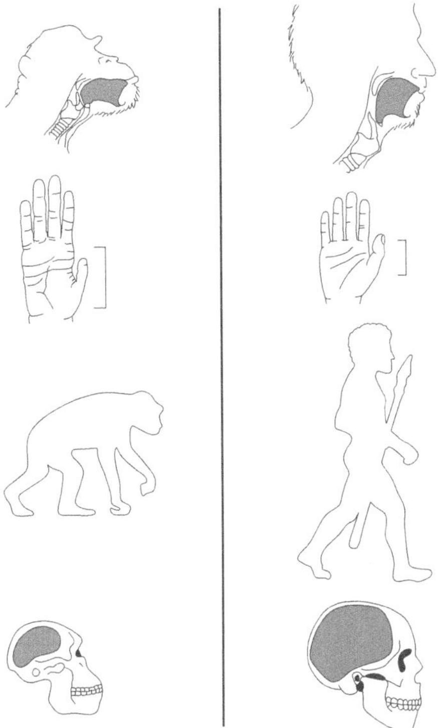

Fig. 2 Anatomical differences between a chimpanzee and a human inspired by Fitch (2000).

language. In the case of humans, that does not seem to be the case since individual and collective abstract thought are possible and are many a time needed to make changes to a current order of things.

The concept of "swarm intelligence" or "collective intelligence" suggests that groups can make decisions that are as good as, or even better than, those made by any single individual. These could be decisions for the well-being of the collective at the cost of the individual that may or may not be purely utilitarian in nature in the conventional sense. When applied to prehistoric humans, however, this idea implies that our ancestors could have benefited as a whole from group decision-making processes. The cooperative nature of early human groups likely played a role in their survival, adaptation, and evolution. Hill et al., (2011) postulate that early human groups often engaged in cooperative hunting and gathering. Collective decisions about where to hunt, when to move to a new territory, or how to share resources would have been critical as Henrich & McElreath (2003) claim that learning from others in the group, a process termed "social learning", was key for the transmission of knowledge and skills. This cumulative cultural evolution may also have enabled early humans to innovate and build upon previous generations' knowledge. The emergence of complex communication among early humans should have facilitated collective decision-making and the transference of knowledge.

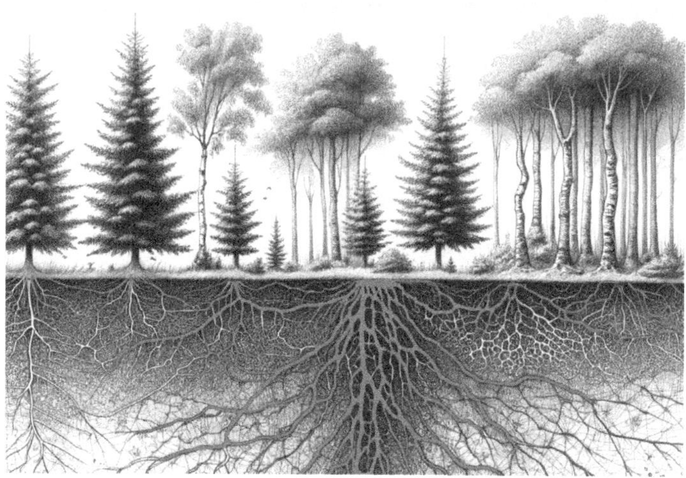

Fig. 3   An illustration of trees communicating through roots called "mycorrhizal network".

Sharing information and experiences within a group would have allowed for more informed and adaptive decisions as Dunbar (1996) had noted. Groups that could effectively harness the power of collective intelligence would have had survival and reproductive advantages. Tomasello and Carpenter (2007) purport a "shared intentionality" that could have potentially driven the evolution of social and cognitive traits that favor group cooperation and decision-making. Marx & Engels (1970) emphasized the primacy of the material conditions of society—the modes of production and labor. Language can be seen as evolving in tandem with changes in these material conditions, serving as a tool to meet the communicative needs of society as it interacts with and modifies its environment. Engels (1950), in particular, linked the development of labor (handwork) with the evolution of the human brain and language. As humans engaged in more complex labor tasks, they needed more sophisticated means of communication, leading to the development of a formal and complex language.

Intelligence a Function of Language or Not

Researchers have long debated the complex relationship between thought and language. The primary evolutionary function of language, as suggested by philosophers and linguists, extends beyond communication, serving to enhance abstract cognitive processes. Language is theorized to have evolved mainly as a cognitive tool, similar to the role of mathematical notations in computations. Contradicting this, Fedorenko & Varley (2016) demonstrated through fMRI (functional magnetic resonance imaging) studies that separate brain pathways are involved in thinking and language processing, indicating that language is not essential for all forms of thought. This finding challenges the Sapir-Whorf hypothesis, which suggests language shapes or is influenced by thought.

Scott-Phillips & Kirby (2010) observed that in controlled experimental settings, the process of "iterated learning"—a chain of communication where each person's output becomes the next person's input—can mimic the natural evolution of languages, becoming more structured and complex. This iterative process suggests that linguistic structures emerge not only from individual cognitive biases but also through cultural transmission across generations. Further experiments revealed that interaction between participants, as opposed to solitary learning, fosters more efficient and shared linguistic conventions. They observed that languages tend to evolve compositionally, where larger expressions derive their meanings from their constituent parts, a key characteristic of natural languages.

Pinker (2003) introduced the "cognitive niche" theory, positing that humans have evolved to solve problems through computational mechanisms (like thinking and reasoning), aided by language as a tool to enhance these abilities. Chomsky also explored the interplay between language and thought, considering language a medium for expressing and shaping thought, though not adhering to the strong Whorfian perspective of language determining thought (Rieber, 1983). The development of herd intelligence in prehistoric humans likely benefited significantly from the advent of spoken language, promoting a social culture conducive to technological advancement.

Complex language systems—spoken, gestured, or written—are pivotal in human evolution and societal development. According to Engels (1950), the development of these systems far surpasses simpler communication methods in other species, like the "Wood Wide Web" of mycorrhizal fungi networks described by Wohlleben (2016). In early human societies, such as hunter-gatherer communities, communication played a crucial role. These societies, often characterized by egalitarian social structures and common ownership (Woodburn, 1982; Engels, 1884; Lee & Daly, 1999; Boehm, 1999), required communication methods that could adapt to their mobile and decentralized nature.

The shift from foraging to farming, known as the Neolithic Revolution, marked a significant societal transformation (Childe, 1936). This transition led to surplus production, population growth, and the emergence of private property and class structures. Patterson (2019) highlights the absence of hierarchies in certain animal societies, introducing the concept of 'heterarchy'—a system with multiple centers of power (McIntosh, 1999). This perspective suggests a decentralized approach to understanding cultural evolution, emphasizing the roles of individual agency and local interactions (Pauketat, 2001).

While many anthropologists uphold the idea of a sort of "primitive communism", some scholars argue that the concept might oversimplify the complexities of early human societies. Some suggest that even in hunter-gatherer societies, there might have been nuances of inequality or structures of power. The rise of centralized institutions and the codification of authority were pivotal in the formation of the state. This centralization was often justified by religious or cosmological principles, Service (1975) claims. Morgan (1877) and Tylor (1871), among the early anthropologists, proposed a unilinear model where societies evolve from 'savagery' to 'barbarism' to 'civilization'. Each stage was characterized by specific technological innovations and socio-political organization.

The evolution of societal structures, whether from a centralized or decentralized perspective, is intimately tied to the development and manifestations of human intelligence. The intricate processes that underlie the organization and transformation of societies are rooted in the cognitive capabilities of human beings. The ability to devise innovative solutions to challenges, a marker of intelligence, has driven societal developments. For instance, the development of agriculture required an understanding of seasons, plant biology, and soil conditions. This required not just manual labor but cognitive advancements in pattern recognition, experimentation, and prediction (Mithen, 1996). The rise of centralized societies, with their complex hierarchies and bureaucracies, demanded a high degree of social intelligence. Managing large groups, navigating interpersonal dynamics, and understanding the nuances of power and influence are all cognitively intensive tasks (Dunbar, 1998). Centralized views of societal evolution emphasize the role of codified laws, scriptures, and administrative records. The development and sophistication of language—both as a tool for communication and abstract thought—have been pivotal in facilitating these centralizing processes (Pinker, 1994). In contrast, decentralized societies often place a premium on oral traditions and communal memories. The cognitive demands of memorizing vast amounts of information, from genealogies to navigational knowledge, highlight the adaptive nature of human intelligence in various societal contexts (Ong, 2013).

It can be concluded, although hesitantly, that perhaps the movement from a decentralized egalitarian society formation to a more centralized unequal society formation a transformation in group intelligence may have occurred. A movement to a centralized way of living became a natural outcome as is seen in modern societies with the development of the state but the decentralized way of intelligent social organization embedded in the genetic coding cannot help but exert its influence. Language did play a deciding role in the transference of intelligence through individuals in a group and also down to newer generations. It seems that language like other cognitive functions was developing alongside the development of the individual and society. The further sophistication of language became a necessary function of intelligence packaging. Nonhuman animals and other biological species may have a system of communication that is sophisticated enough for that particular species but in the context of transforming nature, humans are still incomparable. Along the course of centralizing societies, the ability to organize in a decentralized fashion is seen as unfathomable. However, when put to the test, group intelligence comes to life in the absence of a central control system owed particularly to the social forces that are created through interactions.

A Social Turn

The preamble to collective intelligence is social interaction without which the idea of individual intelligence also comes into question. While humans cannot be seen as a product of history they also cannot be seen only as changers of history. Marx (1852) claimed that history is made by humans, but it is not made as it is pleased. It is, therefore, not made under self-selected circumstances, but under circumstances existing already, that are in place and transmitted from the past. The transition from a decentralized to a centralized society due to agrarian activities has been discussed in the previous section. The production of surplus, its storage, and the final move to industrial society perhaps needed a centralized social interaction. This brought with it the dawn of capital and its accumulation. Capitalism, as an economic system based on private ownership of the means of production and the pursuit of profit, predates industrialization. With its development already begun during the late Middle Ages and Renaissance period, the full potential of capitalism was unleashed with the onset of the Industrial Revolution, argues Landes (1998). Marx (2005) famously claimed that society does not comprise only individuals, but is an expression of the sum of interrelations and the relations within which individuals stand. Historical development is driven by productive forces (like tools and machinery) and the relations of production (how societies are organized around production). The 'social' automatically becomes a product of these material conditions. And indeed the social was redefined to be centrally governed through social hierarchies.

The evolution of democratic thought and the dynamics of crowd behavior present a fascinating journey, as explored by various scholars over time. In the mid-18th century, democracy was viewed skeptically, seen as a primitive, unsustainable form of government in a modern context, as noted by Innes and Philip (2013). This perception, however, shifted dramatically by the mid-nineteenth century, with democracy adapting and flourishing in modern settings.

Groupthink

Gustave Le Bon's seminal work in the late-19th century, a period marked by the ascent of modern democracies and the increasing significance of public opinion, delves into the complexities of crowd behavior. He argues that crowds form a distinct psychological entity, different from the sum of its individual members. In such an entity, individual identities become submerged, often leading to behaviors and actions that individuals would not typically engage in alone. The anonymity within a crowd can dissolve personal accountability, freeing members from social norms and leading to unexpected, sometimes irrational actions. Le Bon (2002) emphasizes that emotions and a sense of uniformity dominate crowd behavior, overshadowing individual rationality and logic.

In a similar vein, Charles Mackay's investigations reveal how societal forces—economic, religious, or cultural—can drive collective irrationality, often resulting in economic collapses, unjust persecutions, and misplaced passions (Mackay, 2012). He underlines the importance of individual critical thinking in mitigating the dangers of such 'groupthink', a term coined by Janis (1972). Janis expanded on this concept, identifying it in scenarios where there's a high pressure

Fig. 4 An illustration of 'groupthink' and how crowds converge toward emotional rhetoric.

for uniformity, leading to a degradation in the quality of decision-making within a group. He suggests that crowds can be more influenced by emotional rhetoric and symbolism than by rational or factual arguments (Figure 4).

Conversely, Surowiecki (2004) posits that groups, under appropriate conditions, can make superior decisions compared to even the most intelligent individuals. He argues that a diverse collection of independently deciding individuals can make better, more accurate decisions. This diversity ensures a comprehensive coverage of a problem's solution space. For this to be effective, it's crucial for individual opinions within the group to remain independent and uninfluenced by peers. Decentralized decision-making allows for multiple perspectives, leveraging localized knowledge for better outcomes.

However, Ariely (2008) and others highlight the susceptibility of humans to cognitive biases and emotional contagion, akin to swarm behavior. Emotional contagion, as described by Hatfield et al., (1993), refers to the automatic and often unconscious process of mimicking and absorbing emotions from others, leading to mood convergence within groups. Such phenomena underline the challenges in ensuring rational, independent thought in group settings.

These insights offer a nuanced understanding of the dynamics of group behavior and decision-making. They underscore the delicate balance between the benefits of collective wisdom and the pitfalls of herd mentality, emphasizing the importance of fostering environments where independent, rational thinking can flourish alongside healthy, collective decision-making.

Intelligence in the Herd

The transition, therefore, is to be made from "a mentality of the group" to "an intelligence of the group". Swarm intelligence, in the context of modern human societies, becomes a fascinating lens to understand our collective behaviors. From voting patterns to market dynamics to the spread of information on social media, humans display behaviors strikingly reminiscent of natural swarms. When applied to humans, swarm intelligence provides insights into how individuals perceive, process, and act upon information in the context of the broader group. One of the fundamental principles of swarm intelligence is the power of collective wisdom. A single ant might not be particularly smart, but a colony of ants can solve problems that would be insurmountable for an individual. Similarly, while individual humans have their limitations, collectively, they possess an immense reservoir of knowledge, creativity, and problem-solving capacity. Beni and Wang (1993) elaborated on how a group, despite individual limitations, can collectively achieve objectives that seem insurmountable for a single entity. Like ant colonies or bird flocks, human societies often show tendencies towards decentralization and self-organization. From grassroots movements to decentralized digital platforms, these structures are pivotal in shaping the direction of societies, often in ways that traditional hierarchical systems cannot. The propensity towards decentralization and self-organization in societies echoes the structures of swarms. Social insects, such as ants, bees, and termites, exhibit collective intelligence even without centralized control. Ants find the shortest path to food sources and show how simple agents following simple rules can exhibit complex and adaptive behaviors.

Bonabeau et al., (1999) detailed how certain decentralized systems outperform centralized ones. A crucial concept introduced is 'stigmergy' which is the indirect communication among agents through the environment. For example, ants deposit pheromones on paths to food, and the intensity of the pheromone influences the choice of paths by other ants. The environment and context in which we meet someone or experience something play a significant role in our judgments.

Gladwell (2008) discusses the example of orchestras using blind auditions (where a screen hides the performer from the judges) leading to more women being selected, suggesting that unconscious biases were at play when the performer was visible. Cues within the environment could shape collective thought to reach a similar conclusion.

The Social Forces of the Individual

Human social structures, as Fisher (2009) explains, exist in a state on the "edge of chaos", balancing between total order and complete anarchy. This dynamic equilibrium results from rules of interaction among individuals, creating complex patterns that transcend the actions of single members. Societies evolve as complex adaptive systems, responding collectively to changing circumstances. This phenomenon is evident in "swarm behavior", which transforms into "swarm intelligence" when a group collectively solves problems in ways beyond individual capabilities.

The concept of "swarm intelligence" hinges on behavioral algorithms —rules guiding interactions among individuals. To comprehend these patterns, one must identify interaction dynamics and the flow of information between individuals. This understanding is exemplified in the 'Boids' simulation by Craig Reynolds, which mimics bird flocking behavior. By employing simple rules like avoidance, alignment, and attraction, Boids exhibit complex group behavior without centralized control. This is similar to what a colony of bats, a flock of birds, and a school of fish exhibit when moving together as shown in Figure 5. The individual first separates itself to create distance, then observes the movements of the rest, and aligns itself with those movements.

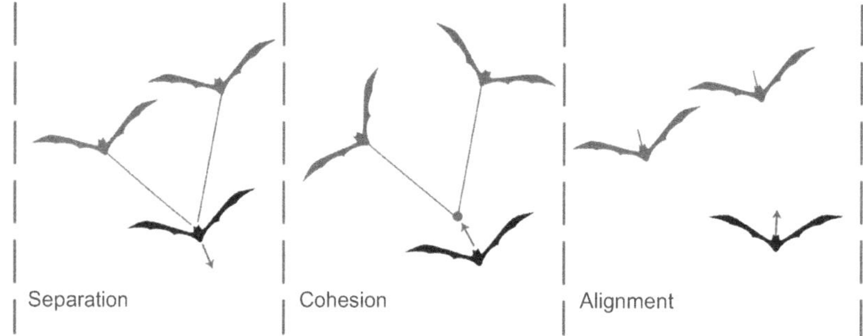

Fig. 5   An illustration of how individual bats align themselves with the rest when flying together.

These principles are mirrored in crowd movements. Individuals in a crowd instinctively avoid collisions, align with nearby people, and move towards average positions, revealing an invisible form of leadership and decision-making. According to Mitchell (1988), echoing Laozi, true leadership is invisible, guiding the group to accomplish tasks as if they were self-initiated.

Social forces, within crowds, act as physical motivators, either attracting individuals towards a goal or repelling them from obstacles. Fisher (2009) suggests that for effective use of swarm intelligence, group members should independently arrive at diverse, original solutions. The key lies in ensuring that these contributions are directed towards a definitive answer, uninfluenced by groupthink or persuasive rhetoric.

In decision-making, the concept of a "quorum response" is crucial, where individuals align with the opinions of informed members, a strategy derived from animal behaviors for cohesive action against threats. Alongside, heuristic rules, as described by Gladwell (2008), allow for rapid, intuitive decision-making based on limited information, a process he terms 'thin-slicing'. Such decision-making is often enhanced in group settings.

Hsieh, Fifić, and Yang (2020) propose a novel method for determining group decision efficiency, finding that decision-making time is generally shorter and outcomes more effective in group settings than individually, supporting Miner's (1984) observations on the superior quality of group decisions.

Human social structures, poised between order and chaos, demonstrate collective intelligence through dynamic interaction patterns. Effective group decisions depend on independent thinking, diversity of thought, and an avoidance of cognitive homogeneity. The essence of good leadership in such systems is its invisible hand, guiding without overt control, leveraging innate human tendencies towards quick, intuitive decision-making, and the collective wisdom of the group.

The Order of Decentralization

Swarm intelligence prioritizes the collective over the individual. Just as a bee sacrifices for the hive, the idea that is promoted is that individuals should work for the collective good. All systems should prioritize the group over the individual. Foster (2000) expands on this idea, elucidating how this collective drive can shape societal structures. A central concept that is examined is the idea of the "metabolic rift"—the separation between human beings and the natural conditions of their existence that result from capitalist production. This rift leads to environmental degradation and the unsustainable extraction of resources. A deep engagement with ecological materialism is necessary to make a return to nature and to also elevate human nature. Swarm intelligence offers such a chance due mostly to a decentralized natural order that is developed and followed.

Wright (2010) examines this dissolution of centralized power structures. Central to this vision is the idea of democratic egalitarianism. The argument is that for any real utopia to be successful, it must be deeply democratic, allowing for genuine participation from its members, and must also work towards reducing systemic inequalities. The introduction of a typology of social empowerment, distinguishing between 'interstitial' strategies that create alternative spaces within the current system, "symbiotic" strategies that involve partnerships with existing central structures, and 'ruptural' strategies that aim for a more direct break from the current system.

While a decentralized society can be imagined after the rule of the collective, this mirrors the decentralized decision-making seen in swarms. The dissolution of centralized power structures and the emergence of self-organized systems are themes common to leveraging the true potential of swarm intelligence. Next we see how this is applied in the field of computer science and Artificial Intelligence (AI).

A Technological Turn

The historical and social dynamics we explored have paved the way for a paradigm shift, echoing the principles of decentralized decision-making observed in natural swarms. Now, as we delve into the realm of AI, we encounter a landscape where decentralized, self-organizing systems offer a departure from traditional cognitive emulation. In the intricate tapestry of AI methodologies, swarm intelligence emerges as a beacon of unconventional problem-solving, utilizing the power of decentralized, autonomous agents engaged in local interactions. This shift not only challenges conventional approaches but also opens a gateway to novel possibilities, inviting us to explore the potential of collective intelligence in the technological domains of computer science and AI.

Decentralization and Artificial Intelligence

The realm of AI has witnessed diverse approaches, each offering unique perspectives on problem-solving. While traditional AI seeks to emulate human cognition, swarm intelligence, in its emphasis on decentralized, simple agents solving complex tasks through local interactions, charts a distinct course.

Traditional AI, exemplified by neural networks, operates on a centralized paradigm. These systems rely on a central controller to manage learning and decision-making processes, a design that can introduce bottlenecks and vulnerabilities. In contrast, swarm intelligence thrives on decentralization, where individual agents operate autonomously based on local information and interactions with neighboring agents. This decentralized approach enhances adaptability and robustness, minimizing susceptibility to single points of failure.

Traditional AI predominantly relies on supervised or unsupervised learning paradigms, training models on extensive datasets to optimize performance. Neural networks adjust weights and parameters to minimize errors. Swarm intelligence, however, often adopts heuristic or rule-based learning. Agents in a swarm follow simple rules, and the emergent collective behavior stems from distributed, decentralized learning, with each agent adapting based on its local environment. While traditional AI excels in tasks with well-defined objectives and clear success criteria, swarm intelligence proves its mettle in addressing complex problems within dynamic and unpredictable environments. The decentralized nature of swarm systems allows them to adapt to changing conditions, making them invaluable for optimization tasks, search problems, and scenarios where flexibility is paramount.

Traditional AI, particularly in neural networks, demands substantial computational resources, especially during the training phase. The models can be resource-intensive, often requiring specialized hardware. In contrast, swarm intelligence, owing to its decentralized and simpler nature, tends to be more resource-efficient. The collective intelligence arises from interactions among individual agents, often necessitating less computational power compared to complex neural network architectures.

Traditional AI models may struggle in dynamic and unpredictable environments due to their reliance on specific datasets and potential lack of adaptability. Swarm intelligence excels precisely in scenarios demanding adaptability and robustness. The decentralized nature of swarm systems enables quick responses to changes in the environment, rendering them suitable for applications in robotics, optimization, and other dynamic fields.

The Impact of Swarm Intelligence

In the ever-evolving landscape of AI, the comparative analysis between swarm intelligence and traditional AI reveals distinctive strengths and applications for each paradigm. While traditional AI approaches have demonstrated effectiveness in various domains, swarm intelligence stands out in situations where adaptability, decentralization, and robustness are paramount. The choice between these paradigms depends on the specific characteristics of the problem at hand and the desired outcomes, emphasizing the versatility and complementary nature of these approaches in shaping the future of AI.

Swarm intelligence has made significant strides in revolutionizing the field of robotics. One exemplary application is swarm robotics, where a large number of relatively simple robots work collectively to achieve complex tasks. Ant-inspired algorithms, for instance, enable robot teams to navigate dynamic environments, allocate tasks efficiently, and respond to changes in real-time. This decentralized approach enhances the adaptability of robotic systems, making them suitable for exploration, search and rescue missions, and environmental monitoring.

The application of swarm intelligence in optimization problems has proven to be particularly fruitful. Metaheuristic algorithms inspired by swarm behavior, such as Particle Swarm Optimization (PSO) (Figure 6) and Ant Colony Optimization (ACO) (Figure 7), excel in solving complex optimization challenges.

Fig. 6   Particle Swarm Optimization (PSO).

These algorithms leverage the collective intelligence of a swarm to efficiently explore solution spaces and converge towards optimal or near-optimal solutions. Industries ranging from logistics and transportation to finance and telecommunications have benefited from the effectiveness of swarm intelligence in optimizing resource allocation, scheduling, and routing problems.

One of the primary challenges faced by swarm intelligence in computer science lies in scalability. As the number of agents within a swarm increases, the complexity of communication and coordination among these agents grows exponentially.

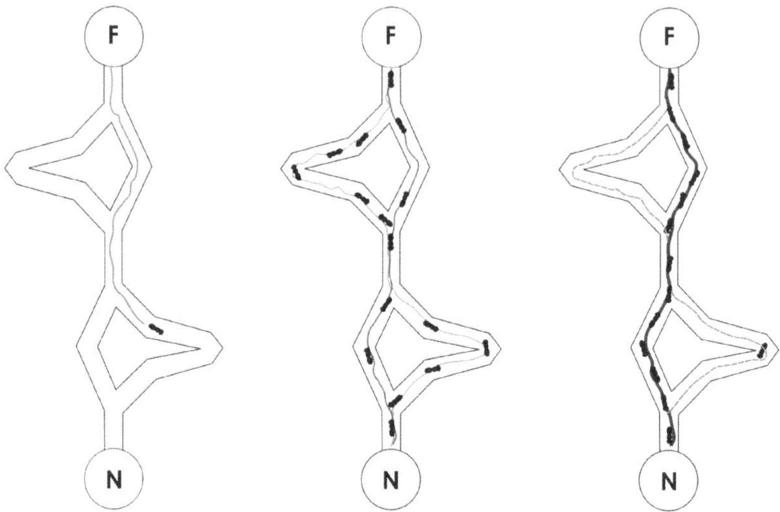

Fig. 7   Ant Colony Optimization (ACO) where F represents Food and N represents Nest
inspired authors below.

The scalability challenge manifests in resource constraints, as the computational overhead required for managing and synchronizing a large number of agents may hinder real-time decision-making. Addressing scalability issues necessitates the development of efficient algorithms and communication protocols that can manage the intricacies of large-scale swarm systems without sacrificing performance.

While swarm intelligence thrives in dynamic and unpredictable environments, ensuring its adaptability poses a significant challenge. Real-world scenarios often involve rapid changes, and swarm systems must be capable of seamlessly adjusting their strategies to cope with evolving conditions. The inherent difficulty lies in designing algorithms that can dynamically reconfigure the behavior of individual agents and the overall swarm without centralized control. Achieving a balance between adaptability and maintaining the integrity of the swarm's collective intelligence is a delicate challenge that researchers must grapple with.

The robustness of swarm intelligence systems is paramount, especially in applications such as robotics and optimization, where reliability is crucial. External factors, such as sensor failures, communication disruptions, or the malfunctioning of individual agents, can compromise the effectiveness of the entire swarm. Developing robust swarm algorithms involves designing mechanisms that can detect and mitigate failures at the individual agent level while preserving the overall functionality of the swarm. Ensuring resilience in the face of unforeseen challenges is an ongoing concern in the application of swarm intelligence in computer science.

Effective communication and coordination are the lifeblood of swarm intelligence. However, the decentralized nature of swarm systems introduces challenges related to information exchange among agents. In scenarios where real-time communication is critical, delays or inconsistencies in the exchange of information can lead to suboptimal performance. Overcoming communication challenges requires the development of robust communication protocols that can handle varying degrees of connectivity and ensure the timely dissemination of critical information within the swarm.

Beyond technical challenges, the application of swarm intelligence raises ethical considerations. In scenarios where swarm systems interact with humans or make decisions that impact human lives, questions of accountability and transparency become paramount. Addressing ethical concerns involves establishing guidelines for the deployment of swarm intelligence in sensitive domains and ensuring that these systems align with human values and ethical standards.

The trajectory of swarm intelligence promises a future replete with innovation and advancements, propelling the field further into the forefront of computer science. Firstly, a compelling avenue lies in the hybridization of swarm intelligence with other AI paradigms. The integration of swarm algorithms with machine learning and deep learning techniques could birth systems that embody the strengths of decentralized decision-making alongside the pattern recognition capabilities inherent in neural networks.

Another direction involves the convergence of swarm intelligence with reinforcement learning. This amalgamation could pave the way for intelligent agents capable of learning and adapting in dynamic environments, with potential applications in robotics, gaming, and autonomous systems. The marriage of swarm intelligence with edge computing and the Internet of Things (IoT) is also on the horizon. Swarm algorithms deployed on edge devices could revolutionize distributed decision-making in IoT networks, offering enhanced real-time processing capabilities, reduced latency, and improved efficiency.

As swarm intelligence becomes more prevalent, emphasis on explainability and interpretability is expected to grow. Transparency in the decision-making processes of swarm algorithms is critical, especially in applications where accountability and understanding are paramount. Lastly, swarm robotics is poised for significant growth in industrial applications. From warehouse automation to collaborative manufacturing, swarms of robots hold the potential to revolutionize industries by efficiently completing tasks that are currently labor-intensive or challenging for single robots.

As the promise of swarm intelligence unfolds, it brings forth ethical considerations that demand careful examination, particularly in the realms of autonomous systems and decision-making. Transparency and accountability are paramount. The decentralized nature of swarm intelligence can present challenges in understanding decision-making processes within a swarm. In critical applications such as healthcare or autonomous vehicles, transparency is essential for building trust and addressing concerns related to accountability. The potential for biases in swarm intelligence algorithms, much like other AI systems, is a growing ethical concern. Efforts are underway to identify and rectify biases in swarm algorithms to ensure fairness and prevent discriminatory outcomes.

Privacy concerns arise in applications involving swarm robotics or IoT, where data collection may be extensive. Striking a balance between the benefits of swarm-based applications and privacy protection is a delicate ethical consideration that requires careful attention. Human-AI collaboration introduces dynamics that necessitate ethical guidelines. Ensuring that swarm systems complement human values, ethical standards, and legal frameworks is essential to prevent unintended consequences. Moreover, the use of swarm intelligence in autonomous weapons raises ethical considerations related to security and safety. Establishing international norms and regulations is critical to prevent misuse and promote responsible AI development in military or security contexts.

In navigating the future directions of swarm intelligence, the field holds immense promise for innovation and advancement. From hybridization with other AI paradigms to applications in swarm robotics and IoT, the potential is vast. However, as this trajectory unfolds, ethical considerations must remain at the forefront. Transparency, fairness, privacy protection, and responsible deployment of swarm-based technologies are essential to harness the full potential of this paradigm while mitigating potential risks. In steering the course through these future directions and addressing ethical considerations, swarm intelligence stands poised to contribute significantly to a more intelligent, collaborative, and ethically sound technological landscape.

References

Ab Wahab, M.N., Nefti-Meziani, S. and Atyabi, A. (2015). A comprehensive review of swarm optimization algorithms. PloS one, 10(5), e0122827.

Almécija, S., Smaers, J.B. and Jungers, W.L. (2015). The evolution of human and ape hand proportions. *Nature Communications*, 6(1), 7717.

Ariely, D. (2008). *Predictably Irrational*. HarperCollins.

Armstrong, D.F. and Wilcox, S. (2007). *The Gestural Origin of Language: Perspectives on Deafness*. Oxford University Press.

Beni, G. and Wang, J. (1993). Swarm intelligence in cellular robotic systems. In Robots and biological systems: towards a new bionics? (pp. 703–712). Berlin, Heidelberg: Springer Berlin Heidelberg.

Boehm, C. (1999). *Hierarchy in the Forest. The Evolution of Egalitarian Behavior*. Cambridge, MA: Harvard University Press.

Bonabeau, E., Dorigo, M. and Theraulaz, G. (1999). *Swarm Intelligence: From Natural to Artificial Systems*. Oxford University Press.

Childe, V.G. (1936). *Man Makes Himself*. London: Watts. 1942. *What Happened in History*. Atlantic Publishers and Distributors.

Chomsky, N. (1980). A review of BF Skinner's Verbal Behavior. *The Language and Thought Series*, 48–64.

Darwin, C. (1906). Mutual affinities of organic beings: Morphology: Embryology: Rudimentary organs. (Chapter 13 in Ambrio Project Encyclopedia (1859) ASU.)

Dunbar, R.I.M. (1996). *Grooming, Gossip, and the Evolution of Language*. Harvard University Press.

Dunbar, R.I. (1998). The social brain hypothesis. *Evolutionary Anthropology: Issues, News, and Reviews: Issues, News, and Reviews*, 6(5), 178–190. Wiley (online).

Engels, F. (1884). *Origin of the Family, Private Property, and the State*. Zurich: Hottingen.

Enard, W., Przeworski, M., Fisher, S.E., Lai, C.S., Wiebe, V., Kitano, T., Monaco, A.P. and Pääbo, S. (2002). Molecular evolution of FOXP2, a gene involved in speech and language. *Nature, 418*(6900), 869–872.

Engels, F. (1950). The part played by labour in the transition from ape to man. http://stars.library.ucf.edu/prism

Fedorenko, E. and Varley, R. (2016). Language and thought are not the same thing: Evidence from neuroimaging and neurological patients. *Annals of the New York Academy of Sciences, 1369*(1), 132–53.

Fisher, L. (2009). *The Perfect Swarm: The Science of Complexity in Everyday Life*. Basic Books.

Fitch, W.T. (2000). The evolution of speech: a comparative review. Trends in cognitive sciences, 4(7), 258–267.

Foster, J.B. (2000). *Marx's Ecology: Materialism and Nature*. Monthly Review Press.

Gladwell, M. (2008). *Blink: The Power of Thinking without Thinking*. USA: Back Bay Books, Little Brown.

Hatfield, E., Cacioppo, J.T. and Rapson, R.L. (1993). Emotional contagion. *Current Directions in Psychological Science, 2*(3), 96–99.

Hauser, M.D., Chomsky, N. and Fitch, W.T. (2002). The faculty of language: What is it, who has it, and how did it evolve? *Science, 298*(5598), 1569–79.

Henrich, J., and McElreath, R. (2003). The evolution of cultural evolution. *Evolutionary Anthropology: Issues, News, and Reviews: Issues, News, and Reviews, 12*(3), 123–35.

Hill, K.R., Walker, R.S., Božičević, M., Eder, J., Headland, T., Hewlett, B., Hurtado, A.M., Marlowe, F., Wiessner, P. and Wood, B. (2011). Co-residence patterns in hunter-gatherer societies show unique human social structure. *Science, 331*(6022), 1286–1289.

Hrdy, S.B. (1999). *Mother Nature: Natural Selection and the Female of the Species*. London: Chatto and Windus.

Hsieh, C.J., Fifić, M. and Yang, C.T. (2020). A new measure of group decision-making efficiency. *Cognitive Research: Principles and Implications, 5*, 1–23.

Innes, J. and Philp, M. (Eds.). (2013). *Re-imagining Democracy in the Age of Revolutions: America, France, Britain, Ireland 1750-1850*. Oxford: Oxford University Press.

Janis, I.L. (1972). *Victims of Groupthink: A psychological study of foreign-policy decisions and fiascoes.* https://doi.org/10.2307/3791464.

Johansson, S. (2005). *Origins of Language: Constraints on Hypotheses*. Amsterdam, Netherlands: John Benjamins.

Johansson, S. (2010). How language did not evolve. *In: Language, Culture and Mind 4. Åbo, 2010.*

Kingsland, S.E. (2018). *Facts or Fairy Tales? Peter Wohlleben and the Hidden Life of Trees*. Ecological Society of America.

Lai, C.S., Fisher, S. E., Hurst, J. A., Vargha-Khadem, F. and Monaco, A.P. (2001). A forkhead-domain gene is mutated in a severe speech and language disorder. *Nature, 413*(6855), 519–23.

Landes, D.S. (1998). *The Wealth and Poverty of Nations: Why Some Are So Rich and Some So Poor*. ISBN 0-393-0-4017-8.

Le Bon, G. (2002). *The Crowd: A Study of the Popular Mind*. Mass. USA: Courier Corporation.

Lee, R.B., Daly, R.H. and Daly, R. (Eds.). (1999). *The Cambridge Encyclopedia of Hunters and Gatherers*. Cambridge University Press.

Mackay, C. (2012). *Extraordinary Popular Delusions and the Madness of Crowds*. New York: Simon and Schuster.

Marx, K. and Engels, F. (1970). *The German Ideology* (Vol. 1). New York: International Publishers Co.

Marx, K. (1852). *The Eighteenth Brumaire of Louis Napoleon, of Die Revolution*. New York.

Marx, K. (2005). *Grundrisse: Foundations of the Critique of Political Economy*. UK: Penguin.

McBrearty, S. and Brooks, A.S. (2000). The revolution that wasn't: A new interpretation of the origin of modern human behavior. *Journal of Human Evolution, 39*(5), 453–563.

McIntosh, S.K. (1999). Beyond chiefdoms: Pathways to complexity in Africa. *(No Title).*

Miner Jr., F.C. (1984). Group versus individual decision making: An investigation of performance measures, decision strategies, and process losses/gains. *Organizational Behavior and Human Performance, 33*(1), 112–24.

Mitchell, S. (1988). (Trans.) *Tao te Ching*. Harper Collins.

Mithen, S. (1996). *The Prehistory of the Mind: The Cognitive Origins of Art and Science*. USA: Thames & Hudson Ltd.

Morgan, L.H. (1877). *Ancient Society: Researches in the Lines of Human Progress from Savagery through Barbarism to Civilization*. Chicago: Charles H. Kerr & Co.

Nakamura, T., Gehrke, A.R., Lemberg, J., Szymaszek, J. and Shubin, N.H. (2016). Digits and fin rays share common developmental histories. *Nature, 537*(7619), 225–28.

Ong, W.J. (2013). *Orality and Literacy*. Routledge.

Patterson, K. (2019). Is there a social hierarchy in the heifer herd of Finca Paraíso? University of South Florida.

Pauketat, T.R. (2001). Practice and history in archaeology: An emerging paradigm. *Anthropological theory, 1*(1), 73–98.

Pinker, S. (1994). *The Language Instinct: How the Mind Creates Language*. Harper Collins.

Pinker, S. (2003). Language as an adaptation to the cognitive niche. *Studies in the Evolution of Language, 3*, 16–37.

Rieber, R. (1983). The psychology of language and thought: Noam Chomsky interviewed by Robert W. Rieber. *Dialogues on the Psychology of Language and Thought*. New York: Plenum.

Scott-Phillips, T.C. and Kirby, S. (2010). Language evolution in the laboratory. *Trends in Cognitive Sciences, 14*(9), 411–17.

Service, E.R. (1975). Origins of the state and civilization: The process of cultural evolution.

Skinner, B.F. (1957). *Verbal Behavior*. New York: Appleton-Century-Crofts.

Surowiecki, J. (2004). *The Wisdom of Crowds*. USA: Anchor.

Tomasello, M. and Carpenter, M. (2007). Shared intentionality. *Developmental Science, 10*(1), 121–25.

Tylor, E.B. (1871). *Primitive Culture*. London: J. Murray.

Wohlleben, P. (2016). *The Hidden Life of Trees: What they Feel, How They Communicate—Discoveries from a Secret World* (Vol. 1). Vancouver: Greystone Books.

Woodburn, J. (1982). Egalitarian societies. *Man*, 431–51.

Wright, E.O. (2010). *Envisioning Real Utopias*. UK/New York: Verso Books.

Young, R.W. (2003). Evolution of the human hand: The role of throwing and clubbing. *Journal of Anatomy, 202*(1), 165–74.

2

Swarm Intelligence: Applications and Implementations in Autonomous Systems

Sasidhar Bhimana[1*] and *Saravanan Ravindran*[2]

Introduction to Swarm Intelligence (SI)

Swarm intelligence (SI) is a collective behavior exhibited by groups of simple agents, such as ants, bees, and birds, which can achieve complex tasks that would be difficult or impossible for a single individual (Dorigo et al., 2019). The collective behavior of these organisms is characterized by decentralized decision-making, self-organization, adaptive responses to environmental changes, and emergent properties that are not present in individual organisms (Beni & Wang, 2004). SI algorithms emulate these features to solve complex optimization, control, classification, clustering, routing, and prediction problems in diverse domains, such as engineering, robotics, biology, economics, social sciences, and humanities. There are two main categories of SI algorithms: swarm-based algorithms and swarm-inspired algorithms (Dorigo & Gambardella, 1996). Swarm-based algorithms involve the simulation of a population of individuals (agents) that interact with each other and their environment to achieve a collective goal. Examples of swarm-based algorithms include ant colony optimization (ACO), particle swarm optimization (PSO), artificial bee colony (ABC), and firefly algorithm (FA) (Kennedy & Eberhart, 1995; Karaboga & Basturk, 2007; Yang, 2010). Swarm-inspired algorithms, on the other hand, extract specific mechanisms or principles from natural swarms and

[1,2] Bharath Institute of Higher Education and Research, Chennai-600073, Tamil Nadu, India.
Email : rssrmcp@gmail.com
* Corresponding author: bhimanasasidhar@gmail.com

incorporate them into conventional optimization or machine learning algorithms. Examples of swarm-inspired algorithms include artificial immune systems (AIS), bacterial foraging optimization (BFO), and grey wolf optimizer (GWO) (Dasgupta & González, 2002; Passino, 2002; Mirjalili et al., 2014).

The success of SI algorithms is attributed to their ability to efficiently explore a large search space, converge to optimal or near-optimal solutions, and handle multiple objectives or constraints simultaneously (Altshuler et al., 2018). The collective intelligence of the swarm enables the sharing and exchange of information, the exploitation of promising regions, and the avoidance of suboptimal regions. Furthermore, the decentralized and distributed nature of the swarm allows for scalability, robustness, fault-tolerance, and adaptivity to dynamic or uncertain environments.

In summary, the concepts of swarm intelligence can be introduced by explaining the collective behavior exhibited by natural swarms and how SI algorithms emulate these features to solve complex problems. The different categories of SI algorithms, such as swarm-based and swarm-inspired algorithms, can be discussed along with their applications in various domains. The advantages of SI algorithms, such as their ability to efficiently explore search spaces and handle multiple objectives, can also be highlighted.

Overview of Swarm Intelligence

The purpose of providing an overview of swarm intelligence is to highlight its collective behavior exhibited by groups of simple agents, such as ants, bees, and birds. Swarm intelligence algorithms emulate these features to solve complex optimization, control, classification, clustering, routing, and prediction problems in various domains. By understanding the principles and applications of swarm intelligence, researchers and practitioners can harness its potential to address challenges in engineering, robotics, biology, economics, social sciences, and humanities.

Importance of Autonomous Systems

Studying and developing autonomous systems are pivotal for various reasons (Dorigo et al., 2019). Firstly, they significantly enhance efficiency and productivity by automating tasks, leading to increased output, reduced costs, and improved overall performance. Secondly, in hazardous environments, autonomous systems operate where human presence is risky, minimizing the potential for injury or loss of life. Thirdly, their unmatched precision and accuracy, especially in fields like manufacturing and healthcare, eliminate human errors. Additionally, these systems are adaptable, capable of scaling based on tasks and real-time conditions, handling complex situations effectively. Lastly, they foster exploration and innovation, delving into new frontiers and conducting experiments that might be unsafe for humans. In essence, autonomous systems drive progress by improving efficiency, ensuring safety, enhancing precision, adapting to change, and enabling exploration and innovation.

Objectives of the Chapter

- Provide an overview of SI, explaining how it draws inspiration from collective behavior in natural swarms and how SI algorithms emulate these features.
- Explore different categories of SI algorithms such as swarm-based and swarm-inspired algorithms along with their applications in various domains.
- Highlight key advantages of SI algorithms such as their ability to efficiently explore search spaces and handle multiple objectives.
- Delve into practical applications of SI in autonomous vehicles, drones, robotics, industrial automation, energy grids, smart cities, IoT ecosystems, precision agriculture, and more.
- Demonstrate SI's versatility in solving complex optimization, control, classification, clustering, routing, and prediction problems.
- Conduct a case study on applying swarm optimization techniques in an industrial automation context.
- Discuss challenges faced in SI implementations and lessons learned from real-world examples.
- Analyze successful SI deployments globally across industries.
- Explore future directions and emerging trends in SI research.
- Summarize research findings on SI and its profound impact on problem-solving across diverse domains.

Swarm Intelligence and Autonomous Vehicles

SI significantly influences the development of autonomous vehicles (Dias & Stentz, 2001). Researchers, inspired by social insects like ants and bees, have created algorithms for autonomous vehicles (Brooks & Flynn, 1989). SI reveals that individual agents need not be highly intelligent; by following simple rules and local cues, they can collectively exhibit intelligent behaviors (Brambilla et al., 2013).

This concept translates into autonomous vehicles as decentralized systems (Hinchey et al., 2007). Similar to how ants cooperate, these vehicles interact locally, sharing information to optimize traffic flow and coordinate movements (Mastellone et al., 2008). SI grants efficiency, scalability, and adaptability (DeLoach & Kumar, 2008). It allows vehicles to explore vast spaces, converge to optimal solutions, and handle multiple objectives (Chalkiadakis et al., 2007). The collective intelligence enables information sharing, promising region exploitation, and suboptimal region avoidance (Wagner & Lindenbaum, 1998).

In essence, SI offers a framework for advanced autonomous vehicle systems, allowing them to navigate complexities, optimize traffic, and coordinate actions effectively.

The Role of Swarm Intelligence in Autonomous Vehicles

SI plays crucial roles in advancing autonomous vehicles (Wagner & Bruckstein, 2001):

Cooperative Behavior: Similar to ants or bees in a colony, swarm intelligence enables vehicles to collaborate, optimizing traffic flow and navigating intricate environments by coordinating actions toward shared objectives (Klos & van Ahee, 2008).

Decentralized Control: Operating without a central system, each vehicle relies on local data and interactions with neighbors. This decentralized approach enhances scalability, robustness, and adaptability, especially in uncertain settings (Felner et al., 2006).

Adaptive and Self-Organizing Systems: Vehicles learn and adapt by observing swarm behavior. This allows individual vehicles to optimize performance, align actions, and achieve collective goals, enhancing overall efficiency (Arkin & Balch, 1997).

Efficient Exploration and Optimization: Swarm intelligence algorithms like particle swarm optimization enhance vehicle behavior. They efficiently explore vast possibilities, converging on optimal solutions and managing multiple objectives or constraints simultaneously (Yang et al., 2013).

Fault Tolerance: Swarm intelligence imparts fault tolerance. When a vehicle fails or faces obstacles, others adjust dynamically, compensating for losses, ensuring mission continuity, and task completion (Sasidhar & Ravindran, 2023).

In summary, SI equips autonomous vehicles with collective intelligence, adaptability, and efficiency, enabling them to navigate intricate and dynamic environments adeptly.

Case Study: Swarm Optimization of Enzyme Production

The focus of the case study is on optimizing α-amylase production using swarm intelligence-based algorithms, specifically particle swarm optimization (PSO) (Wagner & Bruckstein, 2001). The study aims to enhance the recovery and activity of the enzyme by utilizing PSO to find optimal conditions for enzyme production (Sasidhar & Ravindran, 2023). The researchers compare PSO with other optimization techniques like genetic algorithm (GA) (Wagner et al., 1998) and explore the parameters that lead to maximum amylase activity. The study highlights the advantages of SI-based optimization in complex optimization problems and its potential for global perspective and ease of operation.

Genetic Algorithms for Enzyme Production Optimization

Genetic algorithms (GA) excel in enzyme production optimization (Wagner & Bruckstein, 2001):

Mimics Natural Evolution: GA mirrors natural evolution, exploring diverse solutions and adapting to changing environments. This flexibility suits complex problems with multiple optima, ensuring robust optimization (Wagner et al., 1998).

Reproduction, Crossover, and Mutation: GA employs reproduction, crossover, and mutation operations. These processes select good solutions, create new ones by exchanging genetic information, and introduce random changes, enhancing effective exploration of the search space (Wagner & Bruckstein, 2001).

Global Perspective: GA searches globally for optimal solutions, preventing entrapment in local optima. It explores multiple regions simultaneously, increasing the likelihood of finding the global optimum (Wagner & Bruckstein, 2001).

Flexibility and Ease of Operation: GA handles various variables and constraints, offering flexibility. It's easily customized for diverse problems and doesn't require gradient information, making it suitable for nonlinear, non-differentiable objective functions (Wagner & Bruckstein, 2001).

Higher Memory Capability: GA maintains a solution population, retaining valuable solutions across generations. This memory capability preserves essential information, preventing premature convergence (Wagner & Bruckstein, 2001).

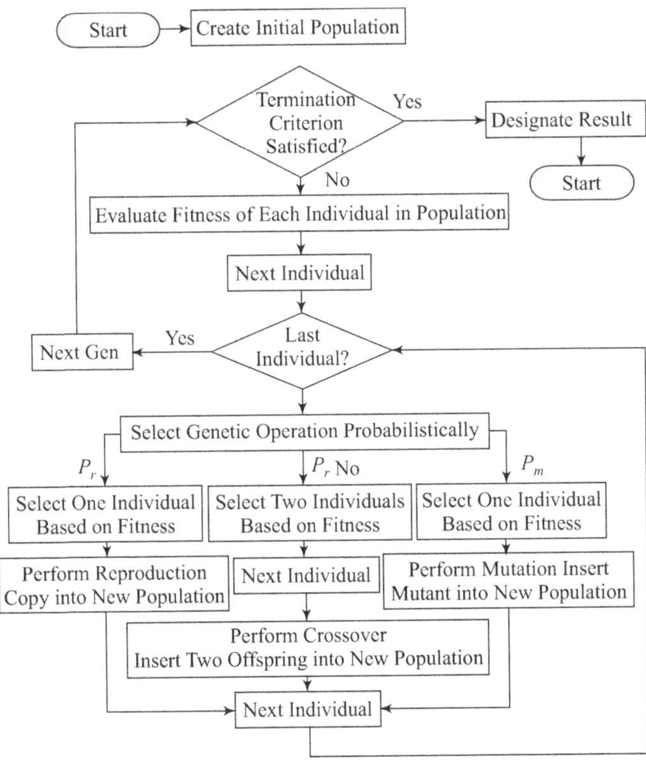

Fig. 1   Flow chart of a simple Genetic Algorithm (GA).

Although PSO has merits like simplicity and fewer parameters, GA's ability to mimic natural evolution, explore the search space globally, and maintain diversity make it the preferred choice for enzyme production optimization (Figure 1).

Particle Swarm Optimization for Enzyme Production Optimization

PSO stands as a powerful tool in optimizing enzyme production within the framework of SI. Inspired by the collective behavior of bird flocking or fish schooling, PSO operates with a population of particles representing potential solutions to the optimization problem.

In the realm of enzyme production, PSO refines parameters such as temperature, pH, substrate concentration, and fermentation time. Each particle assesses its fitness based on criteria like enzyme production yield or activity. Crucially, PSO particles communicate their best-known solutions (personal best) and the swarm's best solutions (global best). This sharing of information efficiently explores the search space, guiding the particles toward optimal or near-optimal solutions (Figure 2).

Through iterative updates informed by personal experiences and shared knowledge, the swarm systematically explores the parameter space. This collaborative approach pinpoints the optimal conditions for maximizing enzyme production. PSO's prowess lies in its ability to efficiently navigate parameter spaces, exploit promising regions, and converge towards optimal solutions. Harnessing the collective intelligence of the swarm, PSO significantly enhances the efficiency and effectiveness of enzyme production processes.

Particle Swarm Optimization vs Genetic Algorithms for Enzyme Production Optimization

PSO presents distinct advantages over Genetic Algorithms (GA) and other optimization methods in the realm of enzyme production optimization (Sasidhar & Ravindran, 2023):

Easier Implementation: PSO boasts a simpler structure and requires fewer parameter adjustments, making it more accessible and user-friendly for researchers and practitioners compared to GA (Sasidhar & Ravindran, 2023).

Minimal Parameter Complexity: PSO demands fewer tuned parameters, reducing the optimization process's complexity and time consumption in contrast to GA (Sasidhar & Ravindran, 2023).

Enhanced Memory Capability: PSO integrates individual memory components, aiding particles in remembering previous best positions. This prevents revisiting explored solutions, facilitating faster convergence toward optimal solutions (Sasidhar & Ravindran, 2023).

Overcoming Local Optima: PSO addresses the challenge of converging towards local optima, a common GA issue. Utilizing a global best guide, PSO influences

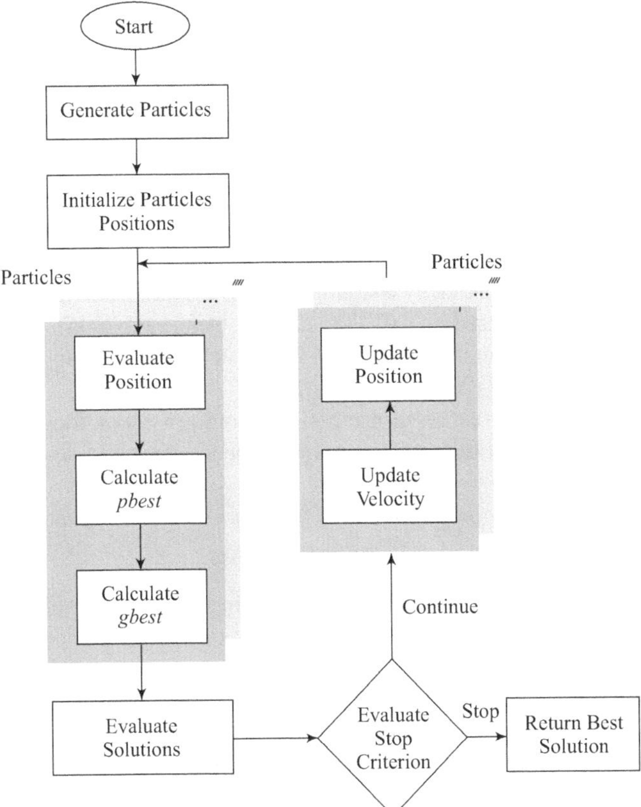

Fig. 2 Simplified illustration of a sequence of events in the Particle Swarm Optimization (PSO).

particle movement, enabling effective exploration of the search space and discovery of superior solutions (Sasidhar & Ravindran, 2023).

Adaptability and Exploration: PSO incorporates an adaptive mutation operator, fostering exploratory capability. Initially applied to the entire population, mutations decrease over time to prevent premature convergence. This adaptability promotes exploration across diverse regions, ensuring the discovery of varied solutions (Sasidhar & Ravindran, 2023).

In summary, PSO excels due to its easier implementation, minimal parameter complexity, superior memory capability, and capacity to overcome local optima compared to GA and other techniques in enzyme production optimization.

Swarm Intelligence in Navigation and Decision-Making for Autonomous Vehicles

SI revolutionizes autonomous vehicle navigation and decision-making through collective agent behavior (Brooks, 1989). SI operates on simple rules, fostering complex behaviors at the swarm level (Brambilla et al., 2013). In autonomous vehicles, SI optimizes traffic flow, coordinates movements, and prevents collisions (Mastellone et al., 2008). SI algorithms enable efficient communication and cooperation among vehicles, ensuring adaptability to changing traffic conditions and real-time decision-making (DeLoach & Kumar, 2008).

One notable application is in traffic management, where swarms optimize flow, reduce congestion, and minimize travel time (Chalkiadakis et al., 2007). Through coordinated movements and shared information, vehicles collectively decide, benefiting the entire system (Hettiarachchi & Spears, 2005). Additionally, SI enhances decision-making by enabling vehicles to learn from their environment, adjusting algorithms based on real-time data (Felner et al., 2006). Observing and interacting with surroundings, autonomous vehicles gather insights, enhancing performance and safety (Arkin & Balch, 1997).

In summary, SI presents a promising paradigm for autonomous vehicle navigation and decision-making, leveraging collective behavior to achieve shared objectives.

Real-World Examples and Success Stories

There are real-world examples of SI being applied to autonomous vehicles (Brooks, 1989). One example is the use of SI in optimizing traffic flow (Sasidhar & Ravindran, 2023). Swarms of autonomous vehicles can coordinate their movements and communicate with each other to optimize traffic patterns, reduce congestion, and improve overall efficiency (Chalkiadakis et al., 2007).

Another example is the use of SI in search and rescue operations (Dias & Stentz, 2001). Swarms of autonomous drones can work together to search large areas and locate missing persons or objects (Mastellone et al., 2008). By collaborating and sharing information, the drones can cover more ground and increase the chances of a successful rescue operation.

These examples demonstrate how SI can be leveraged in the context of autonomous vehicles to solve complex problems and improve overall performance.

Swarm Intelligence in Drones and Robotics

SI is widely applied in drones and robotics, encompassing various functionalities:

Synergetic Mission Planning: Swarms facilitate efficient mission planning, enabling collaboration among drones or robots for complex tasks (Su et al., 2009).

Emergency Detection: Swarm robotics provide decentralized sensing, allowing rapid responses to emergencies, such as natural disasters (Ren & Sorensena, 2008).

Patrolling: Drone or robot swarms ensure surveillance and security in specific areas (Deneubourg et al., 1990).

Fault Tolerance Cooperation: Swarm systems compensate for individual failures, ensuring continuous operation (Drogoul & Ferber, 1992).

Target Tracking: SI aids in tracking and following moving targets or individuals (Weitzenfeld, 2008).

Exploration and Mapping: Swarms explore hazardous terrains, providing valuable data for various applications (Benda et al., 1985).

Collaborative Cleaning: Multiple agents work collectively for efficient area cleaning tasks (Haynes & Sen, 1986).

Control Architecture for Autonomous Drones: SI designs coordinated and intelligent control systems for autonomous drones (Su et al., 2009).

Formation Generation and Keeping: Drones or robots maintain specific formations for coordinated tasks (Deneubourg et al., 1990).

Target Search: Swarm systems efficiently search large areas and locate specific targets through collaboration (Drogoul & Ferber, 1992).

These applications illustrate the versatility of SI in enhancing the capabilities of drones and robotics.

Integration of Swarm Intelligence in Drone Systems

SI seamlessly integrates into drone systems through decentralized decision-making, self-organization, and adaptive responses to environmental changes. In swarm-based drone setups, autonomous drones collaborate for diverse tasks, including mission planning, emergency detection, patrolling, fault tolerance cooperation, target tracking, and search operations (Su et al., 2009; Ren & Sorensena, 2008; Deneubourg et al., 1990; Drogoul & Ferber, 1992; Weitzenfeld, 2008; Benda et al., 1985; Haynes & Sen, 1986). Emulating natural swarm behavior, these systems enhance efficiency and effectiveness compared to single drones by sharing information, exploring promising areas, and avoiding suboptimal regions. This decentralized approach ensures scalability, robustness, fault tolerance, and adaptability in dynamic or uncertain environments. Communication among drones is pivotal, sharing information like pheromone trails or sensor data for swift identification of optimal solutions. Parallel swarm algorithms, such as parallel ant colony optimization or parallel particle swarm optimization, utilize communication to enhance search process efficiency. In essence, integrating SI involves deploying decentralized groups of simple robotic agents that collaborate, communicate, and adapt to accomplish tasks effectively.

Enhancing Drone Navigation and Coordination with Swarm Intelligence

SI transforms drone navigation and coordination by embracing decentralized and collaborative principles. In these systems, drones function collectively, communicating and collaborating to achieve shared objectives, offering numerous benefits (Su et al., 2009; Ren & Sorensena, 2008; Deneubourg et al., 1990; Drogoul & Ferber, 1992; Weitzenfeld, 2008; Benda et al., 1985; Haynes & Sen, 1986). SI seamlessly integrates numerous drones, efficiently handling intricate missions through task distribution among multiple units. Individual drones adapt and compensate for others' failures, ensuring functionality even if some drones are unavailable due to the decentralized nature of SI. Swarm-based navigation enables dynamic path adjustments based on real-time information, with drones collectively responding to changing environmental conditions or mission requirements. SI optimizes navigation and coordination, with drones collaboratively exploring, mapping areas, searching for targets, or conducting tasks like surveillance, enhancing efficiency and reducing mission time. SI equips drones to respond to unforeseen situations, sharing information for collective decisions and adjusting actions, demonstrating adaptability. In summary, SI enhances drone navigation and coordination through decentralized decision-making, collaboration, and adaptability, leveraging SI for efficient and effective task completion.

Robotics Applications and Implementations of Swarm Intelligence

SI finds diverse applications in robotics, including multi-robot coverage, swarm robotics for tasks like mission planning and emergency response, and shaping the control architecture of autonomous drones for various tasks (Su et al., 2009; Ren & Sorensena, 2008; Deneubourg et al., 1990; Drogoul & Ferber, 1992; Weitzenfeld, 2008; Benda et al., 1985; Haynes & Sen, 1986). The applications illustrate the versatility of SI in solving complex problems across domains such as optimization, control, classification, clustering, routing, and prediction.

Case Study: Swarm Optimization Coordinate Industrial Robots for Efficient Manufacturing

This case study demonstrates the application of particle swarm optimization (PSO) to coordinate industrial robots for enhanced manufacturing efficiency (Su et al., 2009). PSO represents each potential assembly schedule and robot coordination pattern as a 'particle' navigating the search space. Through iterative movements guided by each particle's experiences and those of neighbors, the swarm converges towards optimal solutions that minimize production time. Compared to other optimization methods, PSO achieves high-quality solutions with fewer function evaluations, efficiently navigating the complex scheduling problem. This demonstrates how swarm intelligence can streamline industrial automation by coordinating robots as a self-organizing swarm to optimize complex manufacturing tasks for maximum efficiency in a scalable, decentralized manner.

Lessons Learned and Practical Insights

These swarm optimization techniques in industrial robotics can lead to improved efficiency, reduced production costs, and enhanced productivity in manufacturing processes. Swarm-based algorithms in industrial robotics, like ant colony optimization (ACO) and PSO, offer valuable insights into manufacturing optimization. ACO optimizes production scheduling by employing virtual ants representing tasks communicating through pheromone trails. These trails, indicating scheduling attractiveness, are iteratively updated, leading the swarm to an optimal production schedule. PSO optimizes robotic assembly processes using virtual particles representing assembly tasks. Iteratively adjusting positions based on best solutions, the swarm converges to an efficient assembly sequence, minimizing time and resources. These techniques enhance manufacturing efficiency, decrease costs, and boost productivity.

Extending IoT Ecosystems with Swarm Intelligence

SI significantly extends IoT ecosystems by fostering collaborative and adaptive behavior among IoT devices (Altshuler et al., 2010; Puzis et al., 2013; Altshuler et al., 2016; Aknine & Shehory, 2006; Sariel & Balch, 2005; Michael et al., 2008; Gerkey & Mataric, 2002).

Distributed Decision Making: IoT devices autonomously make decentralized decisions based on local data and interactions, ensuring efficient decision processes (Altshuler et al., 2010; Puzis et al., 2013; Altshuler et al., 2016).

Self-Organization: Devices self-organize, forming groups based on proximity or shared goals (Altshuler et al., 2010; Puzis et al., 2013; Altshuler et al., 2016). This adaptive clustering enhances resource allocation and task distribution.

Optimization and Resource Management: Swarm algorithms like ACO or PSO optimize resource allocation, routing, and scheduling (Altshuler et al., 2010; Puzis et al., 2013; Altshuler et al., 2016). This leads to energy efficiency, reduced congestion, and superior system performance.

Fault Tolerance and Resilience: Devices collaborate to compensate for failures, ensuring continuous operation (Altshuler et al., 2010; Puzis et al., 2013; Altshuler et al., 2016). SI enables dynamic rerouting and behavioral adjustments in response to disruptions.

Adaptive Sensing and Data Collection: Devices collectively sense and collect data (Altshuler et al., 2010; Puzis et al., 2013; Altshuler et al., 2016). Coordinated sensing and adaptive sampling, guided by swarm knowledge, enhance data collection efficiency. Leveraging SI enriches IoT ecosystems, enhancing scalability, robustness, adaptability, and efficiency (Altshuler et al., 2010; Puzis et al., 2013; Altshuler et al., 2016). This improvement results in superior performance, heightened reliability, and the ability to tackle complex and dynamic challenges.

Swarm Intelligence and the Internet of Things (IoT)

SI significantly enhances IoT ecosystems in multiple ways (Altshuler et al., 2010; Puzis et al., 2013; Altshuler et al., 2016).

Distributed Decision-making: SI enables decentralized decisions based on local data, optimizing device behavior (Altshuler et al., 2010; Puzis et al., 2013;

Altshuler et al., 2016). For instance, in smart homes, it coordinates devices like thermostats and lights for energy efficiency.

Self-Organization: IoT devices self-organize, allocating resources and coordinating tasks for optimal performance (Altshuler et al., 2010; Puzis et al., 2013; Altshuler et al., 2016). In smart cities, SI optimizes traffic flow, reducing congestion through coordinated vehicle movements.

Adaptability: Devices continuously learn and adjust based on environmental feedback, optimizing performance over time (Altshuler et al., 2010; Puzis et al., 2013; Altshuler et al., 2016). For example, in agricultural IoT, it adjusts irrigation schedules based on real-time weather and crop conditions.

Fault-Tolerance: Swarm intelligence ensures robustness; if a device fails, others seamlessly take over tasks (Altshuler et al., 2010; Puzis et al., 2013; Altshuler et al., 2016). In healthcare IoT, it distributes tasks among devices, ensuring continuous monitoring and timely responses to critical events.

In summary, SI empowers IoT ecosystems with distributed decision making, self-organization, adaptability, and fault tolerance, enhancing their functionality and performance (Altshuler et al., 2010; Puzis et al., 2013; Altshuler et al., 2016).

Collaborative Swarm-Based Algorithms for IoT Device Coordination

Collaborative swarm-based algorithms offer dynamic solutions for coordinating IoT devices (Altshuler et al., 2010; Puzis et al., 2013; Altshuler et al., 2016).

Distributed Task Allocation: Devices act autonomously, making local decisions and optimizing tasks collectively by sharing information (Altshuler et al., 2010; Puzis et al., 2013; Altshuler et al., 2016).

Resource Optimization: Algorithms dynamically allocate tasks, ensuring efficient use of limited resources like processing power and energy, enhancing overall system efficiency (Altshuler et al., 2010; Puzis et al., 2013; Altshuler et al., 2016).

Adaptive Routing and Communication: Swarm algorithms enable self-organization, adapting communication patterns based on network conditions, ensuring scalability and fault-tolerance (Altshuler et al., 2010; Puzis et al., 2013; Altshuler et al., 2016).

Collaborative Sensing and Data Fusion: Devices collaborate in sensing tasks, enhancing accuracy by collectively processing sensor data, vital for applications like environmental monitoring (Altshuler et al., 2010; Puzis et al., 2013; Altshuler et al., 2016).

Self-Healing and Fault-Tolerance: Swarm algorithms empower devices to autonomously detect and respond to failures, ensuring system functionality and recovery (Altshuler et al., 2010; Puzis et al., 2013; Altshuler et al., 2016). In essence, these algorithms offer a decentralized, adaptive approach, enhancing the efficiency, scalability, and robustness of IoT systems (Altshuler et al., 2010; Puzis et al., 2013; Altshuler et al., 2016).

Swarm Intelligence in Communication and Resource Management in IoT Ecosystems

SI revolutionizes communication and resource management in IoT ecosystems (Altshuler et al., 2010; Puzis et al., 2013; Altshuler et al., 2016).

Communication Optimization: Swarm algorithms emulate decentralized decision-making, enhancing communication protocols (Altshuler et al., 2010; Puzis et al., 2013; Altshuler et al., 2016). Result: reduced congestion, improved data rates, and superior network performance.

Resource Allocation: Mimicking natural swarm properties, algorithms dynamically distribute resources (bandwidth, processing power, energy) based on device needs (Altshuler et al., 2010; Puzis et al., 2013; Altshuler et al., 2016).

Outcome: Efficient resource usage and enhanced system performance. Task Distribution: Swarm algorithms coordinate task assignment, ensuring load balance and fault tolerance (Altshuler et al., 2010; Puzis et al., 2013; Altshuler et al., 2016).

Result: Efficient task execution in IoT networks.

Adaptive Routing: Algorithms adapt routing paths, enhancing reliability, scalability, and efficiency (Altshuler et al., 2010; Puzis et al., 2013; Altshuler et al., 2016). Useful for dynamic or disrupted network topologies. In essence, SI, inspired by natural swarms, ensures efficient, scalable, and robust IoT systems (Altshuler et al., 2010; Puzis et al., 2013; Altshuler et al., 2016).

Case Study: Swarm Optimization in IoT Resource Allocation

Swarm optimization in IoT resource allocation demonstrates the power of collective intelligence (Altshuler et al., 2010; Puzis et al., 2013; Altshuler et al., 2016). By leveraging swarm-based algorithms like ant colony optimization (ACO) or particle swarm optimization (PSO), IoT networks can efficiently manage resources (Altshuler et al., 2010; Puzis et al., 2013; Altshuler et al., 2016). These algorithms emulate populations of interacting agents, representing IoT devices, to achieve optimal resource allocation. Factors like resource availability, energy consumption, quality of service, and network conditions are considered. Through coordinated actions and information exchange, these algorithms explore resource allocation options and converge toward optimal solutions. Additionally, swarm-inspired algorithms like artificial immune systems (AIS) or bacterial foraging optimization (BFO) extract natural swarm principles for IoT resource challenges (Altshuler et al., 2010; Puzis et al., 2013; Altshuler et al., 2016). However, it's essential to balance exploration and exploitation, address scalability, and manage communication complexities in these designs. This case study highlights how swarm intelligence enhances resource utilization, improves network performance, and ensures scalability in IoT ecosystems (Altshuler et al., 2010; Puzis et al., 2013; Altshuler et al., 2016).

Future Potential and Challenges of SI in IoT

Indeed, the future of SI in the Internet of Things (IoT) holds immense promise, yet it comes with significant challenges (Altshuler et al., 2010; Puzis et al., 2013; Altshuler et al., 2016). Researchers can explore advanced algorithms and models to enhance SI's efficiency and adaptability in IoT applications. Optimizing communication and coordination among swarm agents, especially in uncertain real-world IoT environments, remains a critical area for improvement (Altshuler et al., 2010; Puzis et al., 2013; Altshuler et al., 2016). Application-focused research is vital, requiring collaboration with industries and government agencies to address practical IoT challenges, from disaster response to urban planning (Altshuler et al., 2010; Puzis et al., 2013; Altshuler et al., 2016). However, challenges persist. Balancing exploration and exploitation is essential to prevent swarms from getting trapped in suboptimal solutions, necessitating diverse and adaptable approaches (Altshuler et al., 2010; Puzis et al., 2013; Altshuler et al., 2016). Scalability poses a significant hurdle. While swarms can comprise millions of agents, managing computational and communication overheads in large-scale systems is complex (Altshuler et al., 2010; Puzis et al., 2013; Altshuler et al., 2016). Efficient algorithms for decision-making and resource allocation are crucial yet challenging to design and optimize. Beyond technical hurdles, ethical and social concerns arise (Altshuler et al., 2010; Puzis et al., 2013; Altshuler et al., 2016). Privacy, security, and control issues may emerge as swarm systems become pervasive. Addressing these concerns involves creating transparent algorithms and collaborating with policymakers to establish appropriate regulations. In essence, while SI offers transformative potential in IoT, addressing challenges through advanced algorithms, real-world applications, scalability solutions, and ethical considerations is crucial (Altshuler et al., 2010; Puzis et al., 2013; Altshuler et al., 2016).

Efficiency Revolution in Manufacturing and Supply Chain Management

Swarm optimization techniques in industrial robotics have transformative potential for revolutionizing efficiency in manufacturing and supply chain management (Kraus et al., 2003; Work et al., 2008; Parker, 1998; Rehak et al., 2008; Mosafi et al., 2022).

Improved Resource Allocation: Swarm optimization algorithms dynamically adjust resource allocation, minimizing idle time and maximizing productivity in manufacturing processes (Kraus et al., 2003).

Enhanced Coordination and Collaboration: Decentralized coordination enables efficient task allocation, cooperative manipulation, and synchronized movement, reducing cycle times (Work et al., 2008).

Adaptive and Flexible Production: Swarm techniques facilitate agile production systems that quickly respond to demand changes, optimizing scheduling and adapting to unforeseen events (Parker, 1998).

Optimal Routing and Logistics: Swarm algorithms optimize material routes, minimizing delays and optimizing inventory management, leading to cost savings and improved customer satisfaction (Rehak et al., 2008).

Predictive Maintenance and Quality Control: The algorithms predict equipment failures by analyzing data, improving product quality, and reducing unexpected disruptions (Mosafi et al., 2022).

In summary, swarm optimization techniques promise a significant shift toward efficient, agile, and adaptive manufacturing and supply chain processes.

Enhancing Manufacturing Efficiency through Swarm Algorithms

Swarm algorithms offer a multifaceted approach to enhancing manufacturing efficiency (Kraus et al., 2003; Manisterski et al., 2008; Mosafi et al., 2022).

Optimizing Production Processes: Swarm-based algorithms like ACO and PSO optimize resource allocation, task scheduling, and production cost reduction by simulating collective problem-solving behavior (Somin et al., 2020).

Improving Quality Control: Swarm-inspired algorithms such as AIS and BFO enhance fault detection, diagnosis, quality control, and maintenance scheduling by mimicking self-organizing swarm behaviors (Altshuler et al., 2012).

Streamlining Automation: Swarm robotics in manufacturing environments automate tasks, boost productivity, and offer fault tolerance and scalability, vital for complex manufacturing systems (Connaughton et al., 2008).

Harnessing the collective intelligence and adaptive nature of swarms, these algorithms promise significant advancements in manufacturing efficiency.

Swarm Intelligence for Resource Allocation and Logistics in Supply Chain Management

Swarm intelligence presents a wealth of opportunities in supply chain management (Mataric, 1994; Manisterski et al., 2008; MacKenzie et al., 1997; Chalkiadakis & Boutilier, 2008; Zheng & Koenig, 2008).

Optimizing Routing and Scheduling: Swarm intelligence finds the most efficient routes and schedules for vehicles through algorithms like ACO and PSO (Mataric, 1994).

Inventory Management: Swarm intelligence optimizes stock levels, minimizing stockouts, reducing holding costs, and enhancing overall inventory management (Manisterski et al., 2008).

Demand Forecasting: Swarm algorithms provide accurate demand forecasts by analyzing historical data and adapting to market changes (MacKenzie et al., 1997).

Collaborative Decision-Making: Swarm-based algorithms facilitate communication and information sharing, aiding in coordinating activities, resolving conflicts, and making collective decisions for the entire supply chain (Chalkiadakis & Boutilier, 2008).

Adaptive and Resilient Supply Chains: Swarm algorithms enable real-time monitoring and adjustment, enhancing adaptability, and resilience in supply chain operations (Zheng & Koenig, 2008).

Real-World Deployments and Case Studies in Manufacturing and Supply Chain

Real-world deployments and case studies in manufacturing and supply chain management illustrate the practical applications of innovative technologies. Here are notable examples:

1. **Myerson, Paul. 2012:** Real-World Examples of Lean Supply Chain and Logistics Management. Chap. A. In: *Lean Supply Chain and Logistics Management.* 1st Edn. New York: McGraw-Hill Education. https://www. accessengineeringlibrary.com/content/book/9780071766265/back-matter/ appendix1.
2. **Predictive Analytics in Supply Chain:** N-iX provides case studies showcasing successful implementations of big data and predictive analytics in supply chain operations. https://www.n-ix.com/big-data-predictive-analytics-supply-chain-case-study/
3. **Diverse Supply Chain Case Studies:** SCM Globe features an online library exploring commercial, humanitarian, and military supply chains, including cases like Zara Clothing Company. https://www.scmglobe.com/online-guide/ case-studies/
4. **IoT and Manufacturing:** Machine Metrics presents IoT use cases in manufacturing, including remote monitoring, supply chain optimization, digital twins, and real-time machine monitoring. https://www.machinemetrics. com/blog/iot-in-manufacturing.
5. **Emerging Technologies:** MIT Sloan discusses technologies such as artificial intelligence, blockchain, robotics, 3D printing, and the Internet of Things that enhance supply chains and manufacturing processes. https://mitsloan.mit. edu/ideas-made-to-matter/5-supply-chain-technologies-deliver-competitive-advantage
6. **Industry 4.0 Solution:** *ScienceDirect* presents a case study on applying an Industry 4.0 waste-to-energy solution in a real-world supply chain, showcasing practical implementations. https://doi.org/10.1016/j.susoc.2022.01.008.

These examples demonstrate how cutting-edge technologies are reshaping manufacturing and supply chain management in practical, impactful ways.

Benefits and Impacts of Swarm Intelligence in Efficiency Optimization

Swarm intelligence provides a robust framework for efficiency optimization (Kraus et al., 2003; Work et al., 2008; Parker, 1998; Rehak et al., 2008; Mosafi et al., 2022).

Efficient Exploration: Swarm algorithms excel in exploring vast search spaces, facilitating exploration of promising regions and avoiding suboptimal areas (Kraus et al., 2003).

Optimal Solutions: Swarm algorithms converge to optimal or near-optimal solutions, ensuring adaptivity, and effective handling of dynamic or uncertain environments (Work et al., 2008).

Scalability: Swarm algorithms scale seamlessly to thousands or millions of agents, ideal for optimizing large-scale systems like traffic management or social networks (Parker, 1998).

Robustness and Fault-Tolerance

Decentralization grants swarm systems robustness, enhancing fault tolerance even if some agents fail (Rehak et al., 2008).

Adaptivity: Swarm algorithms dynamically adjust to changing environments, ensuring efficient optimization in dynamic or uncertain conditions (Mosafi et al., 2022).

Pioneering Precision Farming and Crop Monitoring in Smart Agriculture

SI presents groundbreaking opportunities in smart agriculture (Yamashita et al., 2000; Agogino & Tumer, 2008; Altshuler et al., 2019; Premvuti & Yuta, 1990). Deploying autonomous robots equipped with swarm intelligence in fields enables precise crop monitoring, collecting soil data, detecting diseases, and applying interventions (Yamashita et al., 2000). Efficient sensor networks created by SI gather real-time data on temperature, humidity, and soil conditions (Agogino & Tumer, 2008). Swarm algorithms analyze sensor data, unveiling patterns and anomalies, empowering farmers to optimize management practices (Altshuler et al., 2019).

Swarm Intelligence in Smart Agriculture

SI offers remarkable solutions for smart agriculture, including crop monitoring and precision farming (Bhatt et al., 2009; Arai et al., 1989; Fredslund & Mataric, 2001; Gordon et al., 2003; Bendjilali et al., 2009). Swarm-inspired algorithms optimize resource utilization, strategically positioning sensors, irrigation systems, and fertilizers (Bhatt et al., 2009). Swarm robotics replicates natural pollination processes, ensuring efficient crop pollination (Bhatt et al., 2009). Additionally, swarm robotics automates the harvesting process, reducing labor costs and enhancing productivity (Balch & Arkin, 1998).

Swarm Optimization Techniques for Precision Farming

Swarm optimization techniques can revolutionize precision farming, applying to crop monitoring, resource allocation, field mapping, crop pollination, and decision support systems (Wang, 1989; Pfingsthorn et al., 2008; Rekleitis et al., 2003; Harmatia & Skrzypczykb, 2009; Parker & Touzet, 2000). These techniques empower autonomous drones or robots with sensors to monitor crop health in real-time (Wang, 1989). Swarm optimization algorithms optimize resources like water, fertilizers, and pesticides, considering soil conditions, weather data, and crop needs (Pfingsthorn et al., 2008). Swarm systems create detailed field maps, capturing essential data for precision farming practices (Rekleitis et al., 2003).

Crop Monitoring and Resource Allocation using Swarm Intelligence

Employing SI in precision farming for crop monitoring and resource allocation can bring significant advancements (Wang, 1989; Pfingsthorn et al., 2008). Utilizing swarms of autonomous robots or drones enhances data collection on various crop parameters, promoting healthier crops (Wang, 1989). SI enables dynamic allocation of resources such as water, fertilizers, and pesticides, maximizing yield while minimizing waste (Pfingsthorn et al., 2008). Swarm systems are inherently decentralized, offering scalability, robustness, and adaptability in precision farming (Wang, 1989).

Real-World Examples and Success Stories in Smart Agriculture

The real-world applications of SI in smart agriculture are transforming the industry (Yamashita et al., 2000; Altshuler et al., 2005). Swarm robotics, equipped with sensors and cameras, are deployed for meticulous crop monitoring, increasing yields and minimizing resource wastage (Yamashita et al., 2000). Robotic bees and drones emulate natural pollinators, ensuring efficient crop pollination and addressing the pollination crisis (Altshuler et al., 2005). These examples showcase how swarm intelligence enhances crop management, precision agriculture, and pollination, revolutionizing agricultural productivity, sustainability, and resilience (Yamashita et al., 2000; Altshuler et al., 2005).

Revolutionizing Energy Grids and Smart Cities with Swarm Algorithms

Swarm algorithms have the transformative potential to revolutionize energy grids and smart cities in multifaceted ways. In the realm of energy grids, these algorithms, with their decentralized nature, can effectively manage distributed resources (Altshuler et al., 2008; Hollinger et al., 2009). By dynamically allocating resources, balancing loads, and optimizing energy distribution, they enhance energy efficiency, curbing wastage, and bolstering grid stability.

In the context of smart cities, swarm algorithms prove invaluable (LaValle et al., 1997). They optimize traffic flow by orchestrating vehicle movements, thereby reducing congestion and travel time (Efraim & Peleg, 2007). Additionally, they optimize resource allocation within smart buildings, adjusting heating, ventilation, and air conditioning systems based on occupancy and energy demand (Olfati-Saber, 2006).

Moreover, swarm algorithms contribute to the resilience of these systems (Bonabeau et al., 1999). Their collective intelligence enables adaptation to changing conditions, rapid detection and response to faults, and autonomous reconfiguration. This ensures uninterrupted energy supply and efficient resource utilization, marking a significant stride towards sustainable, self-healing urban environments.

Energy Grid Optimization through Swarm Intelligence

SI can be utilized to optimize energy grids by employing decentralized and self-organizing algorithms inspired by the collective behavior of swarms (Beni & Wang, 1991). These algorithms can help in efficiently managing distributed resources, improving the reliability and efficiency of energy grids.

One approach is to use swarm-based algorithms such as ACO or PSO to optimize the allocation and distribution of energy resources (Wagner & Bruckstein, 1994). These algorithms simulate the behavior of swarms to find optimal solutions for resource allocation, load balancing, and demand response in energy grids.

Another approach is to use swarm-inspired algorithms that extract specific mechanisms from natural swarms and incorporate them into optimization algorithms (Bruckstein, 1993). For example, the grey wolf optimizer (GWO) algorithm is inspired by the hunting behavior of grey wolves and can be used to optimize the operation and control of energy grids.

By leveraging SI, energy grids can benefit from the collective intelligence, adaptability, and scalability of swarms (Altshuler et al., 2011). These algorithms can help in optimizing energy generation, transmission, and consumption, leading to more efficient and sustainable energy grids.

Swarm Intelligence in Smart City Development and Infrastructure Management

SI plays a pivotal role in smart city development and infrastructure management through various applications (Altshuler et al., 2006):

Traffic Management: Swarm algorithms optimize traffic flow, coordinating autonomous vehicles to minimize congestion and travel time (Altshuler et al., 2006).

Energy Management: Coordinating energy-consuming devices like streetlights using SI reduces waste and enhances overall energy efficiency in smart cities (Pagello et al., 1999).

Waste Management: Swarm algorithms optimize waste collection by coordinating vehicles based on real-time data, reducing fuel consumption and improving system efficiency (Yanovski et al., 2003).

Infrastructure Maintenance: Swarms of autonomous robots or drones efficiently inspect and report infrastructure damages, enabling timely repairs, and preventive measures (Wagner & Bruckstein, 2000).

Emergency Response: Autonomous drones or robots, guided by SI, cover large areas during emergencies, gathering real-time data and aiding in search and rescue operations (Yanovski et al., 2001).

Case Study: Swarm Optimization in Energy Distribution

Swarm optimization in energy distribution, specifically in power grid systems, showcases the potential of algorithms like ACO and PSO (Adler & Gordon, 1992). These algorithms optimize power distribution by simulating interactions among various grid components, such as generators, transformers, and consumers.

In power grid systems, the challenge lies in balancing demand, minimizing losses, and ensuring stability. Swarm-based algorithms address these complexities by enabling agents to adapt based on local and global feedback (Gordon, 1995). For instance, in load balancing, swarm agents dynamically allocate power generation and distribution resources according to real-time demand. They exchange information to achieve balanced power distribution, minimizing overloads and losses.

Moreover, swarm algorithms enhance fault detection and optimal routing. Agents monitor component performance, detecting faults and rerouting power to ensure uninterrupted supply. By collectively optimizing power flow, swarm optimization improves the efficiency, reliability, and sustainability of electrical power distribution.

Challenges and Future Directions in the Application of Swarm Intelligence in Energy Grids and Smart Cities

In the realm of energy grids and smart cities, applying SI presents challenges and prompts future exploration:

Communication and Coordination: Efficient communication and coordination among swarm agents, like smart meters and sensors, are pivotal. Coordinating these agents in energy grids and smart cities involves optimizing energy distribution, managing demand-response programs, and ensuring grid efficiency (Rouff et al., 2003).

Scalability: As systems expand, SI algorithms must handle increasing complexities. Computational and communication overheads grow substantially with agent numbers. Developing streamlined algorithms for large-scale systems is essential.

Uncertainty and Complexity: Real-world environments are uncertain and dynamic. SI algorithms must adapt to change, handle uncertainties, and make robust decisions. Innovations are needed to cope with the inherent complexities of these environments.

Practical Application: While lab experiments are promising, real-world deployment lags. Collaboration with industry partners is vital. Addressing issues like disaster response, urban planning, and energy optimization requires practical, field-tested solutions.

Ethical and Social Considerations: Advanced swarm systems raise ethical concerns. Privacy, security, and control need careful handling. Transparent algorithms and collaboration with policymakers are key to establishing ethical regulations and standards (Lazer & Friedman, 2007).

A Profound Dive into Successful Swarm Intelligence Deployments

Some successful SI deployments include:

Multi-robot Coverage Problems: Swarm algorithms, such as parallel ant colony optimization (PACO), have been used to solve multi-robot coverage problems (Gordon, 1995; Rouff et al., 2005; Schultz & Parker, 2013).

Swarm-robotics Application Domains: Swarm robotics has been applied in various domains, such as precision agriculture and distributed systems. Large-scale decentralized systems of autonomous robotic agents have proven to be more effective than a single robot in many areas (Rouff et al., 2003; Schultz & Parker, 2013).

Collaborative Swarm of Autonomous Drones: SI has been utilized in the coordination of autonomous drones. The efficiency of a collaborative swarm of drones has been demonstrated in various works, including the analysis of stochastic frameworks and the establishment of analytic complexity bounds (Rouff et al., 2003; Schultz & Parker, 2013).

Analysis of Successful Swarm Intelligence Deployments across Industries

SI has demonstrated remarkable success in diverse industries, leading to innovative solutions.

Robotics: Autonomous robot swarms excel in exploration, surveillance, and search and rescue operations (Dudek et al., 1996). Their ability to collaborate and cover vast areas efficiently has revolutionized these tasks.

Optimization: SI algorithms like ACO and PSO efficiently explore vast search spaces, converging to optimal solutions (Gordon, 1995). Their ability to handle multiple objectives benefits fields like engineering, logistics, and finance, optimizing complex systems effectively.

Biology: Modeling social insects' behavior aids in route optimization, resource allocation, and task management (Lazer & Friedman, 2007). Understanding collective decision-making processes in ants and bees has inspired efficient algorithms.

Economics: Swarm-inspired algorithms enhance financial market analysis (Altshuler et al., 2012). They process extensive data, identify patterns, and predict trends, enhancing trading strategies' efficiency.

Social Sciences: SI explores social dynamics in online networks (Somin et al., 2022). Swarms of agents adapt to interactions, unraveling phenomena like information diffusion, opinion formation, and trend prediction.

Manufacturing and Supply Chain Management

SI has revolutionized manufacturing and supply chain management, providing solutions to various challenges. Here are some notable applications:

Production Scheduling: SI algorithms like ACO and PSO optimize production schedules (Gordon, 1995).

Inventory Management: SI algorithms optimize inventory levels and reorder points by mimicking ants or bees' foraging behavior (Gordon, 1995). This ensures optimal inventory levels, minimizing stockouts, and excess inventory.

Routing and Logistics: Algorithms like Bee Colony Optimization (BCO) optimize transportation routes (Gordon, 1995). This streamlines supply chain logistics, ensuring efficient delivery schedules.

Demand Forecasting: SI algorithms enhance demand forecasting accuracy (Gordon, 1995). This precision enables better production planning and inventory management.

Quality Control: SI is utilized in quality control processes (Gordon, 1995). Swarms of robots or sensors efficiently inspect products, identify defects, and take real-time corrective actions.

Smart Agriculture

In the realm of smart agriculture, SI has wrought transformative changes. Precision agriculture, in particular, has seen significant advancements through swarm robotics systems. These decentralized groups of robotic agents collaborate seamlessly, handling tasks from planting to harvesting crops (Gordon, 1995).

The benefits of employing swarm robotics in precision agriculture are multifold. Firstly, scalability is inherent, allowing multiple robots to collaboratively cover vast farmlands (Gordon, 1995). This collaboration significantly enhances the efficiency of tasks such as planting and harvesting, drastically reducing both time and labor requirements. Additionally, these systems exhibit adaptability, responding collectively to changing environmental conditions such as soil moisture variations or pest infestations (Gordon, 1995).

Moreover, SI provides robustness and fault tolerance to smart agriculture systems (Gordon, 1995). If one robot encounters an obstacle or fails, others compensate, ensuring uninterrupted operations. This redundancy enhances the reliability of agricultural processes.

In essence, the integration of SI in smart agriculture represents a paradigm shift, promising optimized farming practices, increased crop yields, and minimized resource wastage (Gordon, 1995). The collective intelligence and cooperation of robotic swarms are propelling precision agriculture toward higher efficiency, sustainability, and productivity.

Autonomous Vehicles and Drones

SI algorithms have been applied in the field of autonomous vehicles and drones to optimize traffic flow, coordinate movements, and avoid collisions. Here are a few examples:

Multi-robot Coverage: Swarm-based algorithms, such as PACO, have been used to solve the problem of multi-robot coverage (Gordon, 1995). In this scenario, a group of robots is required to explore an unknown environment while avoiding collisions with each other.

Traffic Management: Swarms of autonomous vehicles or drones can optimize traffic flow and reduce congestion by coordinating their movements and avoiding collisions. Efficient algorithms for path planning, decision making, and communication are required in these systems (Gordon, 1995). Swarm intelligence algorithms can help in achieving efficient traffic management in dynamic and uncertain environments.

These examples demonstrate how SI algorithms can be applied to enhance the capabilities of autonomous vehicles and drones, enabling them to work collaboratively, adapt to changing conditions, and optimize their behavior for improved performance (Gordon, 1995).

Energy Grids and Smart Cities

SI algorithms have found diverse applications in energy grids and smart cities, revolutionizing their efficiency and management.

Power Grid Optimization: Algorithms like ACO and PSO optimize power grids by balancing loads, detecting faults, and routing power flow (Gordon, 1995). This enhances grid efficiency and reliability.

Demand Response: Swarm algorithms optimize energy usage and reduce peak demand by coordinating the consumption of numerous users (Gordon, 1995). This leads to efficient energy management and substantial cost savings in smart grids.

Energy Trading: Swarm algorithms enable decentralized energy trading, promoting renewable energy integration and creating resilient energy systems through peer-to-peer transactions (Gordon, 1995).

Traffic Management: In smart cities, swarm algorithms optimize traffic flow and reduce congestion by coordinating vehicles based on local data (Gordon, 1995). This improves traffic efficiency, reduces travel time, and minimizes fuel consumption.

Waste Management: Swarm algorithms optimize waste collection routes by considering real-time data, minimizing travel distance, reducing fuel consumption, and enhancing overall efficiency in waste management systems (Gordon, 1995).

These examples showcase SI's adaptability and decentralized nature, making it ideal for solving intricate optimization and coordination challenges in energy grids and smart cities.

Learned and Best Practices in Swarm Intelligence Deployment

Understanding the intricacies of SI deployment is crucial. Some key best practices to consider when utilizing SI algorithms are:

Hybridization: Integrating SI algorithms with other optimization or machine learning techniques can enhance their performance, leveraging the strengths of different algorithms for more robust solutions (Gordon, 1995; Schultz & Parker, 2013).

Dynamic Parameter Adaptation: SI algorithms often rely on parameters. Adapting these parameters dynamically during optimization enables the algorithm to adjust to changing environments, enhancing its efficiency (Gordon, 1995).

Domain Knowledge Integration: Incorporating domain-specific knowledge empowers swarm algorithms to make informed decisions. Prior knowledge about the problem domain enhances the algorithm's effectiveness in real-world problem-solving (Gordon, 1995).

Convergence Analysis: Rigorous analysis of convergence properties, including behavior, rate, and criteria, provides insights into the algorithm's performance, guiding the optimization process effectively (Gordon, 1995).

Scalability Design: Ensuring scalability is essential, especially as the number of swarm agents increases. Designing algorithms to handle complexity ensures applicability to larger problem instances, maintaining efficiency (Gordon, 1995).

Communication Optimization: Efficient communication and coordination among swarm agents are fundamental. Optimizing communication mechanisms and synchronization overhead enhances overall algorithm performance (Gordon, 1995).

By adhering to these practices, the challenges and limitations of SI algorithms can be overcome, ensuring their effective application across diverse domains.

Impacts and Benefits of Swarm Intelligence in Various Industries

SI has far-reaching impacts across diverse industries, bringing forth several benefits. Here's an overview of its impact and advantages in various sectors:

Robotics: Swarm robotics harnesses collective behavior, enabling scalable and redundant multi-agent systems (Gordon, 1995). This approach finds applications in search and rescue, environmental monitoring, and precision agriculture, enhancing problem-solving and decision-making capabilities.

Optimization: SI algorithms like ACO and PSO efficiently explore vast search spaces, converging to optimal solutions (Gordon, 1995). Their ability to handle multiple objectives benefits fields like engineering, logistics, and finance, optimizing complex systems effectively.

Social Networks: SI adapts to social dynamics, analyzing behaviors, detecting patterns, and making predictions in social networks (Gordon, 1995). This understanding aids in comprehending complex social behaviors, providing insights into social phenomena.

Traffic Flow: Emulating natural swarm behaviors, SI algorithms optimize traffic flow (Gordon, 1995). By coordinating vehicle movements, these algorithms enhance traffic efficiency, mitigate congestion, and improve transportation systems overall.

Complex Problem Solving: Swarm systems pool collective intelligence to tackle intricate tasks (Gordon, 1995). Their ability to produce intelligent, adaptive behaviors benefits fields like engineering, biology, economics, and social sciences, solving challenges that are difficult for individual entities.

In essence, SI's scalability, adaptability, fault tolerance, and problem-solving capabilities make it a valuable asset across diverse industries.

Real-world Paradigms and Lessons in Innovation

Real-world applications of swarm intelligence algorithms offer valuable lessons in innovation. Here are some key paradigms and insights:

Robotics: Swarm algorithms enhance robotic coordination, enabling tasks like exploration and search and rescue. Redundancy boosts efficiency and reliability, showcasing the power of collaborative intelligence in robotics (Pedemonte et al., 2011).

Optimization: Algorithms like ACO and PCO solve complex problems in engineering and logistics. Mimicking swarm behavior, these algorithms efficiently navigate large search spaces, achieving optimal solutions (Felner et al., 2004; Lalwani et al., 2019).

Network Design: Swarm algorithms optimize communication networks by adapting to changing conditions, improving efficiency, and enhancing fault tolerance. Their ability to dynamically respond to network shifts proves invaluable (Gies & Rahmat-Samii, 2003; Chu et al., 2005).

Social Networks: Swarm algorithms adapt to social dynamics in networks. However, challenges like biases and echo chambers hinder exploring new ideas. Addressing these issues is essential for fostering diverse perspectives (Dorigo et al., 2021; Winfield et al., 2014).

Collaboration: Partnership with industry and government is vital for real-world SI applications. Collaboration fosters practical solutions, tackles societal challenges, and establishes necessary regulations and standards (Yang et al., 2018).

While these applications illustrate the potential of SI, ongoing research is essential. Overcoming challenges such as scalability and parameter settings will drive further innovation.

Companies and Industries Embracing Swarm Intelligence for Innovation

SI has spurred innovation across various industries. While specific articles were not provided, numerous research articles delve into this topic. Some notable ones include:

Emerging Technology and Business Model Innovation by J. Lee explores how AI disrupts industries, fostering innovative business models. (https://www.sciencedirect.com/science/article/pii/S2199853122009817).

Artificial Intelligence in Business: From Research and Innovation to Market Deployment investigates AI's impact on research, innovation, and market implementation. https://www.researchgate.net/publication/340699983_Artificial_Intelligence_in_Business_From_Research_and_Innovation_to_Market_Deployment

Artificial Intelligence in Innovation Research: A Systematic Study by M.M. Mariani delves into how organizations use AI for innovation. https://www.sciencedirect.com/science/article/pii/S0166497222001705.

Use of Artificial Intelligence in Terms of Open Innovation by A. Kuzior explores AI's role in open innovation processes and business model innovation. (https://doi.org/10.3390/su15097205)

Modeling Firm Search and Innovation Trajectory Using Swarm Intelligence by R.R. Chen presents a swarm intelligence-based model for studying firm search across innovation topics. https://doi.org/10.3390/a16020072.

Artificial Intelligence and Radical Innovation: An Opportunity by N. Grashof investigates how AI-related knowledge influences the emergence of radical innovations. https://link.springer.com/article/10.1007/s11187-022-00698-3.

These articles showcase the diverse applications of SI in driving innovation across sectors.

Challenges Faced and Lessons Learned in Swarm-Based Innovation

Swarm-based innovation presents exciting possibilities alongside notable challenges. Here's a summary of the key points:

Challenges

Real-world applications of SI algorithms offer valuable lessons in innovation. Here are some key paradigms and insights:

Robotics: Swarm algorithms enhance robotic coordination, enabling tasks like exploration and search and rescue. Redundancy boosts efficiency and reliability, showcasing the power of collaborative intelligence in robotics (Pedemonte et al., 2011).

Optimization: Algorithms like ACO and PSO solve complex problems in engineering and logistics. Mimicking swarm behavior, these algorithms efficiently navigate large search spaces, achieving optimal solutions (Felner et al., 2004; Lalwani et al., 2019).

Network Design: Swarm algorithms optimize communication networks by adapting to changing conditions, improving efficiency, and enhancing fault tolerance. Their ability to dynamically respond to network shifts proves invaluable (Gies & Rahmat-Samii, 2003; Chu et al., 2005).

Social Networks: Swarm algorithms adapt to social dynamics in networks. However, challenges like biases and echo chambers hinder exploring new ideas. Addressing these issues is essential for fostering diverse perspectives (Dorigo et al., 2021; Winfield et al., 2014).

Collaboration: Partnership with industry and government is vital for real-world SI applications. Collaboration fosters practical solutions, tackles societal challenges, and establishes necessary regulations and standards (Yang et al., 2018).

While these applications illustrate the potential of SI, ongoing research is essential. Overcoming challenges such as scalability and parameter settings will drive further innovation.

Lessons Learned

Simplicity: Simple individual agents often lead to efficient and robust collective behavior.

Communication and Coordination: Effective communication and coordination enhance swarm agents' problem-solving capabilities.

Adaptability: Swarms adapting to dynamic environments perform better in real-world applications.

Hybridization: Combining swarm algorithms with other techniques improves performance and overcomes limitations.

Domain Knowledge: Integrating domain-specific knowledge enhances swarm algorithms' efficiency and problem-solving capabilities.

These lessons underscore the need for ongoing research and innovation, addressing challenges while leveraging the strengths of swarm intelligence.

Real-World Examples and Case Studies Highlighting Innovation through Swarm Intelligence

These real-world examples vividly showcase the transformative power of SI:

Multi-robot Coverage: SI efficiently addresses multi-robot coverage in unknown terrains. Through communication and pheromone sharing, robot groups converge on solutions, even in intricate environments (Dorigo et al., 2021).

Traffic Optimization: Ant-inspired algorithms optimize urban traffic flow. Dynamic adjustments to signals and routes minimize congestion, enhancing overall traffic efficiency (Dorigo et al., 2021).

Disaster Response: Swarm robotics aids disaster response. Collaborative drones search for survivors, assess damage, and deliver aid in challenging areas. Their teamwork covers vast regions, providing effective assistance (Winfield et al., 2014).

Urban Planning: SI refines urban planning. Optimizing locations for facilities like bus stops and waste points, considering factors like population density, enhances urban infrastructure efficiency (Yang et al., 2018).

Social Networks: SI fosters exploration in social networks. Agents learning through interactions open avenues for diverse viewpoints. Addressing biases and echo chambers ensures a balanced exploration of ideas (Decker, 2008).

These cases illuminate SI's potential in tackling intricate real-world issues, driving innovation and societal advancement.

Promoting Adaptability and Resilience through Swarm Intelligence

SI is a beacon of adaptability and resilience in problem-solving. Its decentralized nature fosters information exchange, enabling efficient exploration of vast search spaces and converging on optimal solutions. Individual agents collaborate and adapt to their environment, ensuring scalability, robustness, and fault tolerance. Even if some agents falter, the swarm collectively achieves its goal, showcasing remarkable resilience.

Moreover, SI embodies self-organization and adaptability. Agents dynamically adjust based on shared information, allowing rapid responses to challenges and unexpected events. This adaptability grants the swarm resilience against changing conditions.

In essence, SI's collective intelligence and adaptive nature empower it to navigate complexities, making it a potent tool for addressing evolving real-world challenges.

Conclusion

In conclusion, our exploration of SI reveals its profound impact on problem-solving across diverse domains. Swarm intelligence, exhibited by groups like ants and birds, enables simple agents to tackle intricate tasks collectively. Emulating decentralized decision-making and adaptive responses seen in natural swarms, SI algorithms prove invaluable in solving complex issues such as optimization, control, and prediction.

SI algorithms, categorized into swarm-based and communication-enhanced approaches, find applications in engineering, biology, social sciences, and more. However, scalability poses a challenge in large-scale systems, demanding efficient coordination solutions. Future SI research avenues include refining algorithms, enhancing adaptability, addressing ethical concerns, and fostering collaborations with industry and government. This discussion illuminates the transformative potential of swarm intelligence, paving the way for innovative solutions to multifaceted real-world problems.

Summary of Research Findings

In summary, SI exemplifies the collective problem-solving abilities observed in natural organisms like ants and bees. Researchers have successfully applied SI algorithms to artificial intelligence and robotics, capitalizing on their scalability, robustness, and adaptability. Despite their potential, challenges such as premature convergence and scalability limitations persist.

SI finds diverse applications, particularly in robotics, where redundant teams enhance efficiency. Additionally, SI is valuable in optimization, network design, traffic management, and social networks. However, balancing exploration and exploitation, scalability issues, and the need for refined coordination algorithms pose challenges.

Future research should focus on sophisticated SI algorithms, enhancing adaptability, optimizing communication, and addressing real-world uncertainties. Collaboration with industries, ethical considerations, and applying SI to practical problems are pivotal areas for exploration. Overall, SI's study demonstrates nature's complexity, inspiring advancements across science and technology.

Practical Applications of Swarm Intelligence in Autonomous Systems

Here are some practical applications of swarm intelligence in autonomous systems:

1. **Robotics:** Swarm robotics involves multiple robots collaborating for tasks like exploration and surveillance. SI optimizes their behavior and decision-making.
2. **Optimization:** Algorithms like ACO and PSO solve logistics, scheduling, and resource allocation problems.
3. **Network Design:** SI optimizes communication network designs, aiding routing in computer networks and power distribution in smart grids.
4. **Social Networks:** Analyzing social dynamics in online networks, swarms learn and identify patterns and anomalies.
5. **Disaster Response:** Swarms coordinate autonomous systems like drones for search, rescue, and damage assessment in disasters.
6. **Urban Planning:** Optimizes traffic flow, energy use, and waste management, aiding smart city design and transportation systems.
7. **Environmental Monitoring:** Swarms of sensors monitor air quality, water pollution, and wildlife behavior, aiding resource management.
8. **Medical Applications:** Optimizes treatment plans, drug discovery, and disease diagnosis, analyzing vast medical datasets.
9. **Agriculture:** Enhances precision agriculture by monitoring fields, optimizing crop management, pest control, and irrigation.
10. **Industrial Automation:** Optimizes production, logistics, and quality control in manufacturing and warehouse environments, aiding automation.

Future Directions and Emerging Trends in Swarm Intelligence

Here are the future directions and emerging trends in SI:

1. **Development of Sophisticated Algorithms:** Researchers are focusing on enhancing swarm system efficiency and adaptability. Exploring improved communication and coordination among agents and addressing real-world complexity and uncertainty are key areas of development.

2. **Real-World Problem Applications:** Bridging the gap between lab demonstrations and practical implementation is vital. Collaborations with industries for field-ready solutions and partnerships with government agencies for addressing societal challenges like disaster response and urban planning are essential.

3. **Ethical and Social Considerations:** As swarm systems advance, concerns about privacy, security, and control arise. Researchers are working on transparent, accountable algorithms and collaborating with policymakers to establish regulations, ensuring ethical deployment.

The future of SI holds promise in addressing complex challenges.

Final Remarks and Contributions of the Chapter

This chapter provided an in-depth examination of SI, highlighting its theoretical foundations, practical applications, advantages, challenges, and future prospects. Some key contributions of this chapter include:

- Delving into the biological inspiration behind SI and explaining how it has been applied successfully in diverse domains through various algorithms that emulate collective behavior.

- Presenting a case study on optimizing enzyme production using swarm optimization techniques to demonstrate real-world industrial applications.

- Emphasizing SI's versatility in solving complex problems involving optimization, control, resource allocation, routing, and more through decentralized collective computations.

- Discussing practical implementations of SI across different autonomous systems and emerging technologies and how it enhances efficiency, robustness, and adaptability.

- Analyzing challenges in scaling swarm systems and achieving balanced exploration-exploitation while navigating dynamic conditions.

- Outlining future research directions like developing more sophisticated algorithms, addressing real-world uncertainties, and fostering cross-sector collaborations.

- Underscoring the need for transparent, accountable algorithms and ethical guidelines for responsible deployment of advanced swarm systems.

Overall, this chapter bridges the gap between SI theory and diverse applications, underscoring its transformative potential. It provides insights into this bio-inspired paradigm and its profound impacts on innovation, problem-solving, and progress across industries.

References

Adler, F. and Gordon, D. (1992). Information collection and spread by networks of patrolling agents. *Am. Nat., 140,* 373–400.

Agmon, N., Kraus, S. and Kaminka, G. (2008). Multi-robot perimeter patrol in adversarial settings. *In: Proceedings of the IEEE International Conference on Robotics and Automation (ICRA 2008),* Pasadena, CA, USA, 19–23 May 2008 2339–2345.

Agmon, N., Sadov, V., Kaminka, G. and Kraus, S. (2008). The impact of adversarial knowledge on adversarial planning in perimeter patrol. *In: Proceedings of the AAMAS '08: Proceedings of the 7th International Joint Conference on Autonomous Agents and Multi-agent Systems,* Estoril, Portugal, 12–16 May 2008, 1(May), 55–62.

Agogino, A. and Tumer, K. (2008). Regulating air traffic flow with coupled agents. *In: Proceedings of the AAMAS '08: Proceedings of the 7th International Joint Conference on Autonomous Agents and Multi-agent Systems,* Estoril, Portugal, 12–16 May; International Foundation for Autonomous Agents and Multi-agent Systems, Richland, SC, USA, 535–42.

Aknine, S. and Shehory, O.A. (2006). Feasible and Practical Coalition Formation Mechanism Leveraging Compromise and Task Relationships. *In: Proceedings of the IEEE/WIC/ACM International Conference on Intelligent Agent Technology,* Hong Kong, China, 18–22 December, 436–39.

Alami, R., Fleury, S., Herrb, M., Ingrand, F. and Robert, F. (1998). Multi-robot Cooperation in the Martha Project. *IEEE Robot. Autom. Mag., 5,* 36–47.

Altshuler, T., Altshuler, Y., Katoshevski, R. and Shiftan, Y. (2019). Modeling and Prediction of Ride-Sharing Utilization Dynamics. *J. Adv. Transp., 2019,* 6125798.

Altshuler, Y., Bruckstein, A. and Wagner, I. (2005). Swarm Robotics for a Dynamic Cleaning Problem. *In: Proceedings of the IEEE Swarm Intelligence Symposium,* Pasadena, CA, USA, 8–10 June, 209–16.

Altshuler, Y. and Bruckstein, A.M. (2011). Static and expanding grid coverage with ant robots: Complexity results. *Theor. Comput. Sci., 412,* 4661–74.

Altshuler, Y. and Bruckstein, A.M. (2010).The Complexity of Grid Coverage by Swarm Robotics. *In: Proceedings of the ANTS 2010,* LNCS, Nancy, France, 19–23 July, 536–43.

Altshuler, Y. Dolev, S., Elovici, Y. and Aharony, N. (2010). TTLed Random Walks for Collaborative Monitoring. *In: Proceedings of the NetSciCom 2010 (Second International Workshop on Network Science for Communication Networks),* San Diego, CA, USA, 19 March.

Altshuler, Y., Elovici, Y., Cremers, A.B., Aharony, N. and Pentland, A. (2012). *Security and Privacy in Social Networks.* Berlin/Heidelberg, Germany: Springer Science & Business Media.

Altshuler, Y., Fire, M., Aharony, N., Elovici, Y. and Pentland, A. (2012). How Many Makes a Crowd? On the Correlation between Groups' Size and the Accuracy of Modeling. *In: Proceedings of the International Conference on Social Computing, Behavioral-Cultural Modeling and Prediction,* College Park, MD, USA, 3–5 April. Berlin/Heidelberg, Germany: Springer 43–52.

Altshuler, Y., Fire, M., Shmueli, E., Elovici, Y., Bruckstein, A., Pentland, A.S. and Lazer, D. (2012). The Social Amplifier: Reaction of Human Communities to Emergencies. *J. Stat. Phys., 152,* 399–18.

Altshuler, Y., Pentland, A., Bekhor, S., Shiftan, Y. and Bruckstein, A. (2016). Optimal Dynamic Coverage Infrastructure for Large-Scale Fleets of Reconnaissance UAVs. *arXiv. arXiv:1611.05735.*

Altshuler, Y., Pentland, A. and Bruckstein, A.M. (2018). *Swarms and Network Intelligence in Search.* Berlin/Heidelberg, Germany: Springer.

Altshuler, Y., Pentland, A.S. and Gordon, G. (2015). Social Behavior Bias and Knowledge Management Optimization. *In: Social Computing, Behavioral-Cultural Modeling, and Prediction.* Berlin/Heidelberg, Germany: Springer, 258–63.

Altshuler, Y., Puzis, R., Elovici, Y., Bekhor, S. and Pentland, A.S. (2016). On the Rationality and Optimality of Transportation Networks Defense: A Network Centrality Approach. *In: Securing Transportation Systems.* Hoboken, NJ, USA: John Wiley & Sons: Inc., 35–63.

Altshuler, Y., Shmueli, E., Zyskind, G., Lederman, O., Oliver, N. and Pentland, A. (2014). Campaign Optimization through Behavioral Modeling and Mobile Network Analysis. *IEEE Trans. Comput. Soc. Syst.*, *1*, 121–34.

Altshuler, Y., Shmueli, E., Zyskind, G., Lederman, O., Oliver, N. and Pentland, A. (2015). Campaign Optimization through Mobility Network Analysis. *In: Geo-Intelligence and Visualization through Big Data Trends.* Hershey, PA, USA: IGI Global, 33–74.

Altshuler, Y., Wagner, I. and Bruckstein, A. Collaborative Exploration in Grid Domains. (2009). *In: Proceedings of the Sixth International Conference on Informatics in Control, Automation, and Robotics (ICINCO)*, Milan, Italy, 2–5 July.

Altshuler, Y., Wagner, I. and Bruckstein, A. (2008). On Swarm Optimality. *In: Dynamic and Symmetric Environments. Economics*, *7*, 11.

Altshuler, Y., Wagner, I. and Bruckstein, A. (2006). Shape Factor's Effect on a Dynamic Cleaners Swarm. *In: Proceedings of the Third International Conference on Informatics in Control, Automation, and Robotics (ICINCO), the Second International Workshop on Multi-agent Robotic Systems (MARS)*, Setúbal, Portugal, 1-5 August, 13–21.

Altshuler, Y., Wagner, I., Yanovski, V. and Bruckstein, A. (2010). Multi-agent Cooperative Cleaning of Expanding Domains. *Int. J. Robot. Res.*, *30*, 1037–71.

Altshuler, Y., Yanovski, V., Wagner, I. and Bruckstein, A. (2005). The Cooperative Hunters—Efficient Cooperative Search for Smart Targets Using UAV Swarms. *In: Proceedings of the Second International Conference on Informatics in Control, Automation, and Robotics (ICINCO), the First International Workshop on Multi-agent Robotic Systems (MARS)*, Barcelona, Spain, 14–17 September, 165–70.

Altshuler, Y., Yanovsky, V. and Bruckstein, A., Wagner, I. (2008). Efficient Cooperative Search of Smart Targets Using UAV Swarms. *Robotica*, *26*, 551–57.

Altshuler, Y., Yanovsky, V., Wagner, I. and Bruckstein, A. (2006). Swarm intelligence—Searchers, cleaners, and hunters. *In: Swarm Intelligent Systems.* Berlin/Heidelberg, Germany: Springer: 93–132.

Arai, T., Ogata, H. and Suzuki, T. (1989). Collision Avoidance Among Multiple Robots Using Virtual Impedance. *In: Proceedings of the IEEE/RSJ International Conference on Intelligent Robots and Systems*, Tsukuba, Japan, 4–6 September, 479–85.

Arkin, R. and Balch, T. (1997). AuRA: Principles and Practice in Review. *J. Exp. Theor. Artif. Intell.*, *9*, 175–88.

Balch, T. and Arkin, R. (1998). Behavior-based Formation Control for Multi-Robot Teams. *IEEE Trans. Robot. Autom.*, *14*, 926–39.

Benda, M., Jagannathan, V. and Dodhiawalla, R. (1985). *On Optimal Cooperation of Knowledge Sources; Technical Report BCS-G2010-28.* Arlington, VA, USA: Boeing AI Center.

Bendjilali, K., Belkhouche, F. and Belkhouche, B. (2009). Robot formation modelling and control based on the relative kinematics equations. *Int. J. Robot. Autom.*, *24*, 79–88.

Beni, G. and Wang, J. (2004). Swarm Intelligence in Cellular Robotic Systems. *Proc. IEEE*, *92*, 1227–41.

Beni, G. and Wang, J. (1991). Theoretical Problems for the Realization of Distributed Robotic Systems. *In: Proceedings of the IEEE Internal Conference on Robotics and Automation*, Sacramento, CA, USA, 9–11 April, 1914–20.

Bhatt, R., Tang, C. and Krovi, V. (2009). Formation optimization for a fleet of wheeled mobile robots: A geometric approach. *Robot. Auton. Syst.*, *57*, 102–20.

Bonabeau, E., Dorigo, M. and Theraulaz, G. (1999). *Swarm Intelligence: From Natural to Artificial Systems.* New York, NY, USA: Oxford University Press.

Brambilla, M., Ferrante, E., Birattari, M. and Dorigo, M. (2013). Swarm robotics: A review from the swarm engineering perspective. *Swarm Intell.*, *7*, 1–41.

Brooks, R. (1986). A Robust Layered Control System for a Mobile Robot. *IEEE J. Robot. Autom.*, *RA-2*, 14–23.

Brooks, R. and Flynn, A. (1989). Fast, Cheap and Out of Control, a Robot Invasion of the Solar System. *J. Br. Interplanet. Soc.*, *42*, 478–85.

Bruckstein, A. (1989). Why the Ant Trails Look So Straight and Nice. *Math. Intell.*, *15*, 59–62.

Bruckstein, A., Mallows, C. and Wagner, I. (1997). Probabilistic Pursuits on the Integer Grid. *Am. Math. Mon.*, *104*, 323–43.

Candea, C., Hu, H., Iocchi, L., Nardi, D. and Piaggio, M. (2001). Coordinating in Multi-agent RoboCup Teams. *Robot. Auton. Syst., 36*, 67–86.

Chalkiadakis, G. and Boutilier, C. (2008). Sequential decision-making in repeated coalition formation under uncertainty. *In: Proceedings of the 7th International Joint Conference on Autonomous Agents and Multiagent Systems,* Estoril, Portugal, 12–16 May, 347–54.

Chalkiadakis, G., Markakis, E. and Boutilier, C. (2007). Coalition formation under uncertainty: Bargaining equilibria and the Bayesian core stability concept. *In: Proceedings of the AAMAS '07: Proceedings of the 6th International Joint Conference on Autonomous Agents and Multi-agent Systems,* Honolulu, HI, USA, 14–18 May, ACM: New York, NY, USA, 1–8.

Chevallier, D. and Payandeh, S. (2000). On Kinematic Geometry of Multi-Agent Manipulating System Based on the Contact Force Information. *In: Proceedings of the Sixth International Conference on Intelligent Autonomous Systems (IAS-6),* Venice, Italy, 25–27 July, 88–195.

Chu, S.C., Roddick, J.F. and Pan, J.S. (2005). A parallel particle swarm optimization algorithm with communication strategies. *J. Inf. Sci. Eng., 21*, 809–18.

Chu, S.C. Tsai, P.W. and Pan, J.S. (2006). Cat swarm optimization. *In: Proceedings of the PRICAI 2006: Trends in Artificial Intelligence: 9th Pacific Rim International Conference on Artificial Intelligence,* Guilin, China, 7–11 August;Berlin/Heidelberg, Germany: Springer, *Proceedings, 9*, 854–58.

Connaughton, R., Schermerhorn, P. and Scheutz, M. (2008). Physical parameter optimization in swarms of ultra-low complexity agents. *In: Proceedings of the AAMAS '08: Proceedings of the 7th International joint conference on autonomous agents and multiagent systems.* Estoril, Portugal, 12–16 May: International Foundation for Autonomous Agents and Multiagent Systems, Richland, SC, USA, 1631–34.

Dasgupta, D. and González, F.A. (2002). Immunity-Based Systems: A Survey. *IEEE Trans. Evol. Comput., 6*, 252–67.

Decker, M. (2008). Caregiving robots and ethical reflection: The perspective of interdisciplinary technology assessment. *AI Soc., 22*, 315–30.

DeLoach, S. and Kumar, M. (2008). Chapter Multi-Agent Systems Engineering: An Overview and Case Study. *In: Intelligence Integration in Distributed Knowledge Management.* Hershey, PA, USA: Idea Group Inc. (IGI), 207–24.

Deneubourg, J., Goss, S., Sandini, G., Ferrari, F. and Dario, P. (1990). Self-Organizing Collection and Transport of Objects in Unpredictable Environments. *In: Proceedings of the Japan-U.S.A. Symposium on Flexible Automation,* Kyoto, Japan, 9–11 July, 1093–98.

Dias, M. and Stentz, A. (2001). *A Market Approach to Multirobot Coordination; Technical Report, CMU-RI - TR-01-26.* Robotics Institute, Carnegie Mellon University: Pittsburgh, PA, USA.

Dorigo, M., Birattari, M., Blum, C., Clerc, M., and Stützle, T. (2008). Swarm Intelligence. *Nature, 406*, 39–42.

Dorigo, M. and Gambardella, L.M. (1996). Ant Colony System: A Cooperative Learning Approach to the Traveling Salesman Problem. *IEEE Trans. Evol. Comput., 1*, 53–66.

Dorigo, M., Theraulaz, G. and Trianni, V. (2020). Reflections on the future of swarm robotics. *Sci. Robot., 5*, eabe4385. [PubMed]

Dorigo, M., Theraulaz, G. and Trianni, V. (2021). Swarm robotics: Past, present, and future [point of view]. *Proc. IEEE, 109*, 1152–65.

Drogoul, A. and Ferber, J. (1992). From Tom Thumb to the Dockers: Some Experiments with Foraging Robots. *In: Proceedings of the Second International Conference on Simulation of Adaptive Behavior, International Conference on Simulation of Adaptive Behavior,* Honolulu, HI, USA, 7–11 December, 451–59. [Google Scholar]

Dudek, G., Jenkin, M., Milios, E. and Wilkes, D. (1996). A taxonomy for multi-agent robotics. *Auton. Robot. J., 3*, 375–97.

Efraim, A. and Peleg, D. (2007). Distributed Algorithms for Partitioning a Swarm of Autonomous Mobile Robots. *Struct. Inf. Commun. Complexity, Lect. Notes Comput. Sci., 4474*, 180–94.

Felner, A., Shoshani, Y., Altshuler, Y. and Bruckstein A.M. (2006). Multi-agent Physical A* with Large Pheromones. *J. Auton. Agents Multi-Agent Syst., 12*, 3–34.

Felner, A., Stern, R., Ben-Yair, A., Kraus, S. and Netanyahu, N. (2004). PHA*: Finding the Shortest Path with A* in Unknown Physical Environments. *J. Artif. Intell. Res., 21*, 631–79.

Ferrari, C., Pagello, E., Ota, J. and Arai, T. (1998). Multi-Robot Motion Coordination in Space and Time. *Robot. Auton. Syst.*, *25*, 219–29.

Fredslund, J. and Mataric, M. (2001). Robot Formations Using Only Local Sensing and Control. *In: Proceedings of the International Symposium on Computational Intelligence in Robotics and Automation (IEEE CIRA 2001)*, Banff, AB, Canada, 29 July–1 August, 308–13.

Gerkey, B. and Mataric, M. (2002). Sold! Market Methods for Multi-Robot Control. *IEEE Trans. Robot. Autom. Spec. Issue Multi-Robot. Syst.*, *18*, 758–68.

Gies, D. and Rahmat-Samii, Y. (2003). Reconfigurable array design using parallel particle swarm optimization. *In: Proceedings of the IEEE Antennas and Propagation Society International Symposium. Digest.* Held in conjunction with: USNC/CNC/URSI North American Radio Sci. Meeting (Cat. No. 03CH37450), Columbus, OH, USA, 22–27 June; *1*, 177–80.

Gordon, D. (1995). The expandable network of ant exploration. *Anim. Behav.*, *50*, 372–78.

Gordon, N., Wagner, I. and Bruckstein, A. (2003). Discrete Bee Dance Algorithms for Pattern Formation on a Grid. *In: Proceedings of the IEEE International Conference on Intelligent Agent Technology (IAT03)*, Halifax, NS, Canada, 13–17 October, 545–49.

Hagelbäck, J. and Johansson, S. (2008). Demonstration of multi-agent potential fields in real-time strategy games. *In: Proceedings of the AAMAS '08: Proceedings of the 7th International Joint Conference on Autonomous Agents and Multi-agent Systems*, Estoril, Portugal, 12–16 May; International Foundation for Autonomous Agents and Multi-agent Systems: Richland, SC, USA, 1687–88.

Harmatia, I. and Skrzypczykb, K. (2009). Robot team coordination for target tracking using fuzzy logic controller in game theoretic framework. *Robot. Auton. Syst.*, *57*, 75–86.

Haynes, T. and Sen, S. (1986). Evolving Behavioral Strategies in Predators and Prey. *In: Adaptation and Learning in Multi-Agent Systems; Lecture Notes in Computer Science*. Berlin, Germany: Springer, *1042*, 113–26.

Hettiarachchi, S. and Spears, W. (2005). Moving swarm formations through obstacle fields. *In: Proceedings of the International Conference on Artificial Intelligence*, Las Vegas, NV, USA, 27–30 June.

Hinchey, M.G., Sterritt, R. and Rouff, C. (2007). Swarms and swarm intelligence. *Computer*, *40*, 111–113.

Hollinger, G., Singh, S., Djugash, J., and Kehagias, A. (2009). Efficient Multi-Robot Search for a Moving Target. *Int. J. Robot. Res.*, *28*, 201–19.

Karaboga, D. and Basturk, B. (2007). A Powerful and Efficient Algorithm for Numerical Function Optimization: Artificial Bee Colony (ABC) Algorithm. *J. Glob. Optim.*, *39*, 459–71.

Kennedy, J. and Eberhart, R. (1995). Particle Swarm Optimization. *In: Proc. IEEE Int. Conf. Neural Netw.*, *4*, 1942–48.

Kirkpatrick, S. and Schneider, J. (2005). How smart does an agent need to be? *Int. J. Mod. Phys., C 16*, 139–55.

Klos, T. and van Ahee, G. (2008). Evolutionary dynamics for designing multi-period auctions. *In: Proceedings of the AAMAS '08: Proceedings of the 7th International Joint Conference on Autonomous Agents and Multi-agent Systems*, Estoril, Portugal, 12–16 May; International Foundation for Autonomous Agents and Multiagent Systems: Richland, SC, USA, 1589–92.

Krafft, P.M., Zheng, J.I, Pan, W., Della Penna, N., Altshuler, Y., Shmueli, E., Tenenbaum, J.B. and Pentland, A. (2016). Human collective intelligence as distributed Bayesian inference. *arXiv.*, arXiv:1608.01987.

Kraus, S., Shehory, O. and Taase, G. (2003). Coalition formation with uncertain heterogeneous information. *In: Proceedings of the Second International Joint Conference on Autonomous Agents and Multi-agent Systems*, Melbourne, Australia, 14–18 July, 1–8.

Lalwani, S., Sharma, H., Satapathy, S.C., Deep, K. and Bansal, J.C. (2019). A survey on parallel particle swarm optimization algorithms. *Arab. J. Sci. Eng.*, *44*, 2899–23.

LaValle, S., Lin, D., Guibas, L., Latombe, J. and Motwani, R. (1997). Finding an Unpredictable Target in a Workspace with Obstacles. *In: Proceedings of the 1997 IEEE International Conference on Robotics and Automation (ICRA-97)*, Albuquerque, NM, USA, 25 April, 737–42.

Lazer, D. and Friedman, A. (2007). The network structure of exploration and exploitation. *Adm. Sci. Q.*, *52*, 667–94.

Liu, Y.Y., Nacher, J.C., Ochiai, T., Martino, M. and Altshuler, Y. (2014). Prospect Theory for Online Financial Trading. *PLoS ONE*, *9*, e109458.

Lumelsky, V. and Harinarayan, K. (1997). Decentralized Motion Planning for Multiple Mobile Robots: The Cocktail Party Model. *Auton. Robot., 4*, 121–36.

MacKenzie, D., Arkin, R. and Cameron, J. (1997). Multi-agent Mission Specification and Execution. *Auton. Robot., 4*, 29–52.

Manisterski, E., Lin, R. and Kraus, S. (2008). Understanding how people design trading agents over time. *In: Proceedings of the AAMAS '08: Proceedings of the 7th International Joint Conference on Autonomous Agents and Multi-agent Systems*, Estoril, Portugal, 12–16 May; International Foundation for Autonomous Agents and Multiagent Systems: Richland, SC, USA, 1593–96.

Mastellone, S., Stipanovi, D., Graunke, C., Intlekofer, K. and Spong, M. (2008). Formation Control and Collision Avoidance for Multi-agent Non-holonomic Systems: Theory and Experiments. *Int. J. Robot. Res., 27*, 107–26.

Mataric, M. (1994). *Interaction and Intelligent Behavior*. Ph.D. Thesis, Massachusetts Institute of Technology, Cambridge, MA, USA.

Michael, N., Zavlanos, M., Kumar, V. and Pappas, G. (2008). Distributed multi-robot task assignment and formation control. *In: Proceedings of the IEEE International Conference on Robotics and Automation, 2008 (ICRA 2008)*, Pasadena, CA, USA, 19–23 May, 128–33.

Mirjalili, S., Mirjalili, S.M. and Lewis, A. (2014). Grey wolf optimizer. *Adv. Eng. Softw., 69*, 46–61.

Mosafi, I., David, E., Altshuler, Y. and Netanyahu, N.S. (2022). DNN Intellectual Property Extraction Using Composite Data. *Entropy, 24*, 349. [PubMed]

Olfati-Saber, R. (2006). Flocking for Multi-agent Dynamic Systems: Algorithms and Theory. *IEEE Trans. Autom. Control, 51*, 401–20.

Pagello, E., D'Angelo, A., Ferrari, C., Polesel, R., Rosati, R. and Speranzon, A. (2002). Emergent Behaviors of a Robot Team Performing Cooperative Tasks. *Adv. Robot., 17*, 3–19.

Pagello, E., D'Angelo, A., Montesello, F., Garelli, F. and Ferrari, C. (1999). Cooperative Behaviors in Multi-Robot Systems through Implicit Communication. *Robot. Auton. Syst., 29*, 65–77.

Pan, W., Altshuler, Y. and Pentland, A. (2012). Decoding social influence and the wisdom of the crowd in financial trading network. *In: Proceedings of the IEEE 2012 International Conference on Privacy, Security, Risk and Trust and 2012 International Conference on Social Computing (SocialCom)*, Amsterdam, The Netherlands, 3–5 September, 203–09.

Parker, L. (1998). ALLIANCE: An Architecture for Fault-Tolerant Multi-Robot Cooperation. IEEE Trans. Robot. Autom., *14*, 220–40.

Parker, L. and Touzet, C. (2000). Multi-Robot Learning in a Cooperative Observation Task. *Distrib. Auton. Robot. Syst., 4*, 391–01.

Passino, K.M. (2002). Biomimicry of Bacterial Foraging for Distributed Optimization and Control. *IEEE Control Syst. Mag., 22*, 52–67.

Pedemonte, M., Nesmachnow, S. and Cancela, H. (2011). A survey on parallel ant colony optimization. *Appl. Soft Comput., 11*, 5181–97.

Pfingsthorn, M., Slamet, B. and Visser, A. (2008). A Scalable Hybrid Multi-robot SLAM Method for Highly Detailed Maps. *In: RoboCup 2007: Robot Soccer World Cup XI; Lecture Notes in Computer Science*. Berlin, Germany: Springer, 5001, 457–464.

Premvuti, S. and Yuta, S. (1990). Consideration on the Cooperation of Multiple Autonomous Mobile Robots. *In: Proceedings of the IEEE International Workshop of Intelligent Robots and Systems*, Ibaraki, Japan, 3–6 July, 59–63.

Puzis, R., Altshuler, Y., Elovici, Y., Bekhor, S., Shiftan, Y. and Pentland, A. (2013). Augmented Betweenness Centrality for Environmentally Aware Traffic Monitoring in Transportation Networks. *J. Intell. Transp. Syst., 17*, 91–105.

Regev, E., Altshuler, Y. and Bruckstein, A.M. (2012). The cooperative cleaners' problem in stochastic dynamic environments. *arXiv. arXiv:*1201.6322.

Rehak, M., Pechoucek, M., Celeda, P., Krmicek, V., Grill, M. and Bartos, K. (2008). Multi-agent approach to network intrusion detection. *In: Proceedings of the AAMAS '08: Proceedings of the 7th International Joint Conference on Autonomous Agents and Multi-agent Systems*, Estoril, Portugal, 12–16 May. International Foundation for Autonomous Agents and Multi-agent Systems: Richland, SC, USA, 1593–96.

Rekleitis, I. Dudek, G. and Milios, E. (2003). Experiments in Free-Space Triangulation Using Cooperative Localization. *In*: *Proceedings of the IEEE/RSJ/GI International Conference on Intelligent Robots and Systems* (*IROS*), Las Vegas, NV, USA, 27–31 October.

Ren, W. and Sorensena, N. (2008). Distributed coordination architecture for multi-robot formation control. *Robot. Auton. Syst.*, *56*, 324–33.

Rouff, C., Truszkowski, W., Rash, J. and Hinchey, M. (2003). Formal approaches to intelligent swarms. *In*: *Proceedings of the 28th Annual NASA Goddard Software Engineering Workshop*, Greenbelt, MD, USA, 3–4 December, 51–57.

Rouff, C.A., Truszkowski, W.F., Rash, J. and Hinchey, M. (2005). *A Survey of Formal Methods for Intelligent Swarms*. Greenbelt, MD, USA: NASA Goddard Space Flight Center.

Sariel, S. and Balch, T. (2005). Real-time auction based allocation of tasks for multi-robot exploration problem in dynamic environments. *In*: *Proceedings of the AAAI-05 Workshop on Integrating Planning into Scheduling*, Pittsburgh, PA, USA, 9–10 July, 27–33.

Sasidhar, B. and Ravindran, S. (2023). Evolutionary and Swarm Intelligence in Optimization of α-Amylase From Bacillus velezensis Sp. Research square: posted 18 Aug, 2023. https://doi.org/10.21203/rs.3.rs-3270983/v1.

Savarimuthu, S., Purvis, M. and Purvis, M. (2009). *Tag Based Model for Knowledge Sharing in Agent Society*; Discussion Paper 2009/01; Department of Information Science, University of Otago: Dunedin, New Zealand.

Sawhney, R., Krishna, K., Srinathan, K. and Mohan, M. (2008). On reduced time fault tolerant paths for multiple UAVs covering a hostile terrain. *In*: *Proceedings of the AAMAS '08: Proceedings of the 7th International Joint Conference on Autonomous Agents and Multi-agent Systems*, Estoril, Portugal, 12–16 May; International Foundation for Autonomous Agents and Multi-agent Systems: Richland, SC, USA, 1171–74.

Schultz, A.C. and Parker, L.E. (2008). Multi-Robot Systems: From Swarms to Intelligent Automata. *In*: *Proceedings from the 2002 NRL Workshop on Multi-Robot Systems*. Berlin/Heidelberg, Germany: Springer Science & Business Media.

Simon, H. (1981). *The Sciences of the Artificial*, 2nd Edn. Cambridge, MA, USA: MIT Press.

Somin, S., Altshuler, Y., 'Sandy' Pentland, A. and Shmueli, E. (2022). Beyond preferential attachment: Falling of stars and survival of superstars. *R. Soc. Open Sci.*, *9*, 220899. [PubMed]

Somin, S., Altshuler, Y., Gordon, G., Pentland, A. and Shmueli, E. (2020). Network Dynamics of a financial ecosystem. *Sci. Rep.*, *10*, 1–10.

Somin, S., Altshuler, Y. and Shmueli, E. (2022). Remaining popular: Power-law regularities in network dynamics. *EPJ Data Sci.*, *11*, 61.

Somin, S., Gordon, G. and Altshuler, Y. (2018). Social Signals in the Ethereum Trading Network. *arXiv. arXiv*:1805.12097.

Stone, P. and Veloso, M. (1999). Task Decomposition, Dynamic Role Assignment, and Low-Bandwidth Communication for Real-time Strategic Teamwork. *Artif. Intell.*, *110*, 241–73.

Su, H., Wang, X. and Lin, Z. (2009). Flocking of Multi-agents with a Virtual Leader. *IEEE Trans. Autom. Control*, *54*, 293–07.

Svestka, P. and Overmars, M. (1998). Coordinated Path Planning for Multiple Robots. *Robot. Auton. Syst.*, *23*, 125–52.

Systems, Estoril, Portugal, 12–16 May 2008; International Foundation for Autonomous Agents and Multi-agent Systems: Richland, SC, USA, 1695–96.

Systems, Estoril, Portugal, 12–16 May 2008; International Foundation for Autonomous Agents and Multi-agent Systems: Richland, SC, USA, 55–62.

Trajkovski, G. and Collins, S. (2009). *Handbook of Research on Agent-based Societies: Social and Cultural Interactions*. Hershey, PA, USA: Idea Group Inc. (IGI).

Visser, U., Ribeiro, F., Ohashi, T. and Dellaert, F. (2009). *RoboCup 2007: Robot Soccer World Cup XI; Lecture Notes in Computer Science*. Berlin, Germany: Springer, *5001*.

Wagner, I., Altshuler, Y., Yanovski, V. and Bruckstein, A. (2008). Cooperative Cleaners: A Study in Ant Robotics. *Int. J. Robot. Res.* (*IJRR*), *27*, 127–51.

Wagner, I. and Bruckstein, A. (2000). ANTS: Agents, networks, trees, and subgraphs. *Future Gener. Comput. Syst. J.*, *16*, 915–26.

Wagner, I. and Bruckstein, A. (2001). From Ants to A(ge)nts: A Special Issue on Ant—Robotics. *Ann. Math. Artif. Intell. Spec. Issue Ant Robot, 31*, 1–6.

Wagner, I. and Bruckstein, A. (1994). *Row Straightening via Local Interactions*; Technical report CIS-9406. Center for Intelligent Systems, Haifa, Israel: Technion.

Wagner, I., Lindenbaum, M. and Bruckstein, A. (1998). Efficiently Searching a Graph by a Smell-Oriented Vertex Process. *Ann. Math. Artif. Intell., 24*, 211–23.

Wang, P. (1989). Navigation Strategies for Multiple Autonomous Mobile Robots. *In: Proceedings of the IEEE/RSJ International Conference on Intelligent Robots and Systems* (*IROS*), Ysukuba, Japan, 4–6 September, 486–93.

Weitzenfeld, A. (2008). A Prey Catching and Predator Avoidance Neural-Schema Architecture for Single and Multiple Robots. *J. Intell. Robot. Syst., 51*, 203–33.

Wellman, M. and Wurman, P. (1998). Market-Aware Agents for a Multiagent World. *Robot. Auton. Syst., 24*, 115–25.

Winfield, A.F., Blum, C. and Liu, W. (2014). Towards an ethical robot: Internal models, consequences and ethical action selection. *In: Proceedings of the Advances in Autonomous Robotics Systems: 15th Annual Conference*, TAROS Birmingham, UK, 1–3 September; Berlin/Heidelberg, Germany: Springer, *Proceedings, 15*, 85–96.

Work, H., Chown, E., Hermans, T. and Butterfield, J. (2008). Robust team-play in highly uncertain environments. *In: Proceedings of the AAMAS '08: Proceedings of the 7th International Joint Conference on Autonomous Agents and Multi-agent Systems*, Estoril, Portugal, 12–16 May; International Foundation for Autonomous Agents and Multi-agent Systems: Richland, SC, USA, 1199–02.

Yamashita, A., Fukuchi, M., Ota, J., Arai, T. and Asama, H. (2000). Motion Planning for Cooperative Transportation of a Large Object by Multiple Mobile Robots in a 3D Environment. *In: Proceedings of the IEEE International Conference on Robotics and Automation*, San Francisco, CA, USA, 24–28 April, 3144–51.

Yang, G.Z., Bellingham, J., Dupont, P.E., Fischer, P., Floridi, L., Full, R., Jacobstein, N., Kumar, V., McNutt, M., Merrifield, R., et al. (2018). The grand challenges of science robotics. *Sci. Robot., 3*, eaar7650.

Yang, X.S. (2000). Firefly Algorithm, Stochastic Test Functions, and Design Optimization. *Int. J. -Bio-Inspired Comput., 2*, 78–84.

Yang, X.S., Cui, Z., Xiao, R., Gandomi, A.H. and Karamanoglu, M. (2013). *Swarm Intelligence and Bio-Inspired Computation: Theory and Applications*. Amsterdam, The Netherlands: Newnes, Elsevier.

Yanovski, V., Wagner, I. and Bruckstein, A. (2000). A distributed ant algorithm for efficiently patrolling a network. *Algorithmica, 37*, 165–86.

Yanovski, V., Wagner, I. and Bruckstein, A. (2001). Vertex-ants-walk: A robust method for efficient exploration of faulty graphs. *Ann. Math. Artif. Intell., 31*, 99–112.

Zheng, X. and Koenig, S. (2008). Reaction functions for task allocation to cooperative agents. *In: Proceedings of the 7th International Joint Conference on Autonomous Agents and Multi-agent Systems*, Estoril, Portugal, 12–16 May, 559–66.

Zlot, R., Stentz, A., Dias, M. and Thayer, S. (2002). Multi-Robot Exploration Controlled by a Market Economy. *In: Proceedings of the IEEE International Conference on Robotics and Automation*, Washington, DC, USA, 11–15 May.

3

Federated Learning in Shared Production Scenarios

Kirill Fridman,[1*] *Juergen Grotepass*,[2] *Tatjana Legler*[3]
and *Vinit Hegiste*[4]

Introduction

With digitalization progressing in the manufacturing industry, swarm intelligence will enable a new powerful approach for data-based value creation in supply chains.

Swarm intelligence, a concept inspired by nature's collaborative systems, has gained prominence in optimizing complex processes in manufacturing. Industries continue to generate vast amounts of data and interconnected devices proliferate. But here traditional approaches face limitations in handling the scale and complexity of modern manufacturing environments and solutions are needed.

Federated learning (FL) as the enabler of Swarm Intelligence is a novel paradigm that promises to bridge this gap. This book chapter is dedicated to exploring the potential of federated learning. FL, often described as a decentralized and privacy-preserving machine learning approach, perfectly embodies the ethos of swarm intelligence. We will investigate how this innovative technique, leveraging the collective power of distributed data sources, is bringing about a new era of intelligence and adaptability in manufacturing and supply chain optimization.

[1,2] Huawei Technologies Duesseldorf GmbH, Germany, http://www.huawei.eu

[3] Technologie-Initiative SmartFactory KL e. V., Germany, https://www.smartfactory.de/

[4] Rheinland-Pfälzische Technische Universität Kaiserslautern-Landau (RPTU), Germany, https:// mv.rptu.de/

Email : juergen.grotepass@huawei.com; tatjana.legler@rptu.de; vinit.hegiste@mv.uni-kl.de

* Corresponding author: Kirill.Fridman@huawei.com

By the end of this chapter, you will get a deeper understanding of the possibilities and implications of federated AI in manufacturing based on selected use-cases.

Federated Learning (FL)

Google introduced federated learning (FL) in 2016 (McMahan et al., 2017). FL is a machine learning framework that allows multiple devices to collaboratively learn a model without sharing their private data. This offers ample opportunities in critical domains such as healthcare, finance, etc., where it is risky to share private user information with other organizations or devices.

However, FL also faces several challenges, such as communication efficiency, data heterogeneity, system robustness, and privacy preservation. One of the key challenges is client selection, which refers to the process of choosing a subset of clients to participate in each round of FL training. Client selection can affect the convergence speed, accuracy, fairness, and scalability of FL.

Some of the customer challenges related to client selection are:

- How to select clients that have high-quality and representative data for FL training?
- How to balance the trade-off between communication cost and model performance when selecting clients?
- How to ensure the privacy and security of clients' data and model parameters when selecting clients?
- How to deal with the dynamic and heterogeneous nature of clients (such as availability, reliability, resource constraints, etc.) when selecting clients?
- How to design incentive mechanisms to encourage clients to participate in FL training when selecting clients?
- Lack of trust: for actors to share their data, secure data transmission, storage and access rights are required to protect data from competitors or from the theft of know-how (sensitive production data).

FL can be elucidated by a practical application, as depicted in Figure 1. The Figure illustrates two enterprises, Client1 and Client2, engaged in the production of hardware tools, specifically hammers and screwdrivers. A notable disparity exists in their production scales: Client1 predominantly manufactures screwdrivers, whereas hammers constitute the majority of Client2's output.

If each company were to make a tool classification model on its own, there would be a problem. Client1's model would likely be better at recognizing screwdrivers, and Client2's would be more accurate with hammers. One might think, "Why not combine their data to make a more balanced model?" But here's the catch: neither company wants to directly share its data, mainly because of competition and privacy issues.

This is where FL comes into play. Instead of sharing their data, both companies can use it to train their own models. Once trained, they send the model's weights

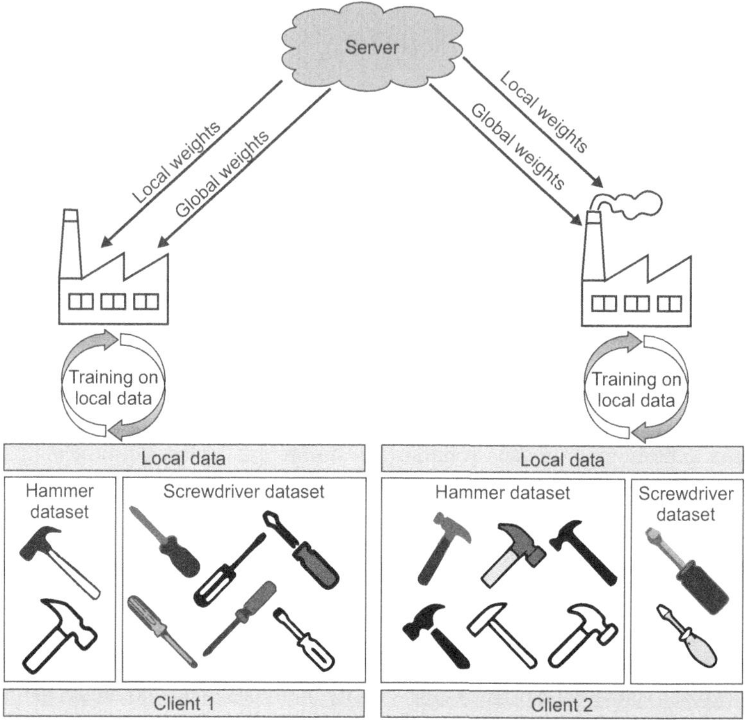

Fig. 1   A visual representation of federated learning with two clients working on classifying hammers.

(not the data) to a neutral server. On this server, algorithms like FedAvg (McMahan et al., 2017), FedProx (Sattler et al., 2019), FedSVGR (Kang & Xiong, 2020), and FedFomo (Karimireddy et al., 2020) are used to merge these weights and create a shared model. This combined model is then sent back to both companies. They can use it as a starting point, train on their data again, and send the updates back. This cycle, known as a "Communication Round", repeats several times. The end result? A model that's good at recognizing both tools for each company.

In the sections that follow, we'll look at how FL can be used for quality checks in manufacturing.

Use-case: Federated Learning for Quality Inspection in Networked Production

Use-case Introduction

This Use Case of "quality inspection in manufacturing ecosystems" is related to the shared production GAIA-X project "smartMA-X" with "SmartFactory-KL" as coordinator and has been demonstrated at Hannover Fair 2023 (Livestream Hannovermesse, 2022).

Two production lines have been interconnected to manufacture customized demo products. Although the products vary in design, shape, material, and process, they share the same need for visual quality inspection. Opposed to the common inspection procedure of processing product and process data on a centralized device on-premise, this solution features edge-based analytics where AI models are deployed locally for each site participating in the shared production scenario.

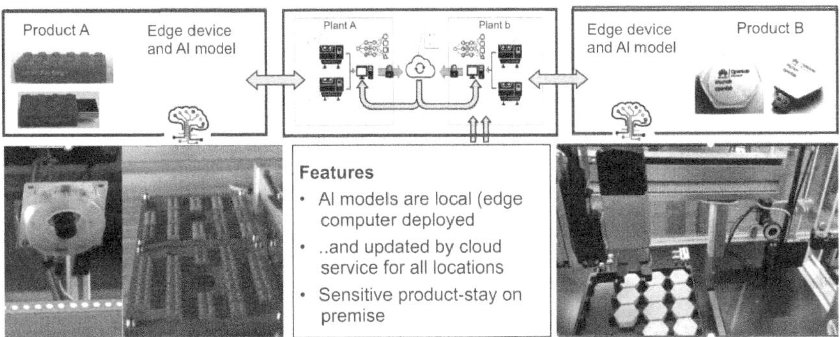

Fig. 2 Collaborative Learning in Networked Production. Different Products – same Needs: Quality inspection as a Service.

Collaborative learning is achieved with all sites sharing their AI model parameters with the cloud model which interpolates them generating an optimized parameter set and being deployed again on all sites locally as a service (Figure 2). This way faults may be detected on all sites, although some might have occurred on one site only before. The significant and decisive advantage is that all sensitive product and process data stay on-premise, with only AI model weights being shared with the cloud model for collaborative learning.

While the data act is motivating data sharing in industrial data spaces no solution has been developed on how to protect partners acting in data ecosystems to generate value—giving answers to industry concerns. This leads to a situation of an innovation obstacle with unequal knowledge in companies to build on such important digital data, holding back digitalization and value creation (Livestream Pitchday, 2022).

The IP protecting value generation approach would support industry partners to share data to take benefit from federated services without disclosing sensitive data. As the shared production scenario is based on the production level 4 paradigm, the service is highly scalable. Furthermore, partners are invited to join offering skills on their legacy machines as brownfield solutions are demonstrated to integrate sensors using 5G connectivity to access edge and cloud-based analytics.

This "Trade Secret/IP Protection as a Service" solution may become a Key Use Case to motivate partners to share data for value generation in a broad range. This enables "swarm learning" to provide better services in different industry segments: connected cars to learn street signs, connected hospitals improving diagnostics and cloud manufacturing for learning skills.

The IP protection service and enabled optimizing services for quality and process enhancement may lead to further innovation cycles for SME industry partners and could become future business cases by themselves.

The initial application of federated learning in the quality inspection of USB sticks was focused on diverse clients having varied error types in USB ports. And their locally trained model was unable to detect the error type of other clients with high confidence. The primary objective was to enable the global federated model to classify all error types without the need for local image data sharing. This goal was successfully achieved, and is shown in the results. In this context, a significant advantage of federated learning emerges, wherein participating clients can derive benefits from the collective model during a federated learning round, rather than relying on their locally trained models, which may not be adept at handling previously unseen error types.

The first use case substantiated the robustness of FL in image classification. The subsequent use case focused on federated object detection, with the primary aim of assessing the global federated model's predictions of bounding boxes for objects similar in type but with slight variations in shape. The findings were intriguing, highlighting the robust nature of the FL model in object detection, effectively defining precise boundaries of bounding boxes over objects. Notably, the locally trained model fell short in producing such precise bounding boxes when objects were placed in an unfamiliar environment or exhibited different shapes.

Dataset

In the field of FL applied to image classification, a compelling use case emerges: the quality inspection of USB products. This premise was thoroughly explored by Hegiste et al., (2022). Figures 2 and 3 offer a comprehensive overview of both the dataset employed and the distribution of samples. Specifically, Figure 3 shows the class types and the number of images associated with each class for every client. Three distinct clients were considered:

1. Huawei USB stick (referred to as Client1)
2. Smart factory USB stick (Blue color, denoted as Client2)
3. WSKL chair at RPTU Kaiserslautern, where a blue USB stick was used (designated as Client3)

For the sake of this investigation, errors were deliberately introduced to the USBs. On Client1's USB, stickers were affixed to simulate scratches. For Client2, the anomaly was a physically damaged USB port, while Client3's USB exhibited rust on its port. It's imperative to underscore that the nature of these errors was diverse across the clients.

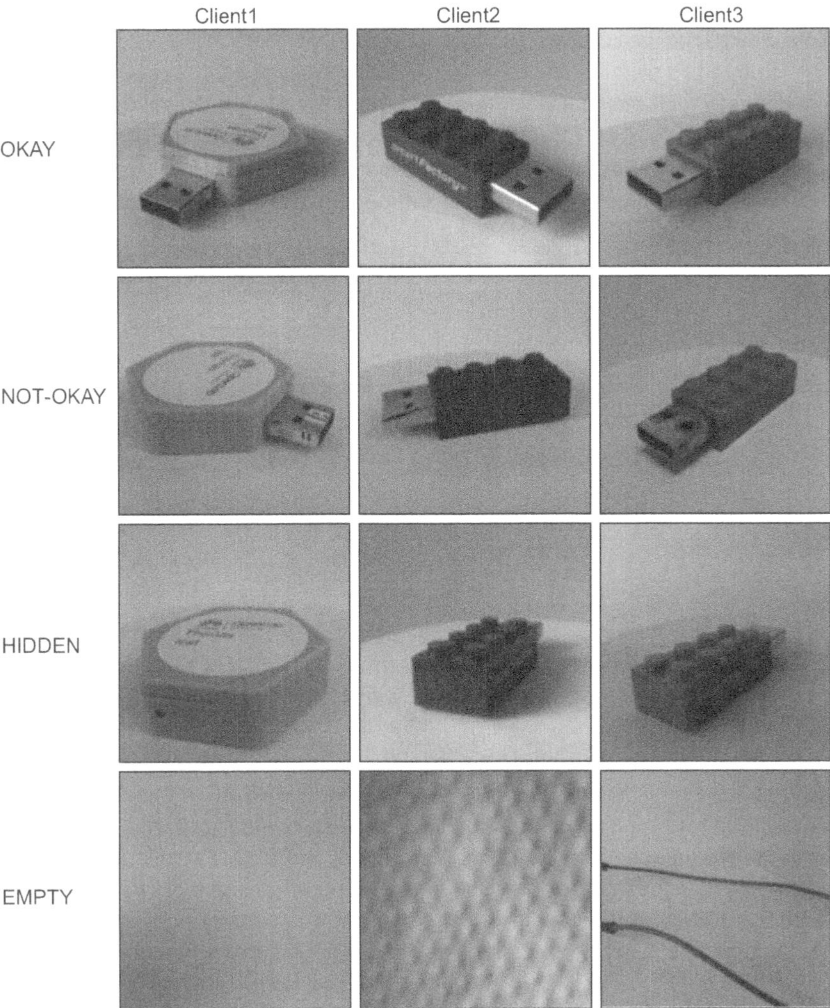

Fig. 3   Three USB manufacturing clients possessing different types of USB port errors

Subsequently, a curated dataset was established encompassing four categories:

1. 'Okay': representing USB ports devoid of any anomalies.
2. 'Not_Okay': denoting USB ports with discernible defects.
3. 'Hidden': an auxiliary category.
4. 'Empty/Not_A_USB': belongs to images that did not feature any USB sticks.

A problem arises when local training is performed. Specifically, Client1's local model exhibits high efficiency in identifying sticker-induced errors as 'Not_Okay'. However, its proficiency diminishes in detecting anomalies in USBs from other clients. The seemingly straightforward solution to this dilemma is the collection of all available data in one place, thereby training a unified USB classification model. Yet, such an approach is connected with data privacy concerns, especially in a manufacturing context.

Enter FL: a paradigm that champions data privacy. It establishes a collaborative framework, facilitates joint ventures between multiple entities or companies. As evident from Figure 4, FL does not mandate uniform dataset distributions across clients. This versatility empowers clients to actively participate in FL rounds, as well as take into account the fact that the resulting global model is adept at assimilating insights from non-IID datasets.

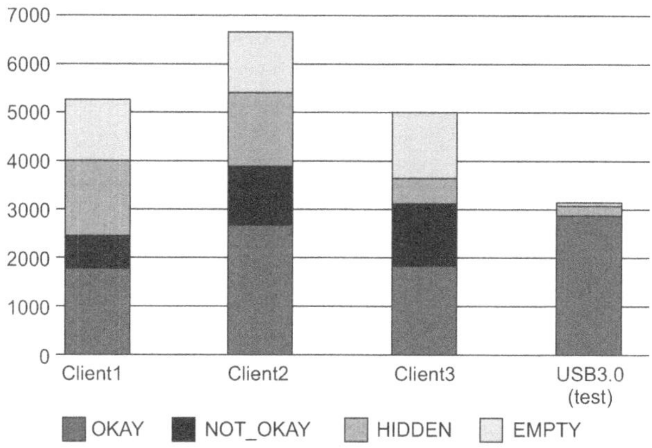

Fig. 4   Image samples distribution per class per client.

Following our work on FL for image classification, we further looked into federated object detection (Hegiste et al., 2023b). The details of this dataset can be seen in Figure 5, and its distribution in Figure 6. Our main goal was to determine if an object in a cabin frame had a windshield, and if so, to accurately mark its location with a bounding box. We worked with four different types of windshields. Figure 6 shows that Client1 makes cabins with blue designs using Windshield Type A. On the other hand, Client2 produces red-designed cabins, but with Windshield Types C and D. It's worth noting that the number of samples from these two clients isn't the same.

Our research focused on comparing models trained locally with a global model trained using FL. We tested these models using images not seen during training. For example, we tested how a model trained on blue cabins responded to images of blue cabins with Windshield Types C and D. We did similar tests for the red cabin model. To push our models further, we even tried combinations like a red windshield on a blue cabin and the other way around. These tests helped us see how well our models performed under different conditions.

System Architecture & Applied Algorithms

The section aims to provide a clear understanding of the system's architecture and the federated machine learning pipeline. The proposed architecture can be

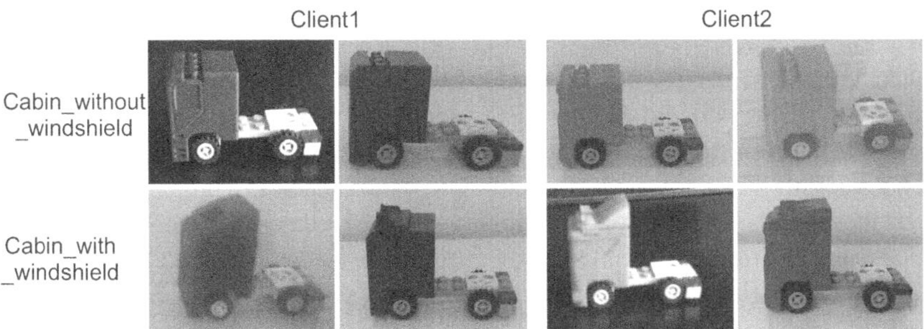

Fig. 5   Cabin with and without Windshield belonging to different manufacturers.

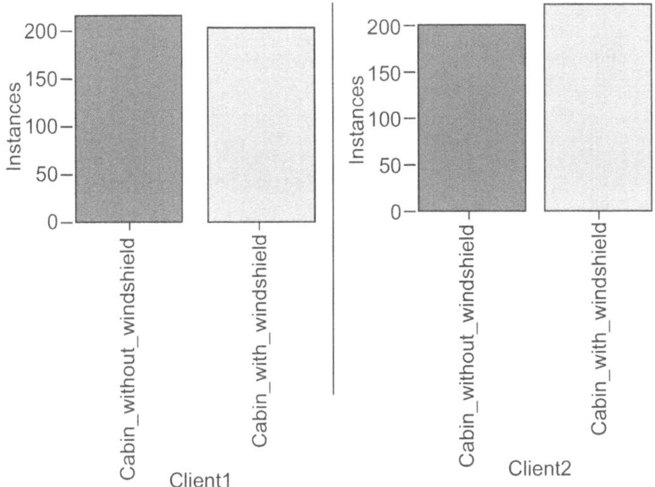

Fig. 6   Dataset distribution for Cabin with and without windshield.

visualized in Figure 7, which illustrates the processes adopted by clients and the server, respectively. While both these process pipelines depict the federated object detection workflow, they are inherently adaptable and apply to federated image classification tasks as well. The only differences lie in the architecture and dataset type.

Client-Side Workflow

The initiation of the FL process starts with the 0th Communication Round (CR). At this stage, no global model exists on the server. Clients, therefore, utilize the pretrained default weights from YOLOv1 (chosen based on the pre-agreed architecture) to train their respective local datasets. Adopting a common set of starting weights is crucial, a fact emphasized in McMahan et al., 2017), which suggests that a global model performs optimally when all participating clients commence from a shared starting point.

Upon completing the training with these weights, clients forward them to a neutral server. Given that this is the 0th CR, the absence of global weights means that the reported mean Average Precision (mAP) is 0. At the server, weights received from all clients undergo an averaging process, neuron-by-neuron. The resultant weights are then stored as the global model weights and disseminated back to the clients. With the conclusion of the 0th CR, as indicated in Figure 7, clients will test these newly acquired global weights against their datasets, recording the mAP. Subsequently, training commences on the local dataset using these global weights. This cycle is reiterated for subsequent communication rounds.

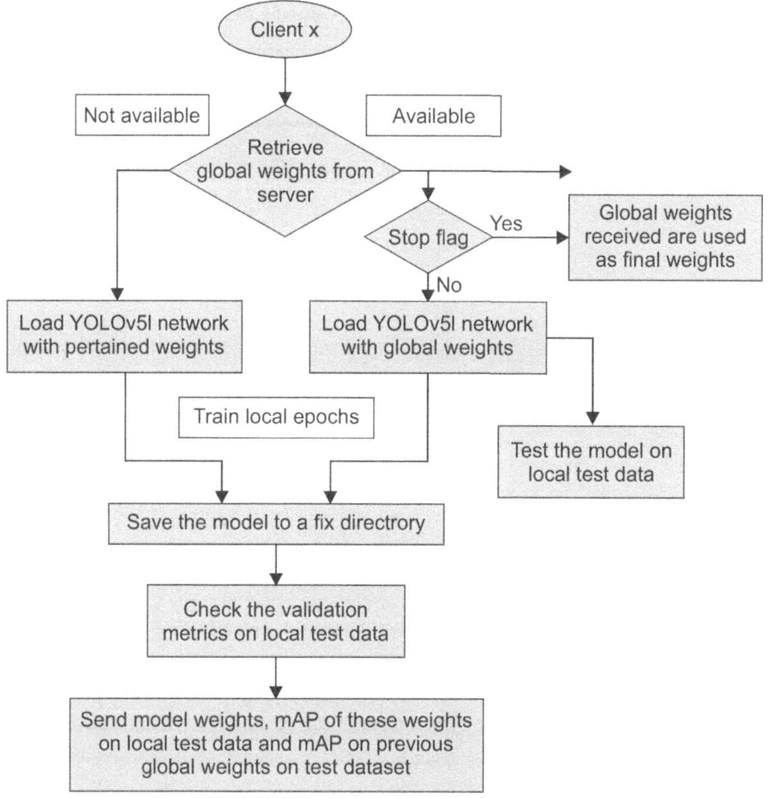

Fig. 7   Client pipeline for Federated Object Detection.

Server-Side Workflow

When focusing our attention to Figure 8, post the 1st CR client-side training, clients are responsible for transmitting both the locally trained model weights and the mAP obtained from testing the previous global model against their local datasets. At the server, mAPs received from all clients undergo an averaging procedure. If this aggregated mAP surpasses a predetermined threshold, the prior global model weights are redistributed to the clients, earmarking them as the finalized federated model—one that exhibits satisfactory performance across varied client datasets. Should the averaged mAP fall short of the threshold, the server persists with the current approach of averaging and sharing global weights. This process endures until a global model meeting the desired mAP threshold is realized.

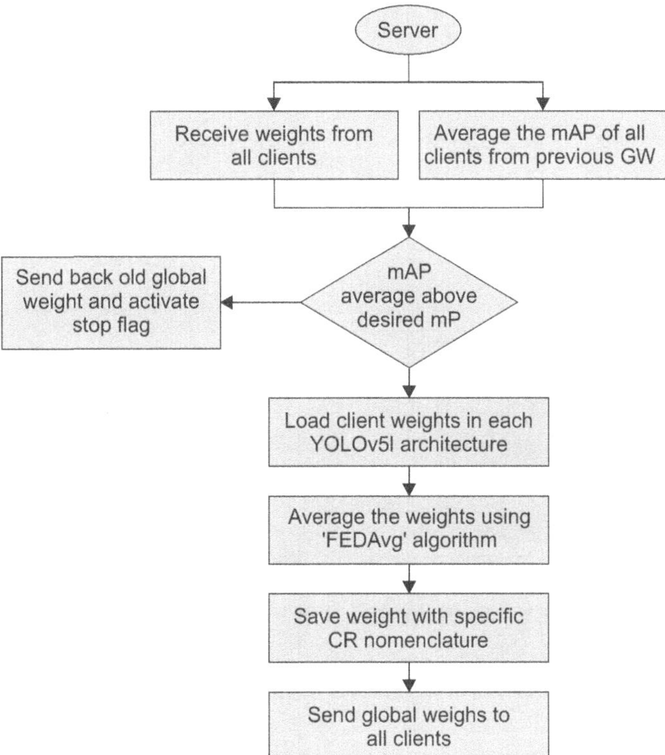

Fig. 8　Server pipeline for Federated Object Detection.

Implementation and Experimentation

The described pipeline, as detailed in the preceding section, is sketched in Figure 9, which provides a graphical representation of the FL procedure, specifically tailored for object detection scenarios. Although the depicted workflow is relevant for both federated image classification (FedIC) and federated object detection (FedOD), the fundamental distinction arises in the metrics and parameters communicated along with the model weights.

For FedIC, the emphasis lies on the loss value, derived from testing the global weights against the dataset. The prime objective in this use case is to minimize this loss. Once the test dataset's loss value drops to the desired threshold, it indicates that the global federated model has reached its optimal performance. The underlying principle is straightforward: a reduced loss signifies a more accurate and effective model.

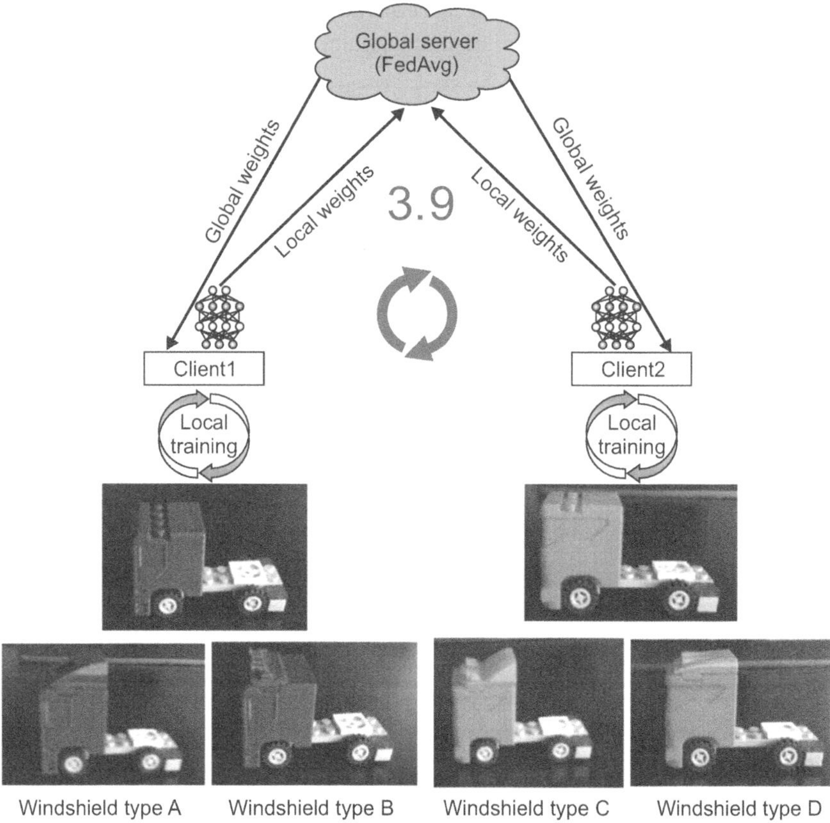

Fig. 9   Federated learning illustration of 2 clients for FedOD (Hegiste et al., 23b).

In contrast, for FedOD, the metric of interest is the mAP. This metric gains prominence since, in object detection scenarios, the accurate classification of objects is only part of the goal. An equally important aspect is the precise positioning of bounding boxes around the detected objects. Consequently, the desired parameter in the FedOD use case is a high mAP, as it ensures not only the correct prediction of object classes but also the accurate placement of bounding boxes.

Federated Image Classification

In the context of the Federated Image Classification (FedIC) pertaining to the USB quality inspection scenario, the global federated model was synthesized adhering to the delineated pipeline. Additionally, three distinct models were systematically conditioned, each tailored to its respective client dataset. An exhaustive examination of the Communication Rounds (CR) and the associated confusion matrix has been documented in Hegiste et al., (2022).

Figure 10 offers a visual representation extracted from an ongoing classification video feed. Specifically:

1. The top-left frame portrays the inference drawn from the global federated model.
2. The top-right presents the output from a model, strictly conditioned on Client1's Huawei USB dataset.
3. The bottom-left frame displays the inference of the model grounded on Client2's blue SmartFactory-KL USB dataset.
4. The bottom-right is the deduced outcome from the Client3 model, which was anchored on its red USB dataset.

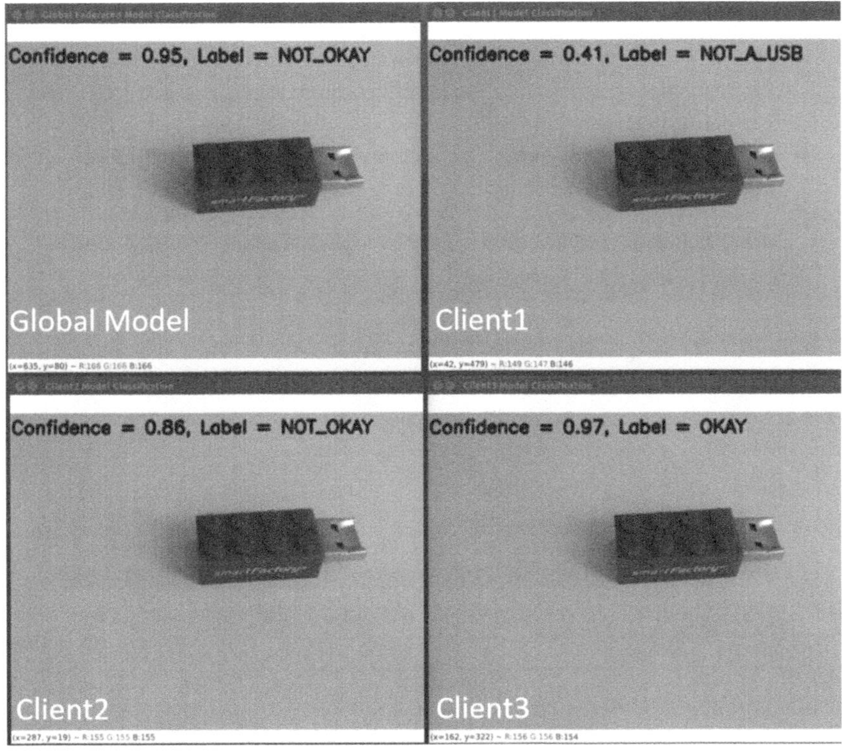

Fig. 10   Live comparison of the global federated model with 3 clients' model trained on their respective dataset.

Within the scope of Figure 10, the subjected test object is a blue USB stick exhibiting a port defect.

Notably:

- Client1's model exhibits a complete inability to identify the object as a USB stick.
- The model from Client3 commits a misclassification, exuding an elevated confidence score. This deviation is attributed to its unfamiliarity with the presented error type, as it was principally trained for the 'rust' error category.
- Both the global model and Client2's model successfully execute a correct classification. Furthermore, the global model's confidence score marginally eclipses that of Client2.

Continuing to Figure 11, which is another visual excerpt from the same live feed, it focuses on a USB port exhibiting rust for classification. Observations include:

- The rust-afflicted red USB stick, innately belonging to Client3's dataset, is accurately identified by Client3's model.
- Contrarily, Client1's model identifies the object as 'Okay'.
- Client2's model, while achieving a correct classification, does so with diminished confidence, likely due to the absence of the rusted USB stick in its training set.
- The global federated model accentuates its robustness, evidenced by its accurate classification complemented by a superior confidence score.

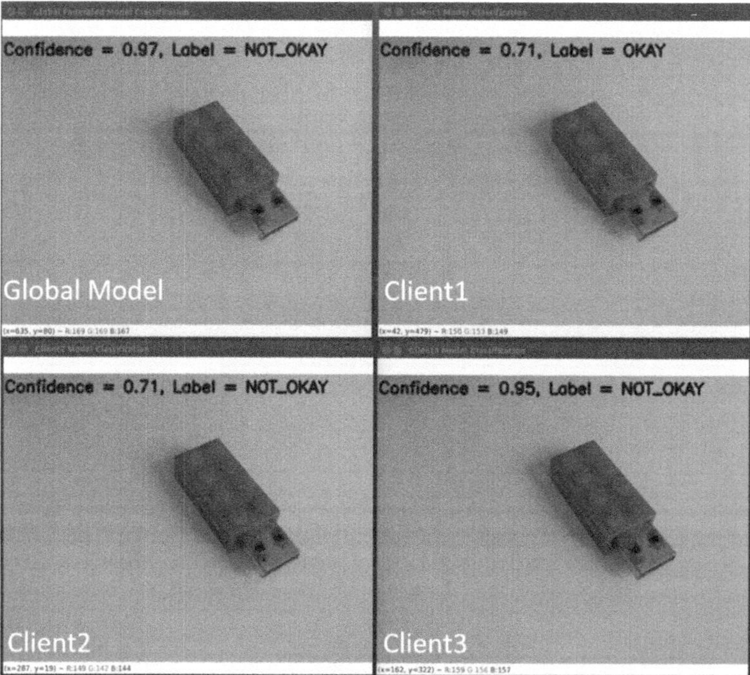

Fig. 11 Live comparison of the global federated model with 3 clients' model trained on their dataset.

Federated Object Detection

In the context of the Federated Object Detection (FedOD) scenario, the models were again subjected to live evaluation, akin to the methodology deployed in the previous use case. Figures 12 and 13 provide a comparative visual representation between the global federated model and the models specifically conditioned on the individual clients' datasets.

Fig. 12   Live comparison of the global federated model with 2 clients' models trained on their respective dataset, evaluating on all both types of Cabin with and without windshields.

Figure 12 presents a scene incorporating four cabins, bifurcated equally to represent each client's type, and showcasing both the classifications: 'Cabin_with_windshield' and 'Cabin_without_windshield'. Some key observations include:

- Both Client1 and Client2 models, predominantly trained on blue and red cabins, respectively, only identify their specific cabin types. Notably, they exhibit an inability to detect cabins of the opposite hue and design.
- It's important to note that the live testing video was recorded under conditions that weren't strictly controlled. Factors such as ambient light penetration from windows, variance in camera angles, and orientation disparities of the objects might influence detection efficacy.

- Furthermore, the bounding boxes delineated by the Client1 and Client2 models do not seamlessly encase the subject. This inconsistency can be ascribed to the fact that the windshields depicted in the live video differ from those the models were initially trained upon.
- Contrasting this, the global federated model demonstrates superior performance by not only accurately classifying the cabins but also rendering impeccable bounding boxes over the detected objects.

Such findings underscore the robustness and adaptability of federated learning models in scenarios with varying environmental conditions and object variations.

Detailed results for this experiment can be found in Table 1, which presents the average precision (AP) values for different IoU thresholds ranging from 0.50 to 0.95, as well as the mean average precision (mAP) at an IoU threshold of 0.5. The FedOD model achieved an AP@ [.50:.05:.95] of 0.93, and an mAP of 1.0, showcasing its robust performance across a wide range of IoU thresholds.

Table 1   mAP metrics comparison of Blue cabin, Red cabin, and FedOD model on an unseen Test dataset.

Model	*Training Dataset*	*Test Dataset*	*mAP*	*AP@ [.50:.05:.95]*	*AP*	*AR*
Client1 (Blue cabin)	Blue colored cabins without windshield and cabin with windshields of type A and B	Blue colored cabins with windshield type C and D and Red colored cabins with windshield type A and B.	0.42	0.35	0.35	0.36
Client2 (Red cabin)	Red colored cabins without windshield and cabin with windshields		0.49	0.42	0.42	0.43
Global Federated model	–		1.0	0.93	0.93	0.96

Within Figure 13, an intriguing experiment was conducted, involving cabins with opposing windshield colors. The outcomes were rather remarkable. The same global federated model, which had proven its mettle previously, continued to demonstrate its efficacy, adeptly classifying this novel combination of attributes.

The results of this experiment are presented in Table 2. The mAP and AP@ [.50:.05:.95] values in Table 2 indicate that the FedOD model outperformed the local models in terms of precision. It consistently achieved higher mAP and AP scores, suggesting that the global FedOD model excels in predicting precise bounding boxes, even when confronted with unseen combination types.

A deeper dive into the results reveals the following:

- The model trained specifically on Client1's data correctly identified the blue cabin equipped with a windshield. However, its bounding box delineation

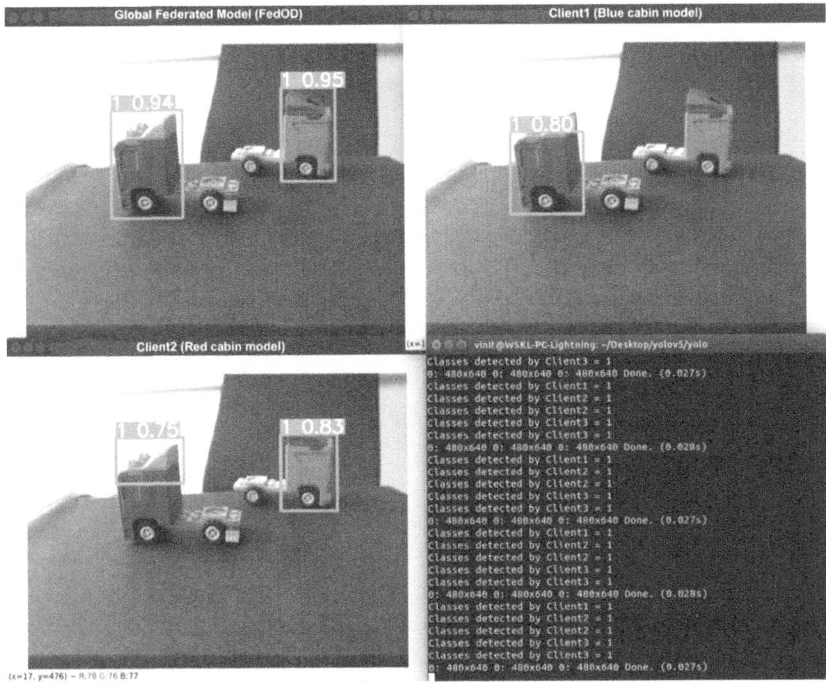

Fig. 13   Live comparison of the global federated model with 2 clients' models trained on their respective dataset, evaluating on all both types of Cabin with different combination of windshields.

Table 2   The centralized trained YOLOv5 models (client1 and client2) trained on their local dataset and FedOD models are tested with Windshield combination not present in their training dataset.

Model	Training Dataset	Test Dataset	mAP	AP@ [.50:.05:.95]	AP	AR
Client1 (Blue cabin)	Blue cabins with windshield of type A and B	Blue cabins with windshield type C and D	0.83	0.70	0.70	0.73
Global Federated model	–	Blue cabins with windshield type C and D	1.0	0.96	0.96	0.98
Client2 (Red cabin)	Red cabins and windshield of type C and D	Red cabins with windshield type A and B	0.97	0.83	0.83	0.86
Global Federated model	–	Red cabins with windshield type A and B	1.0	0.91	0.91	0.93

around the object was less than precise. Furthermore, it failed to recognize the red cabin with a windshield altogether.

- Likewise, the Client2-centric model did manage to classify the red cabin featuring a windshield. Yet, similar to its counterpart, it struggled with the bounding box demarcation due to the absence of this particular combination in its training dataset.
- In contrast, the global federated model showcased a maturity in classifying even such unconventional combinations, thereby solidifying its position as a competitive alternative to models trained using centralized methodologies.

These results highlight the effectiveness of the FedOD model in the cabin quality inspection use-case, demonstrating its ability to accurately detect different cabin and windshield combinations which were not even part of its local training datasets. The superior performance of the FedOD model provides strong evidence for the advantages of FL in collaborative object detection scenarios.

Results and Generated Value

Our experimentation primarily focused on assessing the robustness of the global federated model under various conditions and configurations. Initiating with the FedIC setup, we subjected all four models delineated in the preceding section to novel USB error scenarios, essentially applying one client's error type to another client's USB device.

Figure 14 elucidates one such configuration: a Client2 blue USB bearing the sticker error related to Client1's Huawei USB. The observations were stunning. While Client1's model did identify the stick as 'Not_Okay', its confidence score remained low, possibly due to its familiarity with the error but unfamiliarity with the USB design. In contrast, Client2's model, unknown of this error type, classified the USB as 'Okay', albeit with diminished confidence, suggesting it detected the sticker but couldn't recognize it as an anomaly. Client3's model, unfamiliar with both the USB and error type, confidently misclassified the USB as 'Okay', driven by its similarity to its 'Okay' USB archetype. The global federated model showcased superior adaptability, correctly identifying the anomaly with a formidable 92% confidence. Other tested configurations, like rusted and damaged Huawei USBs, rusted blue USBs, and blue USBs exhibiting sticker and damage, consistently manifested the global federated model's precision, always surpassing 85% confidence in correct classification.

In Figure 15, we focus on a third-party USB, significantly different from any client training dataset, with its port concealed from the camera due to its unique design. Notably, individual client models failed here. Both Client1 and 2 misinterpreted the object, failing to identify it as a USB, while Client3 mistakenly identified it as 'Okay'. In contrast, the global federated model correctly classified the USB with a 60% confidence—considering the strong deviation of this USB's design from the training data. Such outcomes confirm the robustness of the FedIC model, substantiating its efficacy across diverse client scenarios, even in the absence of direct data sharing.

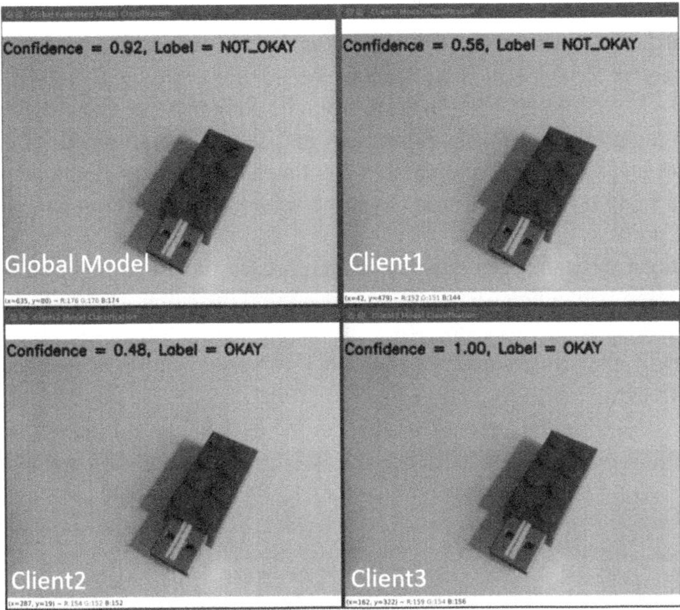

Fig. 14 Live comparison of the global federated model with 3 clients' model trained on their respective dataset, evaluation on a blue USB stick with sticker error (Error belonging to Client1's dataset).

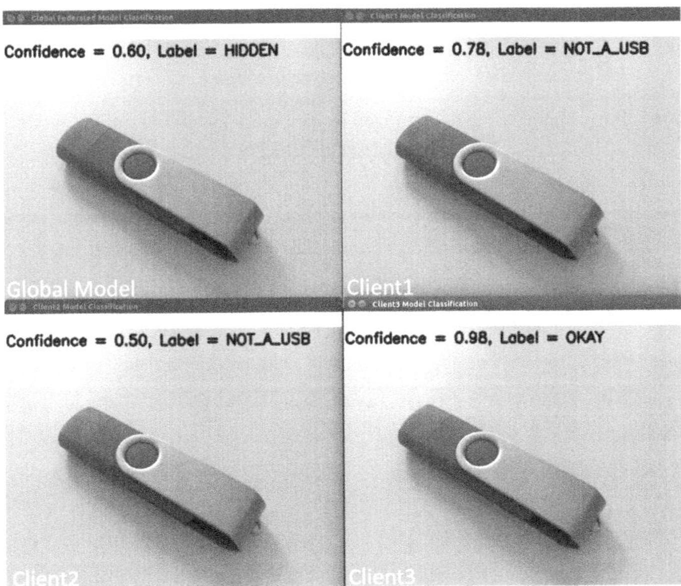

Fig. 15 Live comparison of the global federated model with 3 clients' model trained on their respective dataset, evaluating on a USB which does not belong to the training dataset (ground truth: Hidden).

Moving on to the application of FedOD, this was pragmatically integrated into the quality inspection module of the SmartFactory-KL demonstrator. A notable aspect to highlight is the variance between the training images (Figure 5). Two models were tested using images from the demonstrator's quality inspection module: the global federated cabin model and a centralized trained cabin model. The latter means a model trained on a combined dataset from both clients, using the same YOLOvl architecture and hyperparameters, adopted during the federated learning process.

In Figure 16, the results from the centralized model show accurate object classification with a precise bounding box around the cabin. However, the model produced false positives, especially when capturing the trailer section of a truck. Additionally, the combination of a blue cabin with a red windshield goes undetected by this model.

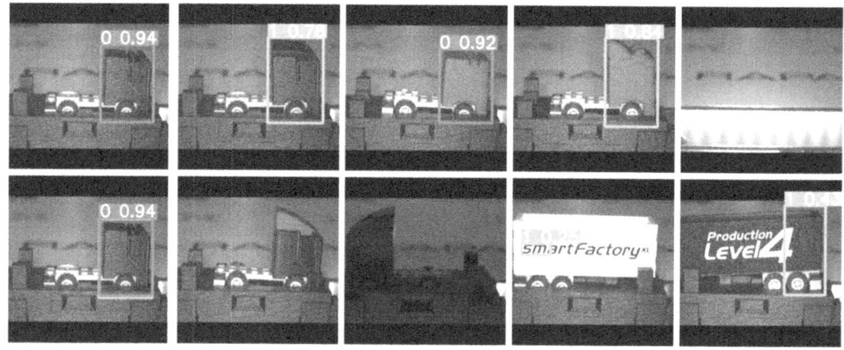

Normal training cabin model

Fig. 16   Performance of the centralized trained model in an unfamiliar environment (Hegiste et al., 2023b).

Federated global cabin model

Fig. 17   Performance of the federated global model in an unfamiliar environment [3].

Comparatively, the global federated model displayed superior performance. As illustrated in Figure 17, the FedOD model showcases its robustness, not only by accurately classifying objects and drawing perfect bounding boxes, but also by avoiding any false positives on unrelated objects.

The primary value is generated through reduced investment costs for on-premise hardware. The secondary value is related to quality and cost improvements by service deployment on demand. Additional value is reduced product carbon footprint at higher quality of products.

Integration in the Industry 4.0 Shared Production Architecture

The shared production scenario is part of the GAIA-X use cases being public-funded on a national level to enable cooperation based on sharing production resources (see Figure 18). GAIA-X is a European initiative designed to create a federated data infrastructure and ecosystem that facilitates the secure sharing of data and resources among organizations, particularly in the context of industry and manufacturing.

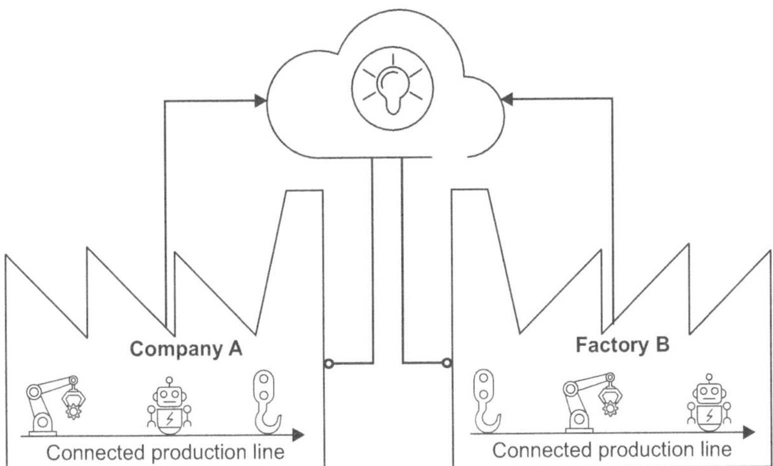

Fig. 18   Shared Production; Cross-Factory and Cross-Company Production; Smart Factory KL Vision 2025: 'Production Level 4'.

This production paradigm offers solutions to utilize production capabilities offered in digital marketplaces whenever needed—not being available on the own shopfloor. Or to offer unused own production capabilities to new customers.

The open and modular approach of production enables companies to work more closely and transparently with each other across company boundaries to customize production.

The control of the production and the value-added networks is data-driven, whereby the ownership rights to the data remain guaranteed.

As in normal value chains, the collection and processing of data can open up potentials for new business models. The difference here is that the potentials could be further increased by the formation of ad hoc value-added networks and new business models could be developed (jointly) in each case.

The future role of AI will be mainly linked to the engineering stage to find and match appropriate digital twins (AAS: asset administration shells) of production services being offered in digital marketplaces.

The AAS is a digital representation of a physical asset, such as a machine, device, or component, in the context of Industry 4.0 (Abdel-Aty et al., 2022). It contains a standardized and structured set of information about the asset, including its properties, behavior, and relationships.

AAS provides a consistent and standardized way to describe assets, making them understandable and accessible in the digital realm. This allows for seamless integration of assets into the Industrial Internet of Things (IIoT) and Industry 4.0 environments.

We will give one possible solution on how we can offer our quality inspection AI-software service to different companies over Gaia-X platform. Therefore, we orientated on the offering of production services in Gaia-X Data Spaces (Jungbluth et al., 2023; Volkman et al., 2023). GAIA-X Data Space, in this context, refers to a virtual space where organizations can securely store, share, and access data and digital assets related to shared production. This data space is designed to enable the seamless exchange of information, resources, and expertise between various stakeholders in the manufacturing and industrial sectors. It aims to promote data sovereignty, data security, and data interoperability, allowing organizations to collaborate while maintaining control over their data. In this case, a potential user of the service can download and use the service with its own dataset in its own production line without any significant efforts. To offer the quality inspection service in Gaia-X the Industry 4.0 complaint way is to describe the quality inspection service (its features, characteristics, properties, and capabilities) with the help of AAS.

With the help of GAIA-X connector, we can connect to the related data space and offer our AAS-based description of the software service. That means it can be found in the Gaia-X service catalog and the service provider can offer the software service via marketplace as shown in Figure 19.

If the customer connects to the related data space, he can browse available software services on the marketplace, select one, that matches his requirements, download it (e.g., as a Docker container) and use it in own production line. All this operation is performed with the help of service generalized description available via AAS. The customer can download the service or, if the service is already running on the customer side, just update the model weights from the global federated model.

Because the quality inspection service is based on the FL algorithm, the customer has also an opportunity to contribute to the service by improving the model quality via an additional round of training on his local dataset. But the main challenge here is to make sure that all customers have similar data classes and use-case. Each FL model is trained for a specific use case, and precise description of

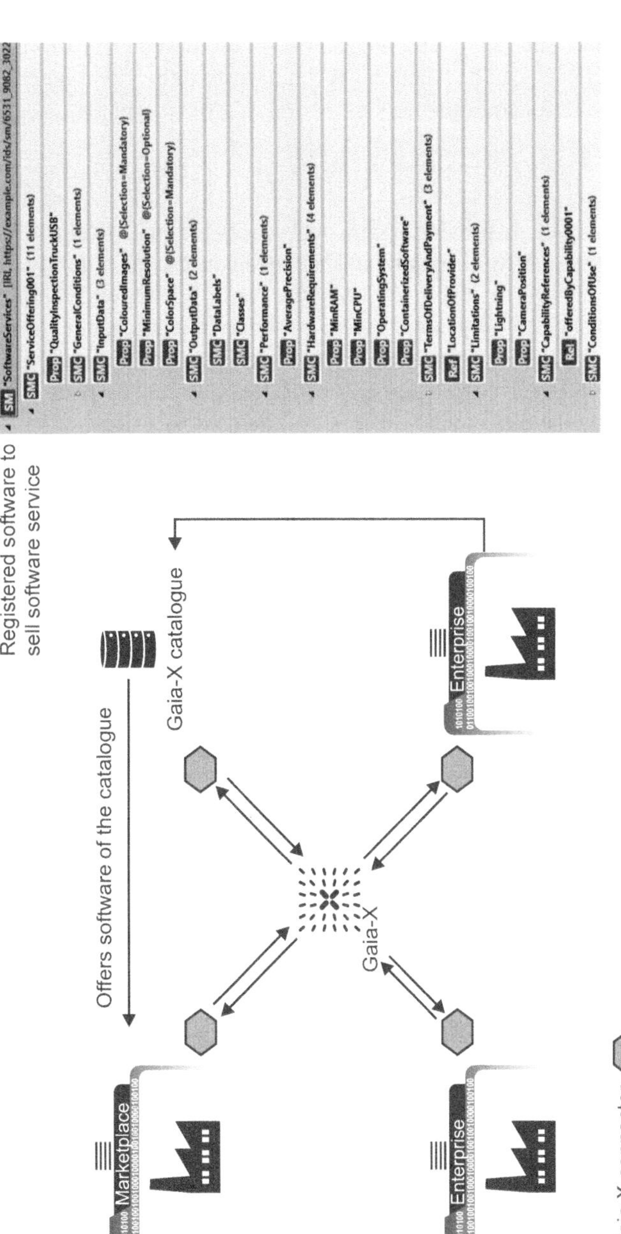

Fig. 19 Industry 4.0 data space for software services.

possible use cases is necessary to offer a product on a marketplace. Otherwise, the model quality can be corrupted by ingesting the wrong local datasets. Although some precautions, such as attempting an automatic assignment of class labels from different clients, can be taken; however, a trustworthy environment must be in place to ensure cooperation.

Conclusion

In our use-case, FL could provide significant benefits for quality control in various domains, such as industrial manufacturing applications. FL's ability to utilize data from multiple sources, while ensuring the data's privacy and confidentiality, mirrors the collaborative dynamics seen in swarm intelligence. FL can enable data-driven machine learning models to leverage the data from multiple sources without compromising the privacy and confidentiality of the data owners. FL can also improve the quality and diversity of the data by incorporating the local knowledge and preferences of different clients.

The integration of FL with object detection yielded significant accuracy and precision enhancements in contrast to isolated client models. Our findings affirm that the global federated model, while avoiding local data sharing, remains resilient even under varying conditions like different lighting, camera angles, and object combinations.

Emphasizing its scalability, FL's potential extends across diverse sectors. Beyond the illustrated use-cases, its benefits are evident in domains where data sharing is constrained by privacy considerations, such as in healthcare or in traffic management for connected vehicles. The cooperative nature of FL, where multiple clients converge to refine a singular global model, optimally positioning FL as a revolutionary tool for enhancing robustness and adaptability in data-driven value creation.

References

Abdel-Aty, T.A., Negri, E. and Galparoli, S. (2022). Asset administration shell in manufacturing: Applications and relationship with digital twin. *IFAC-PapersOnLine*, *55*(10), 2533–38. doi: 10.1016/j.ifacol.2022.08.156.

Hegiste, V., Legler, T. and Ruskowski, M. (2022). Application of Federated Machine Learning in Manufacturing. *In*: *2022 International Conference on Industry 4.0 Technology (I4Tech)*, 1–8. Pune, India: IEEE. dOI: 10.1109/I4Tech55392.2022.9952385.

Hegiste, V., Legler, T. and Ruskowski, M. (2023a). Federated Ensemble YOLOv5: A Better Generalized Object Detection Algorithm. *arXiv*: 2306.17829.

Hegiste, V., Legler, T., Fridman, K. and Ruskowski, M. (2023b). Federated Object Detection for Quality Inspection in Shared Production. *arXiv*: 2306.17645.

Jungbluth, S., Witton, A., Hermann, J. and Ruskowski, M. (2023). Architecture for shared production leveraging asset administration shell and gaia-x (in press).

Kang, J. and Xiong, H. (2020). FSCAVG: Federated Stochastic Cyclic Averaging. *arXiv preprint*. *arXiv*: 2002.10420.

Karimireddy, S.P., Rebjock, Q., Stich, S.U. and Jaggi, M. (2020). FedFomo: Federated FTRL over Momentum for Communication-Efficient Learning on Heterogeneous Data. arXiv preprint. *arXiv*: 2002.07948.

Livestream - Pitchday "Datenraum Industrie 4.0". Retrieved from https://www.plattform-i40.de/IP/Redaktion/DE/Veranstaltungen/2022/09-Pitchday_Stream.html.

Livestream Hannovermesse 2022. Retrieved from https://lnkd.in/eRTg-_p3, start: 1:52:46.

McMahan, B., Moore, E., Ramage, D., Hampson, S. and Aguera y Arcas, B. (2017). Communication efficient learning of deep networks from decentralized data. *In*: *Proceedings of the 20th International Conference on Artificial Intelligence and Statistics*, 1273–82. [Online]. Available: http://proceedings.mlr.press/v54/mcmahan17a.html.

Sattler, F., Wiedemann, S. and Muller, K.-R. (2019). FedProx: Federated Learning with Matched Averaging and Improved Convergence. *In*: *NeurIPS*.

Volkmann, M., Wagner, A., Hermann, J. and Ruskowski, M. (2023). Asset administration shells and gaia-x enabled shared production scenario (in press). *Lecture Notes in Mechanical Engineering* (*LNME*). Springer.

Resilience-X: Towards Resilient Manufacturing Companies
An In-depth Analysis of Requirements and Challenges

Andreas W. Müller,[1*] *Florian Kerber*[2] *and Christoph Legat*[2]

Introduction

In an age characterized by unprecedented global uncertainties and disruptions, the resilience of the manufacturing industry stands as an imperative of utmost significance. The contemporary manufacturing landscape is an intricate web of interconnected processes, technologies, and supply chains that, when subjected to unforeseen challenges, can be vulnerable to disruptions with far-reaching consequences. The capacity of manufacturing enterprises to adapt, recover, and thrive in the face of adversity has become a defining factor in their long-term sustainability and competitiveness.

The manufacturing sector has historically been the backbone of economies, fostering innovation, creating employment opportunities, and driving growth. However, the dawn of the 21st century has ushered in a new era characterized by a profound shift in the dynamics of global trade, the relentless march of technology and innovation as well as an increasing frequency of disruptive events. Natural disasters, geopolitical tensions, supply chain vulnerabilities, and economic

[1] Nuremberg Institute of Technology, Germany.
[2] Technical University of Applied Sciences Augsburg, Germany.
  Email : florian.kerber@tha.de; christoph.legat@tha.de
* Corresponding author: andreas.mueller@th-nuernberg.de

uncertainties have disrupted the manufacturing sector with growing frequency, necessitating a reevaluation of its fundamental resilience.

In this era of global interconnectivity, events in one part of the world can have cascading effects on manufacturing processes and supply chains that span continents. For instance, the COVID-19 pandemic, which began as a localized outbreak, swiftly evolved into a global crisis with profound implications for manufacturing. Lockdowns, travel restrictions, and supply chain disruptions exposed vulnerabilities and underscored the critical need for enhanced resilience strategies.

This book chapter embarks on a comprehensive exploration of the multifaceted concept of resilience within the context of manufacturing. The evolution of manufacturing, intricately intertwined with technological advancements, globalization, and environmental considerations, has heightened the complexity of ensuring resilience. This inquiry into this critical topic aims to provide a roadmap for industry practitioners, policymakers, and scholars navigating the intricate terrain of manufacturing resilience.

The remainder of this chapter is organized as follows: At first, a motivating running example is presented that illustrates challenges in the domain. Afterwards, we examine the understanding of resilience in general and in the manufacturing industry. An overarching reference frame and knowledge scope for aligned understanding is presented subsequently, followed by an investigation of their coverage by current activities in Industry 4.0. Finally, central gaps and possible research directions are explained.

Motivating Example

To illustrate the multifaceted aspects of resilience in the manufacturing industry—the domain considered in this chapter, we consider the example of an SME[1] developing special-purpose production machinery for the automotive industry. Its customers are OEMs[2] that set up new production facilities for the assembly of electric vehicles and in particular the electric drivetrain. The SME is thus integrated into a global supply chain network as a Tier 1 (Figure 1), with relationships to its suppliers and customers, respectively. Within such a network, the partners on these tiers interact collectively towards the OEM.

The advent of electric mobility has had disruptive consequences for the whole automotive industry. New competitors from the US and China challenge the leading OEMs in Europe and North America occupying vacancies in the zero-emissions vehicle market. Moreover, the transition from using internal combustion engines to alternative drive technologies, mainly electric drives, has far reaching consequences in terms of resource allocation and manufacturing technologies. The production of electric drivetrains requires new manufacturing processes, e.g., to assemble battery packs or electric motors. As a result, process technologies such as hairpin

---

[1] Small and medium-sized enterprise.
[2] Original equipment manufacturer.

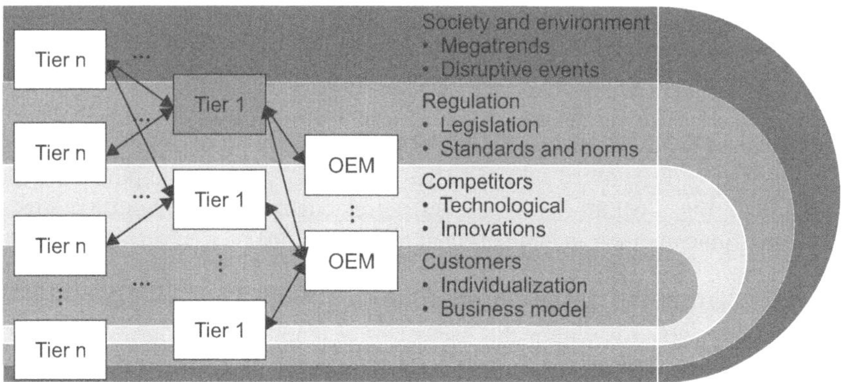

Fig. 1   Globalized supply chain network and influence factors.

soldering, and the resulting quality inspection processes have to be developed. This also comprises a full digital manufacturing twin as part of OEM specifications to be incorporated in digital dataspaces like Manufacturing-X in Germany (VDMA, 2023), Catena-X in Europe (Massoth, 2023; Otto et al., 2022), and Ouranos in Japan.

The additional need for power electronics and control components makes production increasingly dependent on global supply networks and susceptible to shocks such as silicon chip shortages due to trade restrictions. However, current legislative initiatives like the EU supply chain act (European Commission, 2022) cause efforts for supply chain networks. Companies are forced to acquire legal as well as technological expertise and adjust their development and production processes to meet the newly enforced regulatory standards. Typically, the shortage of qualified personnel and expertise required for these processes affects SME suppliers more severely than OEMs as new regulatory and normative requirements are propagated through the supply chain. Consequently, the Tier-1 SME in the example above must invest in human resources to broaden its knowledge base and to develop new partnerships and supplier networks.

New business models like pay-per-use or equipment as a service transform established market principles and foster new competitors even in traditional industries such as manufacturing. Besides, ever increasing requirements for customization cause research and development efforts underpinning the need for modular architectures even for lot size one vendors. In the example above, the Tier-1 SME has to establish standardization concepts to react to changing customer requirements for their product portfolio even during commissioning.

Illustrating some of the most important external influence factors and their effects with the example of the Tier-1 SME, the need for a systematic approach towards comprehensive resiliency mechanisms dedicated to manufacturing companies becomes apparent. The concept presented in the remainder of this chapter will therefore be evaluated from this point of view. The motivating example will be used to identify which particular types of resilience affect a Tier-1 manufacturing

SME, how to adopt the reference frame to develop resilience processes depending on external influences and conditions, and finally how to apply competences to cover these disturbances in a systematic way based on the knowledge scope.

State of Resilience Research and Practice

This section introduces an overview of general approaches and research directions regarding resilience in the manufacturing industry in general.

Common Understanding

Originating from the field of material sciences in the early 19th century, the term and concept of *resilience* has since found its way to a variety of other uses besides engineering. These come from disciplines like ecology, environment, biology, psychology, communities, organizations, or security, considering the interplay of influences from different fields (McAslan, 2010). Today, applications of resilience are typically categorized by main application domains (Florin & Linkov, 2016; Hosseini et al., 2016), which comprise engineering, infrastructure, socio-ecology, and organization as the common denominators relevant in the context of this chapter.

In view of this broad applicability, resilience can be generally understood as:

> *"the intrinsic ability of a system to adjust its functioning prior to, during, or following changes and disturbances, so that it can sustain required operations under both expected and unexpected conditions"* *(Hollnagel et al., 2010).*

Hence, it describes the ability to absorb shocks, adapt to changing circumstances, recover from adversity, and continue functioning effectively. The "resilience curve" (Florin & Linkov, 2016) depicted in Figure 2 illustrates and

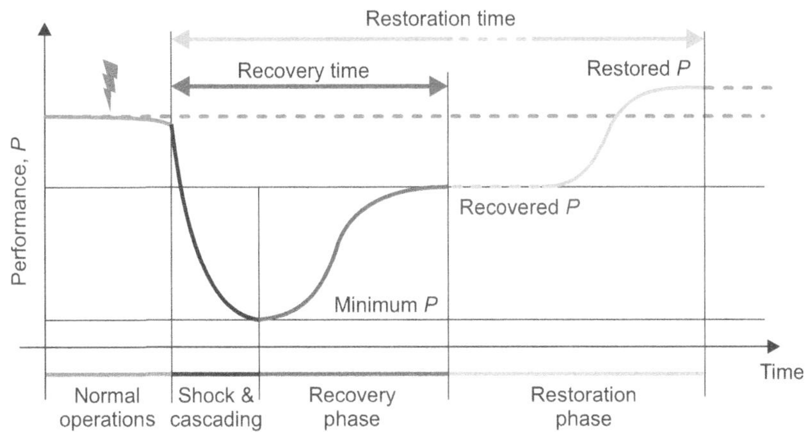

Fig. 2   The resilience curve, taken from (Florin & Linkov, 2016).

emphasizes resilience as a strongly temporal characteristic being subject to various influences during each phase.

It can thus be considered paramount to minimize the overall restoration time, comprising the goals of shortening the shock and cascading, recovery, and restoration phases, respectively. This requires identification, gathering and interpretation of relevant influences to not only cope with the shock but to even achieve an increased performance afterwards.

The definition above further indicates that anything, which can be described, modelled, and understood as a system, along with its (de)compositional and interoperational perspectives, can be subject to a resilience-oriented view as a critical factor. This approach, which is backed by systems theory, is growing in prominence in recent years, up to its extension towards a system-of-systems and cross-domain perspective (Florin & Linkov, 2016).

It is a key characteristic of resilience to form an emergent property of the respective subject, which is positively or negatively affected by various influences and combinations thereof (Florin & Linkov, 2016; Hollnagel et al., 2006). Hence, a complex multifactorial interplay needs to be considered, which varies with the respective domains. For instance, enterprise resilience is often restricted to supply chain management, which in globalized economies naturally has to cope with disruptions. Many conceptual approaches to improve supply chain resilience focus on the adoption of established frameworks such as six sigma techniques, lean management, or agile methods (Thomas et al., 2015). However, there are also quantitative contributions to determine resilience of manufacturing plants based on their technological process cycle and installed equipment (Caputo et al., 2019) and simulation models for supply chain redesign (Carvalho et al., 2012). In civil engineering, the resilience triangle introduced by Tierney & Bruneau (2007) measures a system's performance over time in terms of failure probability, failure consequences, and repair times. In support of the resilience of organizations, ISO 22316 (ISO, 2020) describes the necessary foundational principles, descriptive attributes, and activities for their use, evaluation, and improvement.

Understanding Resilience in the Manufacturing Industry

This section transfers the common understanding of resilience to the considered domain, which is illustrated by the previously introduced example.

Overview

The manufacturing industry, which is focused here, comprises a complex network of activities, from raw material procurement to product distribution. This leads to multiple perspectives, lifecycles and effects becoming interwoven. Generally, a manufacturing company that produces products can appear in the role of customer or supplier. In the case of special-purpose production machinery, the product itself is a machine being tailored to individual requirements. When considering the customer role, one manufacturer can order a machine, which is needed for producing own products, from a supplying manufacturer. For the customer, being resilient means to be able to start or continue operation as quickly as possible (time-to-market). When considering the supplier role, the manufacturing company receives the order for a machine that is needed by another manufacturer. Here, being resilient for the supplier means to deliver the product as quickly as possible (time-to-market). The main lifecycle in such scenarios is formed by the respective manufacturing company since it provides the means for production in the first place. However, the act of ordering induces an interlinking of the main lifecycles of customer and supplier. The products being manufactured are subject to their own lifecycles, which are spawned by the activities of the respective manufacturer as an organization. Hence, each product development initiates a dedicated full lifecycle, with all lifecycles containing the production phase as a logical common meeting point. The time-to-market for each product then depends on a multitude of influences from different fields. This is outlined in Figure 3, with (a) denoting the starting of the lifecycle for a product 'machine' and (b) indicating the respective time-to-market, which can relate to both customer and supplier role. The 'X' shape indicates and emphasizes the logical interconnection and meeting point of the involved lifecycles.

In this context, the time-to-market value is an emergent property and critical factor for the resilience of a manufacturing company, i.e., the ability of a manufacturer to react quickly to, especially external, influences.

When adopting the RAMI 4.0 lifecycle perspective (IEC, 2017) from Industry 4.0 as addressing the manufacturing domain, this maps to a concrete manufacturing company constituting an instance of the type of manufacturing company, i.e., an enterprise that is concerned with manufacturing must have been instantiated (founded) to be able to actually manufacture products. This particular instance of a manufacturing company then initiates the type-instance lifecycles of its ordered

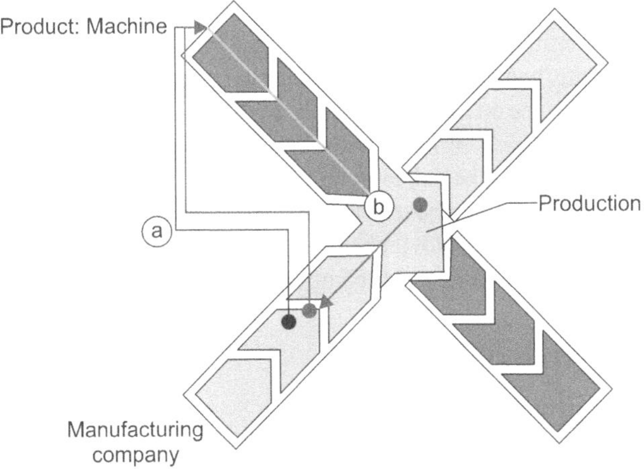

Fig. 3   Interleaving of lifecycles of machinery and manufacturing companies.

products. During this, multiple types of resilience meet on the enterprise level (Bhamra et al., 2011; Florin & Linkov, 2016; Sanchis et al., 2020; Trump et al., 2018) and hence need to be jointly considered (see Table 1):

Table 1   Types of resilience considered relevant for manufacturing companies.

Types of Resilience	*Subject of Consideration*	*Reason*	*Maps to Main Application Domain(s)*
Ecological resilience	Ecosystem	Bidirectional mutual influences of enterprise and ecosystem; likewise applies to data ecosystems.	Socio-ecology, Engineering
Community resilience	Community	Employees of an enterprise form a community, which also interacts with outside communities.	Socio-ecology, Organization
Organizational resilience	Enterprise/ Organization	The entirety of structures, functions, processes constituting an enterprise is potentially subject to shocks.	Organization
Supply chain resilience	Supply chain network	The supply chain as backbone of manufacturing can be subject to shocks.	Infrastructure
Workforce resilience	Human workforce	Employees need to be properly trained to cope with occurring disturbances.	Organization
Manufacturing resilience	Manufacturing system	Production machines, processes and organizations need to be able to properly respond to disturbances.	Engineering

To build overall resilience in the manufacturing domain effectively, it is mandatory to jointly consider and integrate the characteristics and values specific for these types of resilience on a per-enterprise basis as well as from a horizontal supply chain perspective, which forms a network of collaboration (Camarinha-Matos et al., 2022). Conducting the collaborative tasks such that the overall organization or system acts resiliently can benefit from concepts and approaches from the field of collective intelligence (Golovianko et al., 2023). These also address managing the complex interconnections and the knowledge involved.

Transfer to the Domain

Revisiting the automotive SME supplier from Section 2, the competitive market environment demands for a reduced time-to-market of its product, the special-purpose production system for electric drivetrains. The R&D[3] department therefore

---

[3] Research & Development.

adopts agile methods for product development to flexibly react to technological innovations and customer demands alike thus improving its organizational resilience. An accompanying training program for the engineers involved ensures the new processes are skillfully applied in practice to maintain workforce resilience. As a result of the more efficient development process, the machine can be fitted with data analytics algorithms predicting anomalies during productive use. Both supplier and customer thus benefit from this innovation rendering services like predictive maintenance possible and improving machine performance. Resilience is thus fostered both in terms of the organization and workforce as well as manufacturing in general.

Approaching Resilient Systems

This section covers the different aspects to be understood from a technical perspective when addressing the resilience of systems. To this end it first introduces a reference frame for the involved aspects related to the notion of resilient systems. Second, it introduces the corresponding scope of knowledge necessary for their design and operation. In combination, these two form the basis for both per-system and system composition considerations.

Reference Frame

The reference frame intends to clamp the aspects and their interconnections for resilient systems in general.

Overview

Understanding, designing, or engineering resilient systems in any domain requires that any involved subject of resilience can be probed regarding its structure (i.e., time-independent aspects) and both function and behavior (i.e., time-dependent aspects) within the system and regarding the varying influences and conditions in pursue of different goals, such as safety or efficiency (Taha et al., 2021; Wasson, 2015). Influences and conditions are both depending on and comprise aspects such as technical requirements, standards, and legislations, which can be subsumed as conformity requirements.

This requires indicators for phenomena that occur during the system's lifecycle and thus subsequently need to be interpreted for identifying potential countermeasures for the system to maintain its operation. Such indicators are generally required to be in a form that permits simultaneous use for processes, tasks, system models, data acquisition, analytics, and knowledge management. Further, they are required to likewise support perspectives on individual systems, systems (de)composition, and both emergent and reactive behavior. The indicators therefore constitute the basis for the main resilience processes of anticipating situations, monitoring system performance & recognizing situations, responding to situations, and learning from encountered situations. These processes form a loop, in which various patterns, strategies, and models are applied to provide general abilities to systems that can adapt to changing environments (Aven, 2022; Florin & Linkov, 2016; Hollnagel et al., 2010; Trump et al., 2018).

From an application perspective, Constas et al. (2014) and Lezoche & Panetto (2020) list three main purposes of resilience applications, namely, diagnostics ("To which degree is a system resilient to which disturbances?"), evaluation ("To which level are resilience activities successful in achieving their objectives?"), and planning ("What are the consequences of a resilience approach?").

This overall connection of aspects is outlined in conjunction in Figure 4 to form a generic reference frame, with the abbreviations in round brackets introduced forward for later use.

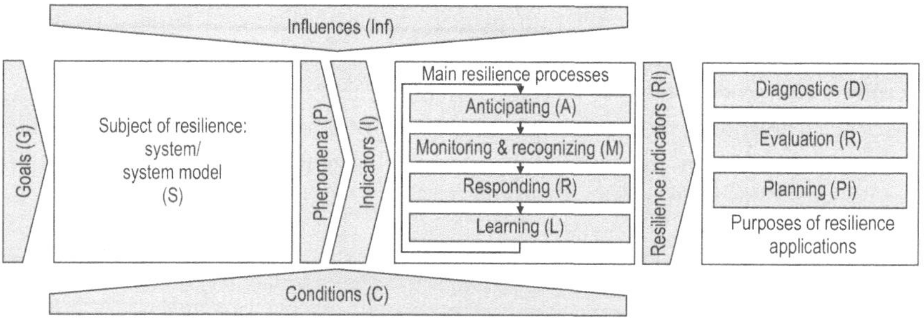

Fig. 4　Reference frame for the overall connection of aspects of resilience.

The reference frame shall be understood as follows: When operating a system in pursuit of given goals, its (de)compositional structures, functions and behaviors (respectively their models) are subject to both less dynamic conditions and more dynamic influences, yielding different phenomena during operation. These phenomena, which also depend on the conditions and influences, need to be well-described to become usable for forming indicators that the main resilience processes can operate on. The resilience process loop takes in the indicators based on the detected phenomena to derive and learn from insights about the system's state of operation, again depending on the conditions and influences. As a result, dedicated resilience indicators are inferred that allow for statements about the degree of resilience of the system with regard to considered aspects. These resilience indicators are then used for diagnosing the system's respective resilience, for evaluating measures influencing the values of indicators, and for identifying and planning suitable measures for improving the system's resilience.

When combining different systems to cover complex tasks and thus forming collaboration networks of systems, the resulting compositions and interconnections lead to the same deliberations from the perspective of collective intelligence (Malone & Bernstein, 2015).

Transfer to the Domain

Consider the example of the special purpose production machine from Section 2 again. The main goal for the supplier company is to meet the commissioning deadline set by its customer, the automotive OEM. The newly adopted agile development process establishes regular customer meetings to react to influences like changing production forecasts, varying throughput times etc. These regular feedback meetings incorporate a dynamic resilience loop during the development phase. For the example of the special purpose production machinery, the SME has to set up a digital toolchain based on digital twin technology, which is introduced in parallel to speed up the development process and make it more traceable at the same time for product traceability. Moreover, the innovation department regularly monitors competitors to spot technological innovations and evaluates them in collaboration with a specialist R&D team. As a result, long-term forecasts of its market performance are provided as part of strategic resilience measures the SME employs.

In a nutshell, the proposed reference frame provides a basis for discussing and aligning necessary tasks and activities towards the design and operation of resilient systems.

Knowledge Perspective on Resilient Systems

This section provides the knowledge scope associated with the reference frame, and outlines implications and connections towards adjacent fields in support of handling the associated knowledge.

Overview

The interactions highlighted in suggest that diverse knowledge must be suitably represented, engineered, managed, and applied collaboratively to yield reliable decision support in addressing disturbances. This requires the convergence and interlinking of heterogeneous expertise.

Typical questions in the manufacturing case may comprise in the following (Florin & Linkov, 2016; Trump et al., 2018):

- Why has something happened, i.e., why has the product been delivered late?
- Does the system meet the requirements, and why (not), i.e., is the delivered product compliant with requirements or has the production met given requirements?
- What are possible remedies, i.e., what can be done to prevent recurring delays?
- What are possible effects and resulting behaviors of individual remedies, i.e., what would happen if a certain remedy is applied?
- Which preventive measures are possible, i.e., which measures are generally possible to ensure short time-to-market?

- How can the system meet given requirements, i.e., how can the system respectively its model be designed for compliance with requirements?

To provide answers to such questions, first describing the statics, i.e., time-independent aspects, and dynamics, i.e., time-dependent aspects, of a system and subsequently capturing and interpreting its current state is considered a prerequisite (Taha et al., 2021; Wasson, 2015). In alignment with Hollnagel et al., (2006, 2010), this involves the competences of information management, communication and coordination, decision-making, and effect control, which simultaneously need to satisfy requirements on coverage of system levels, compositions and interdependencies, lifecycles, and decisions. Additionally, according to Hollnagel et al., (2006, 2010), eight stages of aspirations become important when striving for safe operation of systems, namely on systemic model, complex non-linear accident, systemic ecological system, loss-of-control, integrated wholes and emergence, proactive anticipation, maintenance of control, and constructionist. In total, this spans a universally applicable knowledge scope with regard to the fields of application in Florin & Linkov (2016), Trump et al., (2018). Figure 5 illustrates this scope, with the orderings on the axes indicating the respective levels building on another, i.e., each former is considered prerequisite for the latter.

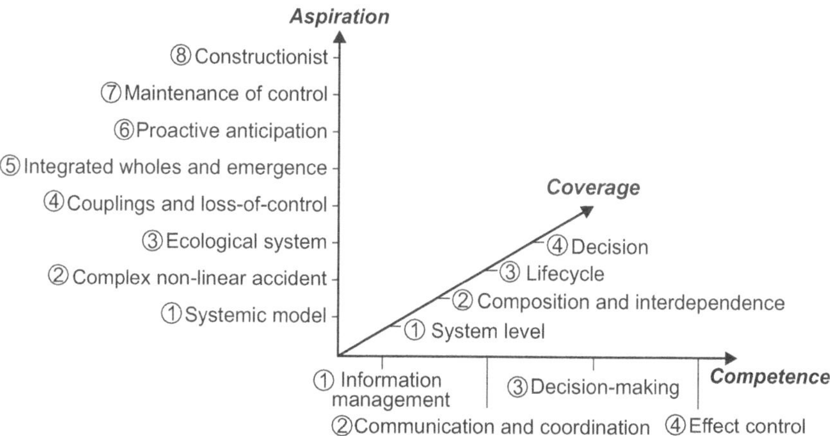

Fig. 5   Scope of resilience-related competences, coverages, and aspirations; derived from (Florin & Linkov, 2016; Hollnagel et al., 2006, 2010; Trump et al., 2018).

The interpretation of this proposed scope of knowledge is then as follows:

- **Axis 'Competence':** This refers to the scope of competences needed to be covered. To suitably manage all necessary information (①) is prerequisite for all further competences, for all abilities along the coverage axis, and for fulfilling the requirements given by the respective perspectives. If information is managed, then it can be communicated and used for coordination purposes (②). If communication is established then informed decision-making (③) becomes possible, and subsequently the control of effects (④) caused by the decisions.

- **Axis 'Coverage':** This refers to the general scope, i.e., types, of knowledge needed to be considered. To gain proper transparency and insights on the system in question, it is prerequisite to describe the system as a (de-) composable entity on a standalone level (①). This permits descriptions of the (de-)composition of the system and in interdependence with other systems as building blocks (②). To understand and manage the dynamics, changes and evolutions of a system and its interdependences with others, covering lifecycle-related information is required (③). Only then describing and making informed decisions becomes possible (④).

- **Axis 'Aspiration':** This refers to the content scope of knowledge required for specific achievements regarding safe and reliable operation. To properly understand aspects of safety and their contexts with regard to resilience, the system in question is considered from different perspectives. First, the model on the systemic level needs to be properly captured (①). If the systemic model is sufficiently described then it is necessary to describe and understand potential and actual accidents with their mechanisms (②). Being able to do that enables modeling the embedding of the system in and its interplay with the surrounding ecological system(s) (③). When understanding this embedding, approaching the perspective of failures and their dynamics, which can lead up to previously uncontrollable effects, becomes feasible (④). That is then the basis to understanding emergent properties when integrating systems (⑤). With this available, it becomes possible to proactively anticipate upcoming situations and prepare for countermeasures (⑥). Such is essential for maintaining control of the system's operation (⑦). With all other perspectives covered, providing a sound constructionist or ontological basis for describing safety-related factors becomes possible (⑧).

This further implies that reaching the highest level in one dimension is not sufficient for reaching resilience for the considered system. Instead, all dimensions need to be addressed.

Implications

In light of the complexity of the knowledge scope it is helpful to identify pivotal considerations in connection to the addressed domain.

Hence, in designing and operating resilient systems, a first step is to address the "information management" competence. This requires identifying, describing and gathering the necessary inputs, outputs, preconditions, used resources, time, and control aspects of the involved functions that contribute to the system and to a certain behavior, along with the respective conditions of applicability, and capabilities for monitoring and control. For this, the Functional Resonance Analysis Method (FRAM) from the field of socio-technical systems offers a structured approach to this identification task and focuses on the dynamics and performance of a system (Hollnagel, 2012). A FRAM-based approach for establishing indicators that can be used in support of this scope is exemplified by Sujan et al., (2021). It is considered transferable to the manufacturing industry domain due to its generic, multi-perspective nature. Additionally, resilience capabilities require dynamic adjustments of systems or processes to recover from undesirable states of operation (Bhamra et al., 2011).

With view to the "information management" competence in its full coverage as comprehensive enabler for the other competences, the perspectives of "proactive anticipation" and "maintenance-of-control" as well as the underlying "ecological system" raise particular interest. This is motivated by the questions stated above and by systems engineering in the manufacturing domain as these perspectives depict key points where changing conditions or influences with their effects on the system are being taken in or reacted on by the resilience processes.

Besides environmental topics, the "ecological system" perspective refers to the space of requirements and constraints the respective system needs to fulfil along its lifecycle. This comprises on the one hand directly system-specific, especially safety- and risk-related, criteria due to required compliance with standards and procedures. This has ever been the subject of resilience research and is backed and exemplified by international standards such as ISO 31000 (ISO, 2022). On the other hand, this refers also and increasingly to the overarching regulatory field, which can be subsumed by "conformity requirements", and which can constitute a strong impact on a system. In that regard, a conformity requirement could be regarded as a disturbance to the system's design or operation, hence posing a risk in the ISO 31000 sense. However, there is little research available in that direction. Sanchis et al., (2020) list regulatory changes as type of disruptive event with respective characteristics.

The "proactive anticipation" and "maintenance of control" perspectives emphasize the need for understanding system dynamics under varying conditions, and for providing suitable up-to-date diagnostic capabilities as pivotal to detect and interpret situations. A typical spectrum of knowledge required for diagnostics is shown in Gao & Liu (2021); Mueller (2012); Mueller et al., (2012). To timely respond on disruptions, simulations of scenarios are a widely employed approach to identify suitable measures (Florin & Linkov, 2016). Given the various simulation methods used in different domains and for different use cases, it is obvious that these exhibit broad demands on understanding and preparing the involved data (Cavalcante et al., 2019; Dong et al., 2022; Kogler & Rauch, 2019; Tan et al., 2020). In Wang et al., (2019), an overview of modeling approaches with regard to simulation of systems is provided, along with specific proposed resilience indicators and involved knowledge, however, for the energy infrastructure domain. In the building energy domain, a multi-scenario simulation for a specific case study was conducted by Katal et al., (2019). An approach to the level-of-detail question for managing resilience-related simulation knowledge is given in Mouelhi et al., (2019), with the notion of cyber-physical systems.

Representing and managing the complex knowledge required for effective assessments on resilience of various subjects requires a high degree of machine-interpretability. For many years, ontologies and concepts from the semantic web, especially knowledge graphs, together with the related standards meet this requirement and offer the needed formal capabilities for knowledge engineering (Blumauer & Nagy, 2020; Pellegrini & Blumauer, 2006; Staab & Studer, 2009; W3C, 2012). In the context of resilience, this has been explored and applied in multiple scenarios (Ameri et al., 2022; Bakirtzis et al., 2022; Dao et al., 2021; Dhakal & Zhang, 2019; Ramzy et al., 2022; Singh et al., 2019; Tian & Li, 2022). A recent example of ontology-based resilience engineering with a combined indicator is given in De Nicola et al., (2023).

As a means for organizing, managing and applying the involved knowledge the widely applicable concept of "digital twin of systems" has emerged in recent years, in a variety of domains and comprising simulations for resilience and simulation-based decision support (Bécue et al., 2020; Brucherseifer et al., 2021; Flammini, 2021; Ivanov et al., 2019; Lektauers et al., 2021; Papacharalampopoulos et al., 2021; Sahlab et al., 2021; Song et al., 2023; Zhou et al., 2021). In the context of Industrie 4.0, this is a subject of ongoing development and standardization in form of the Asset Administration Shell with its domain-specific sub-models (IDTA, 2023; SCI4.0, 2023). These act as redistributable containers of well-structured knowledge.

Eventually, covering the knowledge scope in the addressed domain connects to a variety of approaches in adjacent fields of research.

Transfer to the Domain

A practical implementation of the knowledge scope can be illustrated by means of the example of the special purpose machine for electric drivetrains (see Section 2). In all engineering disciplines, mathematical modeling of systems is used as a prerequisite for system analysis and control. Many different approaches like first principles modeling or energy-based methods (Reddy & Reddy, 2017) like the Port-Hamiltonian framework (Schaft & Jeltsema, 2014) can be used especially for interconnected systems. The mathematical model of the components and their interconnections, in particular the motors to drive the conveyor belt and the robots to mechanically join drivetrain parts as well as the central PLC controlling the machine and interconnecting it to the higher levels of production planning and control, is derived as part of the development process of the machine and stored in a repository for further adjustments and reuse. Operational availability is one of the key selling points of the supplier. Hence, the challenge for the design of the production machine is to compensate external disturbances and uncertainties in the ecosystem it is integrated in. In this example, disturbance observers and impedance controllers for hybrid force-position control make the robot assisted screwing processes for the electric drivetrain robust against sensor failures or environmental influences like changing lighting conditions. As a result, the machinery satisfies key performance indicators like availability, throughput times, and quality requirements based on a physics-based engineering approach.

As a summary, the knowledge scope induces the need for a comprehensive management and joint application of the knowledge on conformity requirements, diagnostic capabilities, and simulation of system behavior as basis for evaluations. With view to the complexity and heterogeneity of systems, this further demands for a standardization and modularization of the technical means to enable and simplify rapid adaptations to varying conditions. Both the fields of ontology and digital twins, along with the Asset Administration Shell concept, offer promising approaches to this challenge in knowledge management and interoperability.

Coverage with View to Industry 4.0

This chapter aims to shed a light on whether current activities and works on comprehensive resilience in the field of Industry 4.0 the focus domain of manufacturing industry described above–address the topics previously outlined by the reference frame and the knowledge scope. By investigating the scope of the selected activities their current coverage and contributions thereto shall be identified and gaps shall be discovered. The selection, which is elaborated in the following subsections, is organized to range in breakdown fashion from the general strategy level to the operational level and features recent related activities with broad scopes in the field. Often, not all aspects are explicitly described or mentioned, hence need to involve interpretations of the selected works. Table 2, using the abbreviations and numberings from Figures 4 and 5, lists the respective coverages as findings.

Strategy on Resilience in Industry 4.0

On an overarching strategic level, Plattform Industrie 4.0 (2022) outlines a classification scheme with resilience phases and focus topics on resilience in alignment with the strategic fields of action for Industry 4.0, which addresses the domain of manufacturing industry considered here. These topics comprise autonomy of all market players, security, i.e., stability, of supply, robust and flexible supply chains, production systems and the role of people, and education, competence, and work organization. For all topics and addressing the types of resilience in Table 1, strategic measures are given that underpin the reference frame and knowledge scope, mapping the ecosystem subject to the concept of data ecosystem.

Design of Value Networks

Focusing on the general design of resilient value networks and hence on the type of supply chain resilience as an essential of the considered domain, acatech (2021) points out network design, data integration, and Industrie 4.0 technologies as the three main fields of action with sets of measures and options for action for each. Among others, these emphasize the need for standardization, for use of software-based risk management systems, and for the use of digital twins for increased transparency. These again underpin the reference frame and knowledge scope.

Table 2   Coverages of the reference frame and the knowledge scope by Industry 4.0 activity ("+" denotes explicit mentioning, "o" denotes "implicitly in scope" to the best of knowledge).

			Considered Current Activities/Works Related to Resilience in Industry 4.0 from Strategy Level (5.1) to Operational Level (5.7)						
			5.1	5.2	5.3	5.4	5.5	5.6	5.7
Reference frame		(Inf)	o	o	o		o	o	o
		(G)	o	o			o		o
		(S)	o	o	+	o	+	+	+
		(P)	o						o
		(I)		o		o		o	o
		(A)	o	o	o	+	o	+	o
		(M)	o	o	o	+	o	+	o
		(R)	o	o	o	+	o	+	o
		(L)	o	o	o	o	o	+	o
		(C)	o	o			o		o
		(RI)				o			
		(D)	o			+	o		
		(E)	o			+	o		+
		(Pl)		o		+	o		o
Knowledge scope	Competence	①	o	o	o	o	o	o	+
		②	o	o		o	o	o	+
		③	o	o		o	o	o	+
		④	o	o		o	o	o	o
	Coverage	①	o	o	+	o	+	o	+
		②	o	o	+	o	+	o	o
		③	o		+		o	o	o
		④	o		o	o	o	o	o
	Aspiration	①	o	o	o	o	+	o	+
		②	o	o	o	o	+	o	o
		③	o	o	o		o	o	+
		④	o	o	o	o	+	o	o
		⑤	o	o	o	o	o		o
		⑥	o	o		o			
		⑦	o	o			o	o	
		⑧	o	o			o		

Sustainability and Resilience in Industry 4.0 and 5.0

On a high level, Aheleroff et al., (2022) outline transitional aspects from Industry 4.0 to Industry 5.0. In that, a proposal for an evolution of the RAMI4.0 to RAMI5.0 is given, along with related challenges. These highlight the importance of better transparency, improved automation of the connected systems involved, decentralization and distribution, and a stronger focus on human workforce as essential to resilience. The outlined concepts apply to the considered domain in full scope and map well to the reference frame and knowledge scope, however, the need for standards-based handling of conformity requirements, indicators, diagnostics and process models can only be assumed.

Smart Resilience Services

Janzen et al., (2021) describe a standards-oriented approach to developing lifelong and collaborative smart resilience services as adaptive IT applications for a proactive, data-driven approach using artificial intelligence and Industry 4.0 standards. In that, requirements' specifications are structured by a 6-step design method to discover involved information objects, involved roles in data handling, required interface services, required internal services, required IoT sensors, and the aspired resilience objective. From a software architecture perspective this results in data-processing services with resilience-related capabilities such as fault pattern recognition, planning of action sequences, representations of data, assets, services, and knowledge. This outlines the range of IT capabilities needed in support of resilience in the domain. While conceptually almost fully covering the reference frame and the knowledge scope from a functional perspective, no aspects related to standards-based handling of conformity requirements, indicators, diagnostics, and process models are given.

Resilience on the Shop Floor

In Weber et al., (2022), an edge computing-oriented resilience concept for the manufacturing industry is introduced. This is based on the concept of assets equipped with RAMI4.0-compatible GAIA-X self-descriptions regarding structure, data, operations, behavior, and dependencies. Provided in semantic web format, these self-descriptions are used to cover challenges from system theory and control, from anomaly detection, and from self-orchestration for automated system reconfigurations. For that, the aspects of information interoperability, stress testing mechanisms, feedback control systems, disturbance-based pattern recognition, and services-based proposition of alternatives are considered. The goal is to achieve the identification and prediction of stress reactions in form of stress scenarios and mechanisms as envisaged foundations for the use of decision support systems and recommender systems. This outlines the range of IT capabilities considered for achieving resilience on the shop floor in the domain. Again, the reference frame and knowledge scope are functionally well-covered but no preparation of and integration with conformity requirements, indicators, diagnostic facilities, and processes in a standards-oriented, reusable form is covered.

Resilience for Smart Manufacturing Services

In Fowler et al., (2023) resilience in the manufacturing context is discussed, including a multilevel categorization of resilience disruptors (macro, meso, micro) to be handled and essential factors for reaching manufacturing resilience, based on the resilience cycle. These factors include and emphasize aspects such as reliability, flexibility, adaptability, configurability, and robustness, all with view to an integrated operation of systems. This outlines the capabilities needed for IT systems in the domain to become resilient themselves, and it maps to the reference frame, especially to the main resilience processes but also as a scope for necessary phenomena and indicators to be considered. Furthermore, this underpins the knowledge scope. Modularity of conformity requirements, indicators, diagnostics, and processes is not addressed but can be assumed to be required.

Smart Manufacturing Based on Swarm Intelligence

In addressing challenges arising with distributed systems that collaborate towards common goals in manufacturing (Guo & Martínez-García, 2021) outline an architecture for real-time processing of sensor data and subsequent self-optimization of the systems involved based on a swarm intelligence approach. In that, resilience can be seen as one of multiple system characteristics to optimized. This approach addresses the distributed nature of the domain as well as that of the interconnected IT systems and it maps to the reference frame especially regarding the aspects of system modeling, phenomena, indicators, and adaptive processes. Further, by the "model fusion" and "data fusion" layers elements of the knowledge scope are addressed with a modularity of all involved entities.

To sum up, from the strategic level to concrete approaches many Industry 4.0 activities on resilience point out and integrate well with the reference frame and knowledge scope, and for many topics there are technical and standardized concepts available. This reaches as far as to employ techniques from the field of collective or swarm intelligence. Within the reference frame gaps in the coverage can be identified especially regarding the reusable modeling and provision of phenomena, conditions, indicators, and resilience indicators, but also regarding the diagnostics, evaluation, and planning processes. Within the knowledge scope the higher aspiration levels also show significant gaps in coverage. Given the variety of applications, however, no overall considerations or standardizations regarding the topics of resilience knowledge itself as outlined by the knowledge scope are addressed.

Gaps and Future Research Directions

As the manufacturing landscape continues to evolve, there are several areas ripe for future research. We identify and discuss essential avenues for in-depth investigation, based on the insights of the previous two sections.

Resilience Language, Metrics, and Assessment

Developing (standardized) metrics and assessment frameworks for resilience measurement is critical to advancing the field. By combining components of different types and origin in systems engineering, phenomena of initially unclear causes can be introduced during system composition. Hence, system resilience is impaired by inhomogeneous interpretation, integration, and use of the knowledge about phenomena and their causal chains. At the same time, changes in the diagnostic knowledge along its lifecycle can result from changes in the system configuration, in the governing influences, such as conformity requirements, or can be based on new insights.

Hence, suitable means to homogeneously describe, capture and interrelate the aspects of resilience given in the reference frame based on the scoped knowledge are required. This induces the need for a standardized domain-specific resilience language and environment, which allows for an overall resilience ecosystem. As a consequence, "knowledge bricks" as redistributable containers of pivotal knowledge along the reference frame and the knowledge scope's aspirations can be identified. In particular, this refers to phenomena in relation to their localization, their interdependencies and effects on system resilience in a redistributable and recombinable form. These are intended to aid in detecting the presence of phenomena depending on the engineered system. Consequently, this must permit a largely automated composition/integration of the diagnostic knowledge into the system engineering tasks to reduce efforts for diagnosing the system upon phenomena detection, and for identifying suitable countermeasures on adverse influences on the system. In covering the reference frame and knowledge scope, this demands for standard knowledge-driven metrics and key performance indicators integrated in an overall knowledge management.

Digital Twins and Simulation

The integration of digital twins and advanced simulation techniques can provide invaluable insights into resilience strategies. Systematic knowledge acquisition and data collection is pivotal during all phases of the life cycle of a system or process. In terms of resilience engineering, simulation and digital twins help understand the normal behavior of systems or processes as well as mechanisms and reactions to disturbances. Simulation models thus provide detailed information in the form of synthetic data that can be used for analysis methods. However, modeling accuracy remains the biggest obstacle when designing simulation models. In many cases, the overall system or process is composed of heterogeneous components the detailed behavior of which is not made available by vendors. Digital twin technology can provide a solution by augmenting multi-physics simulation models with real-life process data.

However, simulation models likewise come with different capabilities that need to be usable interconnectedly, corresponding to the respective configuration of the digitally twinned system. This triggers the need for knowledge bricks for the domain of simulation. Based on standard models these shall support the anticipation of effects and determination of suitable measures in a building block approach to keep up with the changes in the system, influences and behaviors.

Conformity

The dynamically changing conformity landscape of regulations, standards, and sets of requirements imposes a risk to resilient design and operations of systems as well as to companies. Hence, making this knowledge available and usable in machine-interpretable and redistributable form can likewise be a pivotal step towards an automated support of the resilience processes and thus to offering comprehensive decision support.

This motivates the need for knowledge bricks for the domain of conformity, which again permit building block scenarios for knowledge management and application. With the nature of that type of knowledge being strongly driven by natural language, this demands for harmonized modular approaches regarding language models and contents. These can then supplement and influence the resilience and diagnostics knowledge as well as the simulation parts.

Conclusion

In conclusion, this book chapter offers a comprehensive exploration of resilience within the manufacturing industry. By synthesizing existing knowledge and providing insights into the interconnected nature of the manufacturing value chain, the derived reference frame and knowledge scope equip scholars, industry practitioners, and policymakers with a valuable resource for fortifying the resilience of manufacturing enterprises in an ever-changing world. Current activities in the field highlight the importance of modular approaches to keep track of the complex dynamics companies are subject to, but also exhibit gaps regarding the modularity of the involved knowledge. The idea of the availability of standardized re-combinable building blocks for pivotal knowledge raises hope for a simplification of the inherently complex field. As uncertainties and disruptions persist, the pursuit of resilience remains not just a priority but a fundamental imperative for the future of manufacturing.

References

acatech. (2021, Juni 10). Wertschöpfungsnetzwerke in Zeiten von Infektionskrisen – Wie sich Unternehmen für die Zukunft wappnen können. acatech. https://www.acatech.de/allgemein/

publikation-wertschoepfungsnetzwerke-in-zeiten-von-infektionskrisen-wie-sich-unternehmen-fuer-die-zukunft-wappnen-koennen/

Aheleroff, S., Huang, H., Xu, X. and Zhong, R.Y. (2022). Toward sustainability and resilience with Industry 4.0 and Industry 5.0. *Frontiers in Manufacturing Technology, 2.* https://www.frontiersin.org/articles/10.3389/fmtec.2022.951643.

Ameri, F., Sormaz, D., Psarommatis, F. and Kiritsis, D. (2022). Industrial ontologies for interoperability in agile and resilient manufacturing. *International Journal of Production Research, 60*(2), Article 2. https://doi.org/10.1080/00207543.2021.1987553.

Aven, T. (2022). A risk science perspective on the discussion concerning Safety I, Safety II, and Safety III. *Reliability Engineering & System Safety, 217*, 108077. https://doi.org/10.1016/j.ress.2021.108077.

Bakirtzis, G., Sherburne, T., Adams, S., Horowitz, B.M., Beling, P.A. and Fleming, C.H. (2022). An ontological metamodel for cyber-physical system safety, security, and resilience coengineering. *Software and Systems Modeling, 21*(1), Article 1. https://doi.org/10.1007/s10270-021-00892-z.

Bécue, A., Maia, E., Feeken, L., Borchers, P. and Praça, I. (2020). A New Concept of Digital Twin Supporting Optimization and Resilience of Factories of the Future. *Applied Sciences, 10*(13), Article 13. https://doi.org/10.3390/app10134482.

Bhamra, R., Dani, S. and Burnard, K. (2011). Resilience: The concept, a literature review, and future directions. *International Journal of Production Research, 49*(18), Article 18. https://doi.org/10.1080/00207543.2011.563826.

Blumauer, A. and Nagy, H. (2020). *The Knowledge Graph Cookbook.* edition mono/monochrom.

Brucherseifer, E., Winter, H., Mentges, A., Mühlhäuser, M. and Hellmann, M. (2021). Digital Twin conceptual framework for improving critical infrastructure resilience. At - Automatisierungstechnik, *69*(12), Article 12. https://doi.org/10.1515/auto-2021-0104.

Camarinha-Matos, L.M., Rocha, A.D. and Graça, P. (2022). Collaborative approaches in sustainable and resilient manufacturing. *Journal of Intelligent Manufacturing.* https://doi.org/10.1007/s10845-022-02060-6.

Caputo, A.C., Pelagagge, P.M. and Salini, P. (2019). A methodology to estimate resilience of manufacturing plants. *IFAC-PapersOnLine, 52*(13), Article 13. https://doi.org/10.1016/j.ifacol.2019.11.229.

Carvalho, H., Barroso, A.P., Machado, V.H., Azevedo, S. and Cruz-Machado, V. (2012). Supply chain redesign for resilience using simulation. *Computers & Industrial Engineering, 62*(1), Article 1. https://doi.org/10.1016/j.cie.2011.10.003.

Cavalcante, I.M., Frazzon, E.M., Forcellini, F.A. and Ivanov, D. (2019). A supervised machine learning approach to data-driven simulation of resilient supplier selection in digital manufacturing. *International Journal of Information Management, 49*, 86–97. https://doi.org/10.1016/j.ijinfomgt.2019.03.004.

Constas, M., Frankenberger, T., Hoddinott, J., Mock, N., Romano, D., Béné, C. and Maxwell, D. (2014). *A common analytical model for resilience measurement.* Report number, Technical Series Paper No. 2. Resilience Measurement Technical Working Group, Food Security Information Network. Dao, J., Ng, S.T., Yang, Y., Zhou, S., Xu, F.J., and Skitmore, M. (2021). Semantic framework for interdependent infrastructure resilience decision support. *Automation in Construction, 130*, 103852. https://doi.org/10.1016/j.autcon.2021.103852.

De Nicola, A., Villani, M.L., Sujan, M., Watt, J., Costantino, F., Falegnami, A. and Patriarca, R. (2023). Development and measurement of a resilience indicator for cyber-socio-technical systems: The allostatic load. *Journal of Industrial Information Integration, 35*, 100489. https://doi.org/10.1016/j.jii.2023.100489.

Dhakal, S. and Zhang, L. (2019). Ontology-Based Semantic Modeling of Disaster Resilient Construction Operations: Towards a Knowledge-Based Decision Support System. *In*: I. Mutis and T. Hartmann (Hrsg.), *Advances in Informatics and Computing in Civil and Construction Engineering,* S. 789–96. Springer International Publishing. https://doi.org/10.1007/978-3-030-00220-6_95.

Dong, B.-X., Shan, M. and Hwang, B.-G. (2022). Simulation of transportation infrastructures resilience: A comprehensive review. *Environmental Science and Pollution Research, 29*(9), Article 9. https://doi.org/10.1007/s11356-021-18033-w.

European Commission. (2022). Proposal for a DIRECTIVE OF THE EUROPEAN PARLIAMENT AND OF THE COUNCIL on Corporate Sustainability Due Diligence and amending Directive (EU) 2019/1937. https://eur-lex.europa.eu/legal-content/EN/TXT/?uri=CELEX:52022PC0071.

Flammini, F. (2021). Digital twins as run-time predictive models for the resilience of cyber-physical systems: A conceptual framework. *Philosophical Transactions of the Royal Society A: Mathematical, Physical and Engineering Sciences, 379*(2207), Article 2207. https://doi.org/10.1098/rsta.2020.0369.

Florin, M.V. and Linkov, I. (Hrsg.). (2016). *IRGC Resource guide on Resilience.* EPFL International Risk Governance Center (IRGC). https://doi.org/10.5075/epfl-irgc-228206.

Fowler, D.S., Epiphaniou, G., Higgins, M.D. and Maple, C. (2023). Aspects of resilience for smart manufacturing systems. *Strategic Change, jsc., 2555.* https://doi.org/10.1002/jsc.2555.

Gao, Z. and Liu, X. (2021). An Overview on Fault Diagnosis, Prognosis and Resilient Control for Wind Turbine Systems. *Processes, 9*(2), Article 2. https://doi.org/10.3390/pr9020300.

Golovianko, M., Terziyan, V., Branytskyi, V. and Malyk, D. (2023). Industry 4.0 vs. Industry 5.0: Co-existence, Transition, or a Hybrid. *Procedia Computer Science, 217,* 102–13. https://doi.org/10.1016/j.procs.2022.12.206.

Guo, J. and Martínez-García, M. (2021). Key technologies towards smart manufacturing based on swarm intelligence and edge computing. *Computers & Electrical Engineering, 92,* 107119. https://doi.org/10.1016/j.compeleceng.2021.107119.

Hollnagel, E. (2012). FRAM: *The Functional Resonance Analysis Method: Modelling Complex Socio-Technical Systems.* Taylor & Francis Group. http://ebookcentral.proquest.com/lib/thnuernberg/detail.action?docID=906968.

Hollnagel, E., Pariès, J., Woods, D.D. and Wreathall, J. (2010). *Resilience Engineering in Practice: A Guidebook.* Taylor & Francis Group. http://ebookcentral.proquest.com/lib/thnuernberg/detail.action?docID=615608.

Hollnagel, E., Woods, D.D. and Leveson, N. (2006). *Resilience Engineering: Concepts and Precepts.* Taylor & Francis Group. http://ebookcentral.proquest.com/lib/thnuernberg/detail.action?docID=429564.

Hosseini, S., Barker, K. and Ramirez-Marquez, J.E. (2016). A review of definitions and measures of system resilience. *Reliability Engineering & System Safety, 145,* 47–61. https://doi.org/10.1016/j.ress.2015.08.006.

IDTA. (2023, April 18). IDTA: Der Standard für den Digitalen Zwilling. IDTA Deutsch. https://industrialdigitaltwin.org/

IEC. (2017, März). IEC PAS 63088: 2017—*Smart manufacturing—Reference Architecture Model Industry 4.0 (RAMI4.0).* https://www.vde-verlag.de/iec-normen/224330/iec-pas-63088-2017.html.

ISO. (2020, September 15). ISO 22316: 2017—*Security and resilience—Organizational Resilience—Principles and Attributes.* ISO. https://www.iso.org/standard/50053.html.

ISO. (2022). ISO 31000: 2018 *Risk management—Guidelines.* https://www.iso.org/standard/65694.html.

Ivanov, D., Dolgui, A., Das, A. and Sokolov, B. (2019). Digital Supply Chain Twins: Managing the Ripple Effect, Resilience, and Disruption Risks by Data-Driven Optimization, Simulation, and Visibility. *In*: D. Ivanov, A. Dolgui, and B. Sokolov (Hrsg.), *Handbook of Ripple Effects in the Supply Chain,* S. 309–32. Springer International Publishing. https://doi.org/10.1007/978-3-030-14302-2_15.

Janzen, S., Öksüz, N., Sporkmann, J., Schlappa, M., Gerhard, J., Ortjohann, L. and Becker, P.A. (2021). Smart Resilience Services for Industrial Production. *In: VDI Wissensforum GmbH (Hrsg.), Automation, 2021,* S. 621–36. VDI Verlag. https://doi.org/10.51202/9783181023921-621.

Katal, A., Mortezazadeh, M. and Wang, L. (Leon). (2019). Modeling building resilience against extreme weather by integrated CityFFD and CityBEM simulations. *Applied Energy, 250,* 1402–17. https://doi.org/10.1016/j.apenergy.2019.04.192.

Kogler, C. and Rauch, P. (2019). A discrete-event simulation model to test multimodal strategies for a greener and more resilient wood supply. *Canadian Journal of Forest Research, 49*(10), Article 10. https://doi.org/10.1139/cjfr-2018-0542.

Lektauers, A., Pecerska, J., Bolsakovs, V., Romanovs, A., Grabis, J. and Teilans, A. (2021). A Multi-Model Approach for Simulation-Based Digital Twin in Resilient Services. *WSEAS Transactions on Systems and ControL, 16,* 133–45. https://doi.org/10.37394/23203.2021.16.10.

Lezoche, M. and Panetto, H. (2020). Cyber-Physical Systems, a new formal paradigm to model redundancy and resiliency. *Enterprise Information Systems, 14*(8), Article 8. https://doi.org/10.1080/17517575.2018.1536807.

Malone, T.W. and Bernstein, M.S. (2015). *Handbook of Collective Intelligence*. MIT Press. http://ebookcentral.proquest.com/lib/thnuernberg/detail.action?docID=4093115.

Massoth, H. (2023). *Catena-X Operating Model Whitepaper*.

McAslan, A. (2010). *The Concept of Resilience—Understanding its Origins, Meaning, and Utility* (A Strawman Paper). Torrens Resilience Institute, Flinders University.

Mouelhi, S., Laarouchi, M.-E., Cancila, D. and Chaouchi, H. (2019). Predictive Formal Analysis of Resilience in Cyber-Physical Systems. *IEEE Access*, *7*, 33741–33758. https://doi.org/10.1109/ACCESS.2019.2903153.

Mueller, A. (2012). Meeting Challenges of Generic Pervasive Diagnostics. *Proceedings of 17th International Conference on Emerging Technologies & Factory Automation* (*ETFA*).

Mueller, A., Hofmann, I., Oberkampf, H. and Zillner, S. (2012). Knowledge Engineering Requirements for Generic Diagnostic Systems. *Proceedings of the 4th International Conference on Knowledge Management and Ontology Development* (*KEOD 2012*).

Otto, B., Ten Hompel, M. and Wrobel, S. (Hrsg.) (2022). *Designing Data Spaces: The Ecosystem Approach to Competitive Advantage*. Springer International Publishing. https://doi.org/10.1007/978-3-030-93975-5.

Papacharalampopoulos, A., Michail, C.K. and Stavropoulos, P. (2021). Manufacturing resilience and agility through processes digital twin: Design and testing applied in the LPBF case. *Procedia CIRP*, *103*, 164–69. https://doi.org/10.1016/j.procir.2021.10.026.

Pellegrini, T. and Blumauer, A. (Hrsg.) (2006). *Semantic Web*. Springer Berlin Heidelberg. http://link.springer.com/10.1007/3-540-29325-6.

Plattform Industrie 4.0. (2022). *Resilience in the Context of Industrie 4.0* (Whitepaper). BMWK. https://www.plattform-i40.de/IP/Redaktion/DE/Downloads/Publikation/Resilienz.html.

Ramzy, N., Auer, S., Ehm, H. and Chamanara, J. (2022). MARE: Semantic Supply Chain Disruption Management and Resilience Evaluation Framework. *arXiv*: 2205.06499; Nummer *arXiv*: 2205.06499). arXiv. https://doi.org/10.48550/arXiv.2205.06499.

Reddy, J.N. and Reddy, J.N. (2017). *Energy Principles and Variational Methods in Applied Mechanics* (3rd Edn.). Wiley.

Sahlab, N., Kamm, S., Müller, T., Jazdi, N. and Weyrich, M. (2021). Knowledge Graphs as Enhancers of Intelligent Digital Twins. *2021 4th IEEE International Conference on Industrial Cyber-Physical Systems* (*ICPS*), 19–24. https://doi.org/10.1109/ICPS49255.2021.9468219.

Sanchis, R., Canetta, L. and Poler, R. (2020). A Conceptual Reference Framework for Enterprise Resilience Enhancement. *Sustainability*, *12*(4), 1464. https://doi.org/10.3390/su12041464.

Schaft, A. van der and Jeltsema, D. (2014). Port-Hamiltonian Systems Theory: An Introductory Overview. *Foundations and Trends® in Systems and Control*, *1*(2–3), 173–378. https://doi.org/10.1561/2600000002.

SCI4.0. (2023). Industrie 4.0 Komponente und das Konzept der Verwaltungsschale. Standardization Council Industrie 4.0. https://www.sci40.com/german/themenfelder/verwaltungsschale/

Singh, S., Ghosh, S., Jayaram, J. and Tiwari, M.K. (2019). Enhancing supply chain resilience using ontology-based decision support system. *International Journal of Computer Integrated Manufacturing*, *32*(7), Article 7. https://doi.org/10.1080/0951192X.2019.1599443.

Song, J., Liu, S., Ma, T., Sun, Y., Tao, F. and Bao, J. (2023). Resilient digital twin modeling: A transferable approach. *Advanced Engineering Informatics*, *58*, 102148. https://doi.org/10.1016/j.aei.2023.102148.

Staab, S. and Studer, R. (Hrsg.) (2009). *Handbook on Ontologies*. Springer. https://doi.org/10.1007/978-3-540-92673-3.

Sujan, M., Watt, J., Patriarca, R., Costantino, F., Villani, M.L. and De Nicola, A. (2021, Februar 9). Developing Leading Safety Indicators using the Functional Resonance Analysis Method. *Conference on Safety Critical Systems Symposium*, York.

Taha, W.M., Taha, A.-E. M. and Thunberg, J. (2021). *Cyber-Physical Systems: A Model-Based Approach*. Springer Nature. https://doi.org/10.1007/978-3-030-36071-9.

Tan, W.J., Cai, W. and Zhang, A.N. (2020). Structural-aware simulation analysis of supply chain resilience. *International Journal of Production Research*, *58*(17), Article 17. https://doi.org/10.1080/00207543.2019.1705421.

Thomas, A., Pham, D.T., Francis, M. and Fisher, R. (2015). Creating resilient and sustainable manufacturing businesses: A conceptual fitness model. *International Journal of Production Research, 53*(13), Article 13. https://doi.org/10.1080/00207543.2014.975850.

Tian, Y. and Li, W. (2022). GeoAI for Knowledge Graph Construction: Identifying Causality Between Cascading Events to Support Environmental Resilience Research. *arXiv*: 2211.06011; Nummer *arXiv*: 2211.06011). *arXiv*. https://doi.org/10.48550/arXiv.2211.06011.

Tierney, K.J. and Bruneau, M. (2007). Conceptualizing and measuring resilience: A key to disaster loss reduction. *Scopus*. https://api.semanticscholar.org/CorpusID:63244164.

Trump, B.D., Florin, M.-V. and Linkov, I. (Hrsg.) (2018). *IRGC Resource Guide on Resilience* (Volume 2). International Risk Governance Center (IRGC). https://doi.org/10.5075/epfl-irgc-262527.

VDMA. (2023). Vorstudie Datenraum Manufacturing-X.

W3C. (2012, Dezember 11). *OWL 2 Web Ontology Language Direct Semantics* (2nd Edition). https://www.w3.org/TR/2012/REC-owl2-direct-semantics-20121211/

Wang, J., Zuo, W., Rhode-Barbarigos, L., Lu, X., Wang, J. and Lin, Y. (2019). Literature review on modeling and simulation of energy infrastructures from a resilience perspective. *Reliability Engineering & System Safety, 183*, 360–73. https://doi.org/10.1016/j.ress.2018.11.029.

Wasson, C.S. (2015). *System Engineering Analysis, Design, and Development: Concepts, Principles, and Practices*. John Wiley & Sons.

Weber, M., Brinkhaus, J., Dumss, S., Henrich, V., Hoffmann, F., Ristow, G.H., Schickling, C., Trautner, T., Grafinger, M., Weigold, M. and Bleicher, F. (2022). EuProGigant Resilience Approach: A Concept for Strengthening Resilience in the Manufacturing Industry on the Shop Floor. *Procedia CIRP, 107*, 540–45. https://doi.org/10.1016/j.procir.2022.05.022.

Zhou, C., Xu, J., Miller-Hooks, E., Zhou, W., Chen, C.-H., Lee, L. H., Chew, E.P. and Li, H. (2021). Analytics with digital-twinning: A decision support system for maintaining a resilient port. *Decision Support Systems, 143*, 113496. https://doi.org/10.1016/j.dss.2021.113496.

5

Multi-agent Reinforcement-Learning for Solving Flexible Job Shop Scheduling Problems

Felix Schmalzel,[1*] *Martin Springer*[2] *and Thorsten Schöler*[3]

Introduction

In recent decades, manufacturing has transformed significantly from traditional mass production to adaptable manufacturing. This shift is driven by factors such as the demand for shorter product lifecycles and catering to diverse consumer preferences. The industry now emphasizes flexibility and customization. This product variation in combination with the rigidity of traditional manufacturing presents a problem. Modern technologies, such as automation, robotics, and IoT (internet of things), enable real-time monitoring, data-driven decision-making, and seamless integration of complex systems. This increases the need for optimal automatic scheduling solutions. The problem can be characterized as a flexible job shop scheduling problem (FJSSP). It is an NP-hard problem, meaning there is no straightforward method for finding an optimal solution. (MacCarthy and Liu, 1993; Keddis et al., 2013)

Traditional methods have leaned on heuristics that use models and constraints. These methods, while providing some relief, are not without limitations. Creating a fully formulated mathematical model with all constraints is time-intensive. The

[1,2,3] Technical University of Applied Sciences Augsburg, Germany.
Email : Martini.Springer@googlemail.com; thorsten.schoeler@hs-augsburg.de
* Corresponding author: Schmalzel.Felix@gmail.com

optimization is a resource-intensive operation, often done in advance, making it less adaptable to sudden changes (A.P. Muhlemann and Farn, 1982).

Recent developments in the field of machine learning present reinforcement learning as a powerful tool for solving complex problems. Reinforcement learning has showcased remarkable results in a variety of applications. DeepMind's AlphaGO defeating the world champion Go player (Silver et al., 2016) or OpenAI's Dota 2 Bot that is able to play at a high level (Berner et al., 2019). The techniques have matured (Liang et al., 2018) and are employed in production systems used in everyday life by companies like YouTube (Covington, Adams, and Sargin, 2016).

Our approach adopts a multi-agent reinforcement learning (MARL) framework to tackle the FJSSP. Agents represent products within a grid-based factory environment. Unlike traditional methods (e.g., swarm heuristics and genetic algorithms), this grid-based representation offers more flexibility, allowing for modifications to the manufacturing layout (as parameters of the factory model) without the need for additional fine-tuning (of the generalized model's structure). Only the contents of the observations change and not the structure of the observations themselves. We do not use the network to plan the complete manufacturing run but instead infer it for every time step, unlike traditional approaches. With this we hope to be able to quickly adapt to changes such as machine failures, product variations, factory expansions, etc.

We honor a distributed learning approach for the products, yielding a decentralized control algorithm. Every product is represented by an agent. Those agents choose their next action independently. Coordination and cooperation of agents is a result of the learned behavior and their global observations. With this, we are able to find optimal solutions by using many small models, through their collective intelligence. It is important to note that all those models have the same structure and weights.

A similar approach was chosen by Bär et al. (Baer et al., 2019). They used a petri-net to represent the factory. We use a grid to ensure easy changes to the layout without the need to relearn the transitions.

Through the development and testing of a prototype, we compare our solution with a theoretical optimum for a small factory, providing a rigorous evaluation of its effectiveness.

In the following pages, we present our methodology, findings, and implications. By addressing the core challenges faced by manufacturers, this research hopes to increase efficiency and reduce downtime for modern manufacturing.

Problem Definition

In this chapter we want to mathematically formulate the problem with the goal of portraying the challenge the MARL algorithm has to solve. We begin by defining the variables that are present in the problem, as seen in Equation 1.

Equation 1

$$t: \text{Timestep}$$
$$P(t): \text{List of products}$$
$$P_i(t): \text{A single product}$$
$$P_i^{pos}(t): \text{Position at step } t \text{ for product } P_i$$
$$P_i^{proc}(t): \text{List of outstanding processes at step } t \text{ for product } i$$
$$P_i^{counter}(t): \text{Remaining time of the currently running manufacturing process}$$
$$Processes: \text{Set of all processes}$$
$$M: \text{List of all machines}$$
$$M_i: \text{A single machine}$$
$$M_i^{proc}: \text{Set of processes that is supported by this machine}$$
$$M_i^{time}: \text{The time this machine takes to complete a process}$$
$$Actions: \text{All possible actions}$$
$$G: \text{A matrix representing the grid world}$$
$$\varepsilon: \text{Empty field}$$
$$B: \text{Omnidirectional transport belt}$$
$$I: \text{Input module}$$
$$O: \text{Output module}$$
$$valid\ Action(i, t): \text{True if the action is valid}$$
$$Input_{pos}: \text{Position of the Input module}$$
$$Output_{pos}: \text{Position of the Output module}$$

The product $P_i(t)$ is a tuple made up of a position and a list of required processes.

Equation 2

$$P(t) = [P_0(t), P_1(t), P_2(t), ..., P_n(t)]$$
$$P_i(t) = (P_i^{pos}(t), P_i^{proc}(t), P_i^{counter}(t))$$
$$P_i^{pos}(t) = (x, y)$$

The position $P_i^{pos}(t)$ is a simple tuple of two integers. All products have to start at the position of the input module. The position can be manipulated by moving in any of the cardinal directions using the corresponding actions.

Equation 3

$$P_i^{pos}(0) = Input_{pos}$$

$$P_i^{pos}(t) = P_i^{pos}(t-1) + \begin{cases} (0, 1) & \text{if } action_i(t-1) = N \\ (1, 0) & \text{if } action_i(t-1) = E \\ (0, -1) & \text{if } action_i(t-1) = S \\ (-1, 0) & \text{if } action_i(t-1) = W \\ (0, 0) & \text{otherwise} \end{cases}$$

The list of processes a product has to go through ($P_{iproc}(t)$) is given in order. If the product was processed, then the first element of the list is removed.

Equation 4

$$Processes \subset N_0$$
$$P_i^{roc}(0) \neq \varnothing$$
$$p_i^{proc}(t) \subseteq Processes$$
$$P_i^{proc}(t) = \begin{cases} P_i^{proc}(t-1)\backslash\{P_i^{proc}(t-1)_0\} & \text{if } action_i\,(t-1) = Machining \\ P_i^{proc}(t-1) & \text{otherwise} \end{cases}$$

A product cannot do anything whilst it is using a machine. After the machining time, the product can move freely again.

Equation 5

$$P_i^{counter}(t)$$
$$= \begin{cases} M_{curretn}^{time} - 1 & \text{if } action_i\,(t-1) = Machiing \Rightarrow P_i^{pos}(t) = (x, y) \Rightarrow M_{current} = g_{x,y} \\ P_i^{counter}(t-1) - 1 & \text{if } P_i^{counter}\,(t-1) > 0 \\ 0 & \text{otherwise} \end{cases}$$

The machine *Mi* is a tuple made up of a list of supported processes and the time the processing takes. The Processes list contains only the processes that are supported by at least a single machine.

Equation 6

$$Machines = [M_0, M_1, M_2, ..., M_n]$$
$$M_i = (M_i^{proc}, M_i^{time})$$
$$M_i^{proc} \subseteq Processes \wedge M_i^{proc} \neq \varnothing$$
$$\forall p \in Processes\ \exists i: p \in M_i^{proc}$$
$$M_i^{time} > 0$$

To manipulate the products, actions are used. The possible actions include moving in the cardinal directions: *N,E,S,W*, using a machine: *Machining* and doing nothing: *Sleep*. Accordingly, the set of actions is defined as *Actions =* {*N,E,S,W,Machining,Sleep*}.

The grid is made up of modules. These modules include an input module, output module, belts, and all the machines previously defined. An empty field is represented using ε. All products start on the input module. They have to go to the output module after finishing their respective process lists. The belts allow movement. There can only be a single input module and a single output module.

Equation 7

$$G = \begin{pmatrix} g_{0,0} & g_{1,0} & g_{2,0} & \cdots & g_{x,0} \\ g_{0,1} & & & & \\ g_{0,2} & & \ddots & & \vdots \\ \vdots & & & & \\ g_{0,y} & & & & g_{x,y} \end{pmatrix}$$

$$g_{x,y} \in \{\varepsilon, B, I, O\} \cup M$$
$$\exists!x\exists!y\ g_{x,y} = I \Rightarrow Input_{pos} = (x, y)$$
$$\exists!x\exists!y\ g_{x,y} = O \Rightarrow Output_{pos} = (x, y)$$

Not all actions can be used at all times. To use the *Machining* action, the product has to be in a machine that supports the next needed process. For $\{N,E,S,W\}$ the target module cannot be an empty field. The target module also has to be empty of other products. The output module can only be entered if all processes are finished. See Figure 1.

The goal of the simulation is to minimize the makespan (Leung and Pinedo, 2007): Given $M,P(0),G$, find a list of $action_i(t)$ for all *is* and *ts* that optimizes $\min_t\{t|\forall iP_{ipos}(t)=Output_{pos}\wedge P_{iproc}t=\emptyset\}$.

Equation 8

$$G = \begin{pmatrix} \varepsilon & O & \varepsilon \\ I & B & M_0 \\ \varepsilon & M_1 & \varepsilon \end{pmatrix}$$

$$P(0) = [((0,2), [0,1]), (0,2), [0, 1])]$$
$$M = [(\{0\}, 5), (\{1\}, 5)]$$

Simulation

Environment

The foundation of this research lies in the design and implementation of a dynamic and customizable environment based on grids. This environment serves as the simulated factory for the agents to interact, learn, and make decisions. The grid-based structure allows for a structured representation of the workspace, with each cell representing a module that may be traversed by the products. The modular design of these cells enables addition and removal of modules directly within the source code, providing flexibility in constructing diverse scenarios and tasks. In addition to that, new functionalities such as new cells can also be implemented and used in the environment. A picture of a possible layout can be seen in Figure 2.

Input: P, M, G, i, t
Result: $isValid$

$x, y \leftarrow P_i^{pos}(t)$
if $P_i^{counter}(t) > 0$ **then**
  if $action_i(t) = Sleep$ **then**
    $isValid \leftarrow True$
  else
    $isValid \leftarrow False$
  end if
else if $action_i(t) \in \{N, E, S, W\}$ **then**
  $x_2, y_2 \leftarrow P_i^{pos}(t + 1)$
  if $g_{x2,y2} \neq \varepsilon$ **then**
    if $\forall j\, P_j \in P \backslash \{P_i\}: P_j^{pos}(t + 1) \neq P_i^{pos}(t + 1)$ **then**
      if $g_{x2,y2} \neq O$ **then**
        $isValid \leftarrow True$
      else if $P_i^{proc}(t) = \emptyset P$ **then**
        $isValid \leftarrow True$
      else
        $isValid \leftarrow False$
      end if
    else
      $isValid \leftarrow False$
    end if
  else
    $isValid \leftarrow False$
  end if
else if $action_i(t) = Machining$ **then**
  if $g_{x,y} \in M$ **then**
    $M_{current} \leftarrow g_{x,y}$
    if $P_i^{proc}(t)_0 \in M_{current}^{proc}$ **then**
      $isValid \leftarrow True$
    else
      $isValid \leftarrow False$
    end if
  else
    $isValid \leftarrow False$
  end if
else
  $isValid \leftarrow True$
end if

Fig. 1   Algorithm to check if an action is valid.

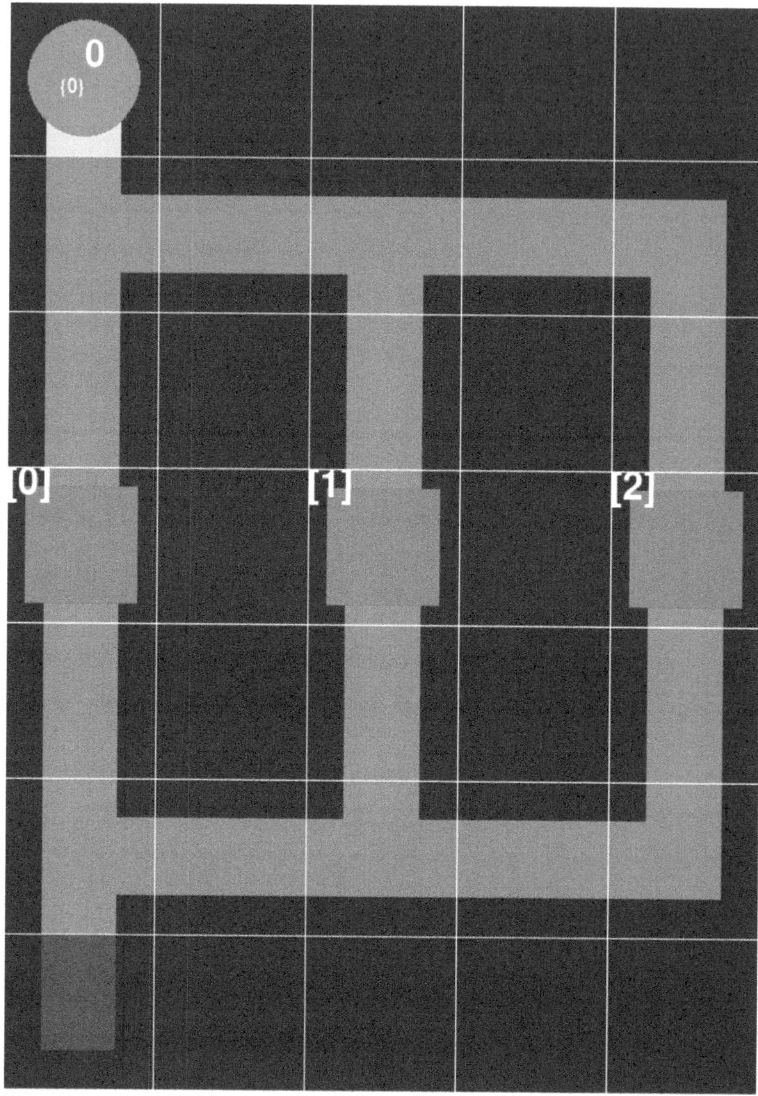

Fig. 2 Exemplary Layout of a possible factory simulation.

The environment not only fosters visualization and interpretation of the agent's actions but also offers opportunities for the study of spatial relationships, as well as coordination among multiple agents. By incorporating diverse modules and altering their configurations, we can simulate various real-world scenarios, each posing unique challenges for the MARL agents to adapt and optimize their strategies. An example of that would be the paths the agents take and whether they can make space for a passing agent or not.

In this research, we present how the dynamic nature of the environment empowers us to explore different multi-agent scenarios, varying complexities, and heterogeneous agent capabilities. The flexibility to manipulate the factory layout through the source code provides the possibility to investigate the performance and robustness of our MARL agents under diverse conditions.

Observations

Equation 9

$$Observation(i, t): \text{Observations for product } P_i \text{ at step } t$$
$$Observation(i, t) = Transitions \oplus Targets(i, t) \oplus Positions(i, t)$$

For the implementation of the reinforcement learning, we adopt global observations that provide the agents with the necessary information about the state of the factory simulation, leaving no internal state withheld. Our focus on a factory setting allows us to leverage a 1D-array of combined and flattened observations that mirror real-world scenarios. The observations consist of three key components described in the following.

Transition Map

The Transition map is an array with a size of $m*n*4$ for grid of shape $m \times n$. In this array, each tile contains 4 bits that signify whether the agent can exit the tile by moving North, East, South, or West. This information is crucial for the agents to know in which ways the product can proceed from any given tile. See Equation 10.

Equation 10

Transitions = [

$tr_{0,0,0}, tr_{0,0,1}, tr_{0,0,2}, tr_{0,0,3}, tr_{1,0,0}, tr_{1,0,1}, tr_{1,0,3}, ..., tr_{x,0,0}, tr_{x,0,1}, tr_{x,0,2}, tr_{x,0,3}, tr_{0,1,0},$
$tr_{0,1,1}, tr_{0,1,2}, tr_{0,1,3}, tr_{1,1,0}, tr_{1,1,1}, tr_{1,1,2}, tr_{1,1,3}, ..., tr_{x,1,0}, tr_{x,1,1}, tr_{x,1,2}, tr_{x,1,3}, ..., tr_{x,y,0},$
$tr_{x,y,1}, tr_{x,y,2}, tr_{x,y,3}]$

$$tr_{x,y,0} = \begin{cases} 1 & \text{if } g_{x,y} \neq \varepsilon \wedge g_{x,y-1} \neq \varepsilon \\ 0 & \text{otherwise} \end{cases}$$

$$tr_{x,y,1} = \begin{cases} 1 & \text{if } g_{x,y} \neq \varepsilon \wedge g_{x+y,1} \neq \varepsilon \\ 0 & \text{otherwise} \end{cases}$$

$$tr_{x,y,2} = \begin{cases} 1 & \text{if } g_{x,y} \neq \varepsilon \wedge g_{x,y+1} \neq \varepsilon \\ 0 & \text{otherwise} \end{cases}$$

$$tr_{x,y,3} = \begin{cases} 1 & \text{if } g_{x,y} \neq \varepsilon \wedge g_{x-1,y} \neq \varepsilon \\ 0 & \text{otherwise} \end{cases}$$

Target Map

The *Target(i,t)* is an array with a size of *m*n* for a grid of shape *m×n*. This map serves as a guide for the agents by indicating the current target they need the product to interact with. For instance, if the product requires process 5 and there are multiple machines capable of performing this process, the corresponding tiles will be marked with a 1 in this array. Furthermore, when the product is fully processed and ready for exiting the factory, a 1 is placed at the output module in the Target map. See Equation 11.

Equation 11

$$Targets(i, t) = [$$
$$target_{0,0},\ target_{1,0},\ target_{2,0},\ ...,\ target_{x,0},$$
$$target_{0,1},\ target_{1,1},\ target_{2,1},\ ...,\ target_{x,1},$$
$$...,\ target_{x,y}]$$

$$target_{x,y}(i, t) = \begin{cases} 1 & \text{if } g_{x,y} = 0 \wedge P_i^{proc}(t) = \varnothing \\ 1 & \text{if } g_{x,y} \in M \Rightarrow M_j = g_{x,y} \wedge P_i^{prox}(t)_0 \in M_j^{proc} \\ 0 & \text{otherwise} \end{cases}$$

Position Map

The *Position(i,t)* is an array with a size of *m*n*2* for a grid of size *m×n*. The first index of this array is set to 1 where the product is located and 0 elsewhere, while the second index is set to 1 where other products are present and 0 elsewhere. This configuration enables the agents to discern the positions of themselves and other products in the environment, aiding them in strategic navigation and coordination. See Equation 12.

Equation 12

$$Positions(i, t) = [$$
$$position_{0,0,0},\ position_{0,0,1},\ position_{1,0,0},\ position_{1,0,1},\ ...,\ position_{x,0,0},$$
$$position_{x,0,1},\ position_{0,1,0},\ position_{0,1,1},\ position_{1,1,0},\ position_{1,1,1},\ ...,$$
$$position_{x,1,0},\ position_{x,1,1},\ ...,\ position_{x,y,0},\ position_{x,y,1}]$$

$$position_{x,y,0}(i, t) = \begin{cases} 1 & \text{if } P_i^{pos}(t) = (x, y) \\ 0 & \text{otherwise} \end{cases}$$

$$position_{x,y,1}(i, t) = \begin{cases} 1 & \text{if } \exists j P_j \in P \setminus \{P_i\} \Rightarrow P_j^{pos}(t) = (x, y) \\ 0 & \text{otherwise} \end{cases}$$

Example Observation

For an example observation of a small grid of shape 3×3 with 2 products, see Equation 13 and Figure 3.

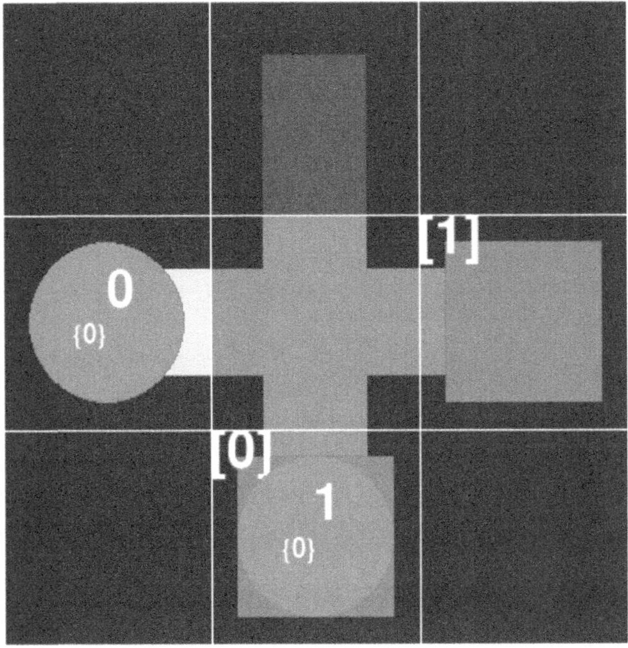

Fig. 3 Layout used for the example observations.

Equation 13

$$t = 5 \Rightarrow$$

$$P(t) = [((0,1), [0,1]), ((1,2), [0,1])]$$

$$M = [([0], 5), ([1], 5)]$$

$$G = \begin{pmatrix} \varepsilon & O & \varepsilon \\ I & B & M_0 \\ \varepsilon & M_1 & \varepsilon \end{pmatrix}$$

$$i = 0 \wedge t = 5 \Rightarrow$$

$$Transitions = [\; 0,0,0,0,\quad 0,0,1,0,\quad 0,0,0,0,$$
$$0,1,0,0,\quad 1,1,1,1,\quad 0,0,0,1,$$
$$0,0,0,0,\quad 1,0,0,0,\quad 0,0,0,0,$$

$$Targets(i, t) = [\; 0,\quad 0,\quad 0,$$
$$0,\quad 0,\quad 1,$$
$$0,\quad 0,\quad 0]$$

$$Positions\,(i, t) = [\; 0,0,\quad 0,0,\quad 0,0,$$
$$1,0,\quad 0,0,\quad 0,0,$$
$$0,0,\quad 0,1,\quad 0,0,]$$

Rewards

The agents are motivated by a set of rewards that drive their learning and decision-making processes. Each reward is designed to encourage efficiency and goal-oriented behavior. It is important to note that the rewards are reset after each step and are not cumulative. This approach ensures that the agents are continually motivated to make optimal decisions and optimize their behavior at each step. Without the reset, the reward value could stay high after using one correct machine, leading to random decision-making as every subsequent step would still be considered beneficial due to the positive cumulative reward. The following rewards are provided to the agents (Equation 14):

$reward_{timestep}$: To promote time efficiency, a slight negative reward of -1 is applied for every time step taken by the agents. This discourages unnecessary delays and encourages the agents to minimize the time required to achieve their objectives.

$reward_{backtrack}$: To discourage repetitive or redundant movements, a negative reward of -1 is imposed if a product returns to the tile it was previously on. This drives the agents to take more purposeful actions and reduce the wear on the modules. It is important to note that the sleep action does not give a penalty.

$reward_{process}$: When a product uses the appropriate machine to undergo the required manufacturing process, a reward of $+20$ is granted.

$reward_{exit}$: When successfully completing the manufacturing process and exiting the production line grants, a reward of $+20$.

Equation 14

$$reward(i, t):\text{Reward given to } W_i \text{ at step } t$$

$$reward(i, t) = reward_{timestep} +$$

$$\begin{cases} reward_{backtrack} & \text{if } P_i^{pos}(t-2) \wedge P_i^{pos}(t-1) \neq P_i^{pos}(t) \\[6pt] reward_{process} & \text{if } \begin{aligned} &action_i(t) = Machining \wedge P_i^{pos}(t) = (x, y) \wedge g_{x,y} \in Machines \\ &\Rightarrow M_{current} = g_{x,y} \wedge P_i^{proc}(t)_0 \in M_{current}^{proc} \end{aligned} \\[6pt] reward_{exit} & \text{if } P_i^{proc}(t) = \varnothing \wedge P_i^{proc}(t) = (x, y) \wedge g_{x,y} = O \\[6pt] 0 & \text{otherwise} \end{cases}$$

Setup

For the implementation and training of the model, we utilized a laptop equipped with an Intel Core i7-7820HK using a NVIDIA GTX1080 Mobile Graphics Card and 64 GB of RAM. The laptop uses Ubuntu 20.04 as its operating system. For running our environment, we utilize the RLlib library. (Liang et al., 2018)

Layout

The used layout is a simple crossing of size 5×5 with two agents/products. The products need to use both machines. The first product uses machine 0, then machine 1. The second product uses machine 1 first, and then machine 0. Both machines have different processing delays. Machine 0 takes 5 timesteps and machine 1 takes 10 timesteps. See Equation 15 and Figure 4. The theoretical maximum can be achieved in 35 steps with a total reward of 87.

Equation 15

$$P(0) = [((0,2), [0,1]), ((0,2), [1,0])]$$

$$M = [([0], 5), ([1], 10)]$$

$$G = \begin{pmatrix} \varepsilon & \varepsilon & O & \varepsilon & \varepsilon \\ \varepsilon & \varepsilon & B & \varepsilon & \varepsilon \\ I & B & B & B & M_0 \\ \varepsilon & \varepsilon & B & \varepsilon & \varepsilon \\ \varepsilon & \varepsilon & M_1 & \varepsilon & \varepsilon \end{pmatrix}$$

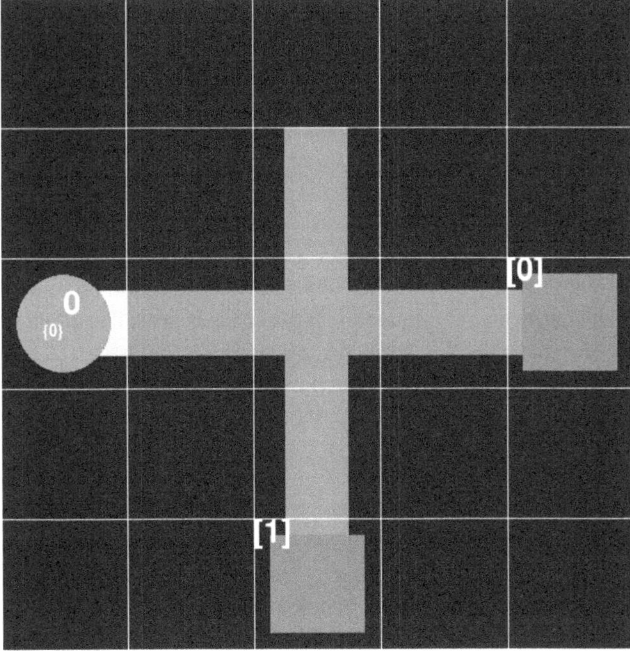

Fig. 4 Layout used in the experiment.

Hyperparameters

For the actual MARL model, we used Deep Q-Learning (DQN) with a shared policy for all agents using the dueling architecture (Roderick, MacGlashan, & Tellex, 2017; Wang et al., 2016). The hyperparameters can be seen in Table 1. Throughout the training we logged the process after every training iteration. Hyperparameters used for training the model are given in Table 1.

Table 1   Hyperparameters used for training the model

Hyperparameter	*Value*
FC Hidden Layers	256×256
Hidden Layer for Value/Advantage FC	4096
FC Activation	Tanh
Learning Rate	0.0005
Batch mode	Truncate Episodes
Grad Clip	40 by Global Norm
Train Batch Size	32
Exploration function	Greedy Epsilon
Epsilon Initial	1.0
Epsilon Final	0.02
Epsilon Steps	100.000

Discussion

In this paper, we have demonstrated that Multi-agent Reinforcement Learning (MARL) presents a formidable alternative to conventional planning methods in the domain of flexible job shop scheduling. Referring to Figure 5, it is evident that our MARL approach reaches the absolute optimal solution with minimal hyperparameter adjustments within a short training duration of approximately 30 minutes on the hardware mentioned in chapter [hardware]. The green line signifies the time at which the max obtainable reward has been achieved. The red dotted line is the max obtainable reward. Utilizing a two hidden neural network layers of 256 neurons, we achieve significant computational efficiency without compromising on performance.

The GPU, VRAM usage is almost constantly 0. CPU Usage sits at about 40% and RAM Usage is about 10 GB (Figure 6 Even with our limited hardware, there is room for improvement proving the computational efficiency.

A distinctive aspect of our MARL framework is the individualistic 'actor' agents, which play a crucial role in deadlock prevention. These agents learn and adapt strategies that inherently avoid blocking scenarios without the need for explicit deadlock handling mechanisms.

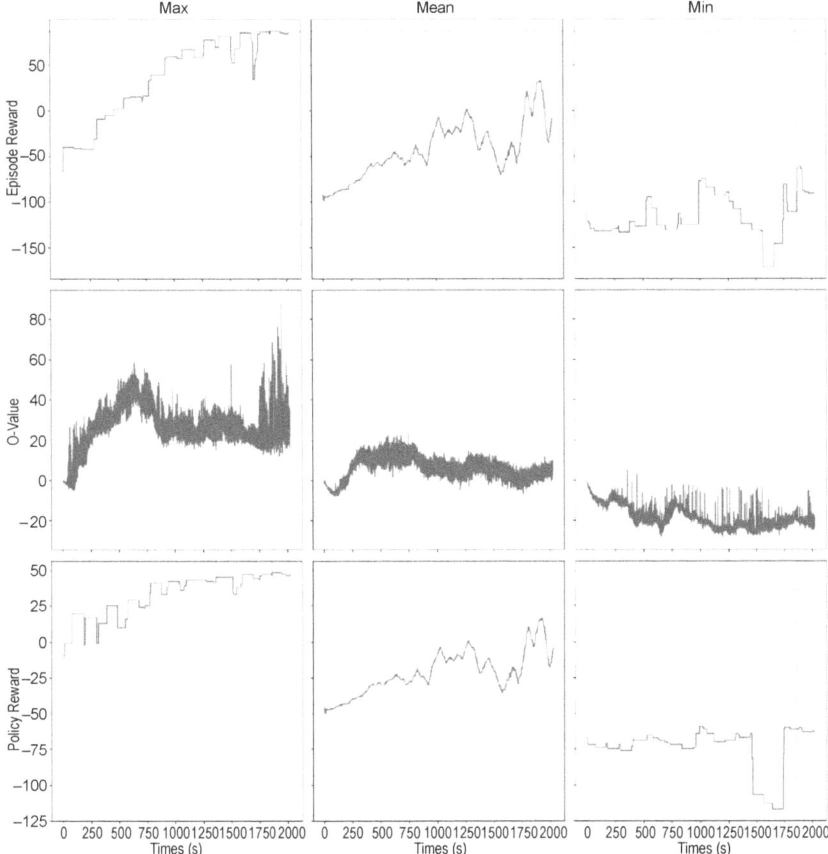

Fig. 5　Episode reward, Q-value, policy reward for the environment.

The ease of environment development in our MARL setup stands in stark contrast to the complexity observed in the methods employed by Bussmann and Schild (2000). This simplicity accelerates the development cycle and reduces the barrier to entry for deploying MARL in practical, real-world job shop scenarios. Moreover, the adaptability of our model to accommodate various factory configurations and operational contingencies underscores the robustness and versatility of MARL as a tool for contemporary and future scheduling challenges.

The results we have gathered provide a solid foundation for optimism about the efficacy of MARL in job shop scheduling and its role in smart manufacturing paradigms.

Societal Impact

The aim of this chapter is to provide insights into the broader implications of the proposed swarm intelligence, beyond the technical sphere, focusing on the societal

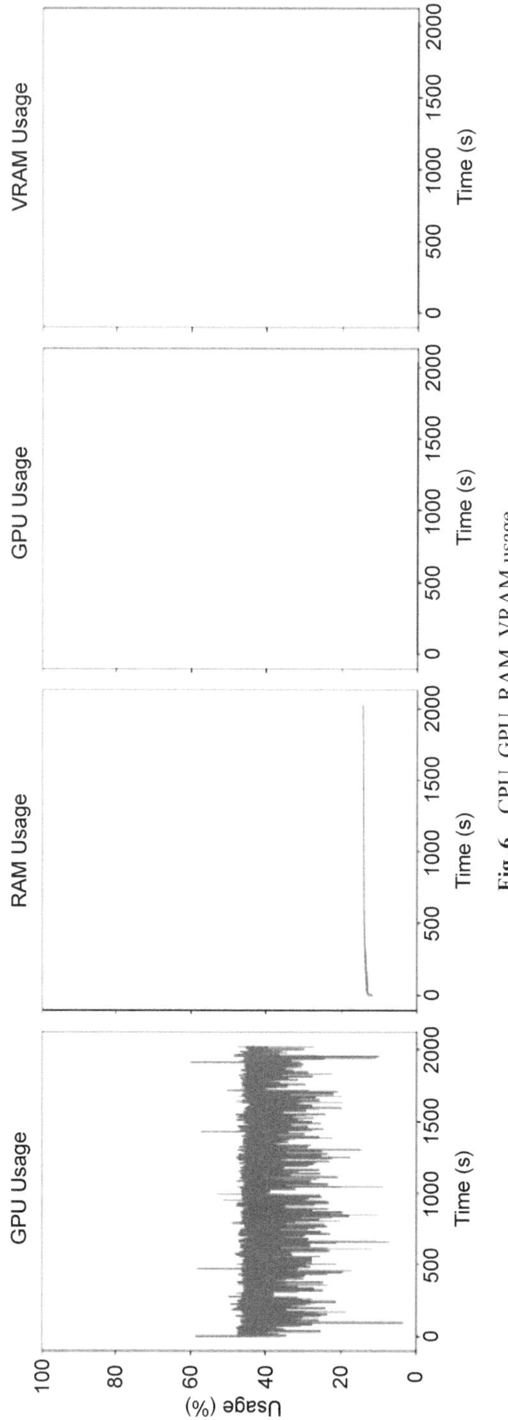

Fig. 6   CPU, GPU, RAM, VRAM usage.

impact. First, by applying this technology to real-word problems, the manufacturing efficiency gets significantly increased. This is achieved by reducing production downtimes and eliminating the need for additional neural network training for new factory layouts. This can reduce the operational costs, which can directly translate into lower prices for the customer and increased competitiveness by the factory in the market. Additionally, by directly reacting to faults in the production line the use of energy is optimized, and the ecological footprint is reduced.

Future Research

As we progress in exploring MARL for factory automation, several promising directions for future research emerge. First, **simulating failures** within the manufacturing process would offer insights into the model's adaptability and resilience. By introducing various failure scenarios, we can evaluate the agents' ability to respond optimally, minimizing downtime, and maximizing productivity. This exploration would provide a more comprehensive understanding of the model's robustness in real-world dynamic environments. Further closing the gap between the proposed simulation and a real-world application as machine failures are a common occurrence.

Additionally, training our model with multiple different factories presents an opportunity to investigate into the **transferability of learned policies across different unknown factory settings**.

Furthermore, **introducing another neural network that proposes factory layouts** for the agent neural network to evaluate and improve upon could lead to further improvements in terms of robustness of the model. This two-step approach would leverage the strengths of both networks, with the proposal network generating improved layouts. Such findings would contribute to the design and optimization of physical factories, enhancing productivity and resource utilization.

Another main goal for our coming research is to **compare our approach with a calculated maximum** derived from a more complex factory.

Overall, by delving into these future research areas, we aim to further enrich the applicability and effectiveness of our MARL approach in the context of factory automation. These explorations have the potential to unlock new avenues for efficiency, adaptability, and innovation in manufacturing processes.

References

Baer, S., Bakakeu, J., Meyes, R. and Meisen, T. (2019). Multi-agent reinforcement learning for job shop scheduling in flexible manufacturing systems. *In: 2019 Second International Conference on Artificial Intelligence for Industries (AI4I)*, 22–25.

Berner, C., Brockman, G., Chan, B., Cheung, V., Debiak, P., Dennison, C., Farhi, D., Fischer, Q., Hashme, S., Hesse, C., Jozefowicz, R., Gray, S., Olsson, C., Pachocki, J., Petrov, M., de Oliveira Pinto, H. P., Raiman, J., Salimans, T., Schlatter, J., Schneider, J., Sidor, S., Sutskever, I., Tang, J., Wolski, F. and Zhang, S. (2019). Dota 2 with large scale deep reinforcement learning. *CoRR.*, abs/1912.06680. http://arxiv.org/abs/1912.06680.

Bussmann, S. and Schild, K. (2000). Self-organizing manufacturing control: An industrial application of agent technology. *In: Proceedings Fourth International Conference on Multi-agent Systems,* 87–94. https://doi.org/10.1109/ICMAS.2000.858435.

Covington, P., Adams, J. and Sargin, E. (2016). Deep neural networks for youtube recommendations. *In: Proceedings of the 10th ACM Conference on Recommender Systems, RecSys '16,* 191–98. New York, NY, USA: Association for Computing Machinery. https://doi.org/10.1145/2959100.2959190.

Keddis, N., Kainz, G., Buckl, C. and Knoll, A. (2013). Towards adaptable manufacturing systems. *In: 2013 IEEE International Conference on Industrial Technology (ICIT),* 1410–15. https://doi.org/10.1109/ICIT.2013.6505878.

Leung, J.Y.-T., Li, H. and Pinedo, M. (2007). Scheduling orders for multiple product types to minimize total weighted completion time. *Discrete Applied Mathematics, 155*(8), 945–70. https://doi.org/10.1016/j.dam.2006.09.012.

Liang, E., Liaw, R., Nishihara, R., Moritz, P., Fox, R., Goldberg, K., Gonzalez, J., Jordan, M. and Stoica, I. (2018). RLlib: Abstractions for distributed reinforcement learning. *In: J. Dy and A. Krause (Eds.), Proceedings of the 35th International Conference on Machine Learning; Proceedings of Machine Learning Research, 80* 3053–62.: PMLR. https://proceedings.mlr.press/v80/liang18b.html.

MacCarthy, B. and Liu, J. (1993). Addressing the gap in scheduling research: A review of optimization and heuristic methods in production scheduling. *International Journal of Production Research, 31,* 59–79. https://doi.org/10.1080/00207549308956713.

Muhlemann, A.P., A.G.L. and Farn, C.-K. (1982). Job shop scheduling heuristics and frequency of scheduling. *International Journal of Production Research, 20*(2), 227–41. https://doi.org/10.1080/00207548208947763.

Roderick, M., MacGlashan, J. and Tellex, S. (2017). Implementing the deep q-network. *CoRR.,* abs/1711.07478. http://arxiv.org/abs/1711.07478.

Silver, D., Huang, A., Maddison, C.J., Guez, A., Sifre, L., van den Driessche, G., Schrittwieser, J., Antonoglou, I., Panneershelvam, V., Lanctot, M., Dieleman, S., Grewe, D., Nham, J., Kalchbrenner, N., Sutskever, I., Lillicrap, T., Leach, M., Kavukcuoglu, K., Graepel, T. and Hassabis, D. (2016). Mastering the game of go with deep neural networks and tree search. *Nature, 529*(7587), 484–89. https://doi.org/10.1038/nature16961.

Wang, Z., Schaul, T., Hessel, M., Hasselt, H., Lanctot, M. and Freitas, N. (2016). Dueling network architectures for deep reinforcement learning. *In: M.F. Balcan and K.Q. Weinberger (Eds.), Proceedings of The 33rd International Conference on Machine Learning: Proceedings of Machine Learning Research, 48,* 1995–03. New York. New York, USA: PMLR. https://proceedings.mlr.press/v48/wangf16.html.

On Joint Exploration and Self-guided Task Allocation for Mobile Robots in an Industrial Environment

Maximilian Schnitzler[1][*] and *Florian Kerber*[2]

Introduction

The future of industrial production is undergoing rapid changes. New consumer-driven demands for product variety require greater flexibility in production. Companies are therefore facing new challenges for their production environments. In the context of Industry 4.0, flexibility is achieved by digitally linking all resources as well as by decentralizing production planning and control (Vogel-Heuser et al., 2020; Zobl, 2022).

For decades, production technology has been driven by adaptability, flexibility, and reconfigurability. Simultaneously, the advent of cyber-physical production systems has increased the capabilities of individual production systems. As a forerunner, the automotive industry has started to integrate alternative manufacturing systems and thus change the layout of their shop floors. Fluid Manufacturing Systems (Fries et al., 2021) aspire to break down all means of production into location-flexible autonomous modules that are digitally linked with each order to dynamically form and dissolve production units according to customer demand, portfolio management, and supply chain restrictions (Bauernhansl et al., 2020; Bauernhansl et al., 2014). As a result, the shop floor layout dynamically morphs and changes both on a macro and micro level affecting both production and logistics (Hoßfeld & Ackermann, 2020; Wiendahl et al., 2007).

[1,2] Augsburg Technical University of Applied Sciences, Germany.
Email : florian.kerber@hs-augsburg.de

[*] Corresponding author: maximilian.schnitzler@hs-augsburg.de

To implement fluid production, more autonomous as well as mobile systems are required. In particular, the potential of mobile robots in industrial applications seems promising. Increased flexibility and performance over classical solutions have emerged as predominant advantages (Mehami et al., 2018; Michalos et al., 2016; Ueda et al., 2001). Mobile robots provide flexible material flow control and possess expanded production capabilities, e.g., in assembly, compared to conventional industrial technologies.

The standard industrial application for mobile robotics to date has been autonomously guided vehicles (AGV). Although AGVs are seemingly flexible, their application in modern industrial plants is based on sequential mapping and localization routing over predefined virtual paths or corridors (Hamner et al., 2010). To accurately navigate along paths on the shop floor, infrastructure such as reflectors are sometimes needed to guide AGVs, which is costly and severely limits flexibility (Ullrich, 2014). Generally, AGVs are used to perform intralogistics tasks such as the transport of goods. The actual vehicle control including the safety functionalities is performed by the vehicles themselves, see Figure 1.

The desire to entrust the system with more complex tasks demands a higher degree of flexibility. Autonomous mobile robots (AMR) can both provide extended flexibility and autonomy (Cupek et al., 2020; Ullrich, 2014). AMR refers to an umbrella term in mobile robotics defining mobile units with largely autonomous decision-making and manipulation capabilities for their respective tasks (Alatise & Hancke, 2020). An autonomous vehicle, a mobile service robot, and a more sophisticated industrial AGV are all considered AMR. In terms of industrial environments, AMR systems could be used to replace rigid assembly lines that allow mobile manipulation during transport. In addition, machining centres for multi-variant products become more viable if they can be automatically set up and loaded by AMRs. Ultimately, such AMR systems with interchangeable manipulators could solve most tasks in manufacturing in a self-organised way, turning a mere logistics device into a highly flexible, self-organized production assistance unit.

However, additional autonomy requires also more sophisticated control architectures. In this chapter, we give an overview over all relevant aspects of autonomous robot systems and their use in industrial production scenarios. Section 2 introduces the technology of AMR systems starting with the more common, yet less performant, AGV systems. Simultaneous localization and mapping is one of the core problems for mobile robot systems. The basic concept of single robot and multi-robot SLAM algorithms is summarised and a detailed description for a visual MR-SLAM is introduced in Section 3. Considering fleets of mobile agents, task allocation and joint exploration strategies have to be considered, especially in industrial applications. Sections 4 and 5 will outline the main approaches and scientific state of the art in these domains, respectively. A short summary and outlook conclude this treatment of AMRs.

The Technology of AGV and AMR Systems

Modern AGV systems consist of mobile transportation units and a central server for coordination and control. This fleet manager receives transport orders from a higher level IT system for production planning, often a manufacturing execution system (MES) instance. Its central functionalities are guidance control, i.e., the management of orders and their allocation to individual AGVs, and mapping of the environment, see Figure 1. The path network editor contains the environment map and a layout of the virtual path network. The individual AGVs possess optical measurement devises like LiDaR sensors and cameras to localize themselves within a pre-recorded map. For high precision navigation e.g., at pickup locations.

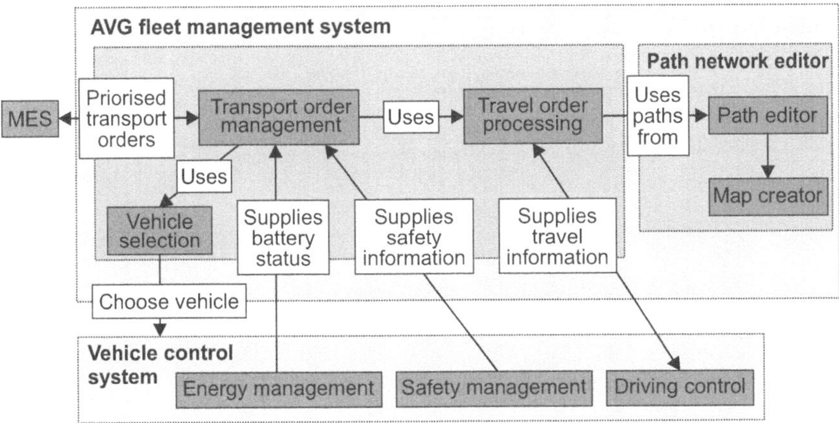

Fig. 1   General control architecture of an AGV fleet management system.

AGVs often require physical guidance structures or markers that can be identified by the installed sensors. The optical sensors also implement safety functions.

The usage of predefined environment representations reduces the flexibility of currently available AGVs with regard to path planning and collision avoidance. Fluid production and logistics with mobile assembly and just in sequence material distribution could thus not be implemented (Roozbahani Hamid, 2017; Unhelkar et al., 2018). Upgrading the AGVs to AMRs provides the aforementioned functionalities for the fleet of robots and extends their capabilities. Additional hardware like manipulators, special purpose end effectors and sensors can be installed on an AMR, promoting it to a mobile production system rather than a logistics device. When working together in teams, they can thus replace conventional machine centres and robotic cells in preproduction and assembly. They can also eliminate the need for conveyor belts which have been the standard in mass production since the beginning of the 20th century. Automotive original equipment manufacturers (OEMs) have started to redesign factories demonstrating that this no longer remains a pure vision (Hoogeveen, 2022; Kathmann et al., 2023; Kiefl et al., 2023).

However, this additional autonomy requires sharing additional information and an increase in computational power. Implementing a centralised architecture, further information regarding motion planning and SLAM has to be transmitted from and to AMRs, see Figure 2. The additional computational demand for path planning, map generation and extended capabilities raises hardware requirements of the central server. Depending on the communication technology and the server hardware, only a limited number of AMRs can be coordinated efficiently with a single server (Ryck et al., 2020). Increasing the number of AMRs thus results in updating the server hardware and adopting multiple communication networks. As with all centralised architectures, a failure of the central unit inevitably leads to the failure of the whole system.

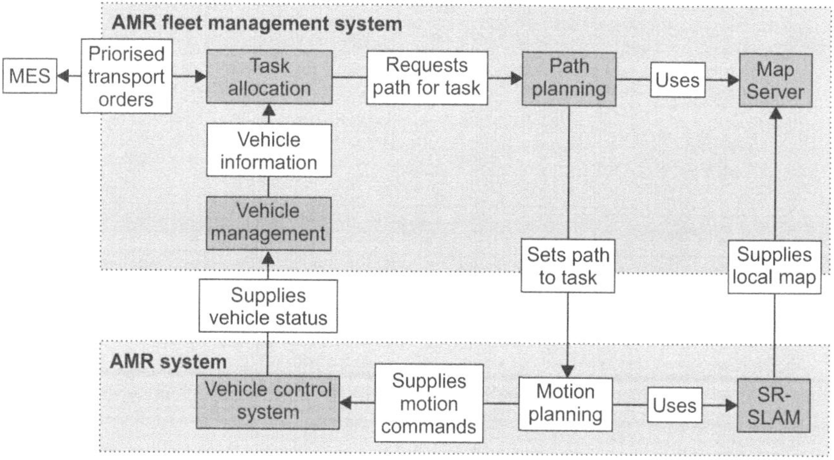

Fig. 2   General control architecture for a centralised AMR fleet management system.

To realise the full potential of the AMR it is therefore advisable to decentralise the system, see Figure 3. By enhancing the computational power of the individual AMR, the central server can be dissolved and its tasks distributed to the fleet of AMRs. This leads to even more autonomy promoting decentralised decision-making such as task allocation and exploration. As each AMR covers its own computing requirements with its dedicated hardware, the negative impact of the number of AMRs on the computing power requirements can be cancelled out. However, both algorithmic and implementation challenges still have to be solved for the coordination and control of such a decentralised AMR fleet. In the following sections, a fully decentralised control architecture for AMRs will be outlined.

Localization and Mapping: Development of SLAM Algorithms

Mobile systems have to navigate autonomously in complex environments to perform scheduled tasks in time while avoiding obstacles. To navigate in any

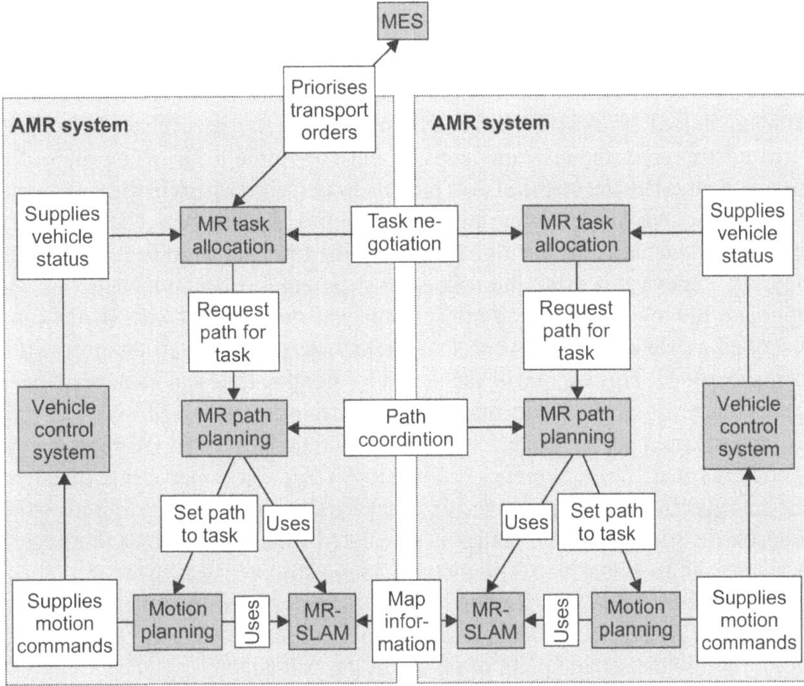

Fig. 3 General control architecture for a decentralised AMR fleet management system.

kind of environment, a robot needs a map of the environment and has to localize itself on the map. While the use of manually recorded maps can provide a detailed understanding of the environment, the effort and cost of creating and maintaining such maps makes them unsuitable for autonomous robot navigation. Indeed, these maps require constant work and control strategies to accommodate changing environments or to extend the covered territory. To avoid the aforementioned drawbacks, it is desirable to let the system generate and continuously update its own map of the environment. This requires the ability to navigate through an unknown environment using a process known as simultaneous localization and mapping (SLAM). Smith, Self, and Cheeseman (Smith et al., 1987) presented a first approach to address the SLAM problem. Since then, research has been conducted to provide robust and accurate SLAM algorithms for single robots (SR-SLAM). They merge odometry and sensor data, e.g. from a light detection and ranging device or a camera (Mur-Artal et al., 2015). The current standards for SR-SLAM are Kalman filters, particle filters, and graph optimization techniques. It should be noted that SLAM turns into a trivial problem when a global reference frame is provided. This can be done via features in the environment with known orientation or with satellite positioning (GNSS). However, installing features may not be practical due to cost, environmental constraints, or possible acquisition errors due to environmental disturbances. Indeed, indoor environments generally do not allow for satellite positioning.

Multi-Robot SLAM

Using multiple robots for faster and more reliable mapping of the environment is called multi-robot SLAM (MR-SLAM). MR-SLAM is an even more complex problem than SR-SLAM. However, reliability and convergence can be improved when all robots in the environment use and contribute to the same map of the environment, speeding up the SLAM process by perceiving parts of the environment individually and then merging this information. However, new challenges arise regarding the creation of the global map and the localisation of each robot in this map. To achieve this goal, the robots must be able to communicate with each other in order to exchange their information. The main task of MR-SLAM can be described as identifying the coordinate transformation between the poses of the robots involved. This is typically achieved by merging the local maps of all robots generated by SR-SLAM. The industrial standard is centralised MR-SLAM which uses a dedicated server to perform the computationally intensive map merging (Abdulgalil et al., 2019; Schmuck & Chli, 2019). Each local map is thus transferred to a central entity. The entity is used to detect common map features. Based on these features, the transformation matrix is calculated which then allows to merge the local maps into a global map. This global map is distributed back to the participating robots. However, this puts a bound on the overall system complexity since the computational resources of the central mapping server are restricted. Besides, the server represents a single point of failure of the system.

Within academia, decentralised MR-SLAM algorithms have been the focus of research recently (H. Chang et al., 2007; Lazaro et al., 2013; Nikolay Atanasov et al., 2016; Rooker & Birk, 2007). Most of these approaches attempt to align and merge individual robot maps to produce a joint map (Birk & Carpin, 2006; Carpin, 2008; Konolige et al., 2003; Li et al., 2014; Rooker & Birk, 2007).

Decentralised MR-SLAM systems distribute raw data, local maps or features to merge the various local maps into one global map. This is organized independently by the participating robots and does not require a central authority to merge or distribute the map. Decentralised MR-SLAM comes with advantages in regard of scalability and the elimination of the central entity as a single point of failure. However, decentralised control requires faster communication and higher bandwidths to take full advantage of the system's improved scalability. For larger robot fleets, a decentralised MR-SLAM structure is beneficial to enable the necessary up-scaling of computational power. As a result, each and every robot becomes self-sufficient and the whole system is merely limited by the required bandwidth and the algorithmic complexity of the control algorithms.

A Decentralised MR-SLAM Algorithm Based on Direct Sparse Odometry

The general approach to decentralised MR-SLAM tries to fuse maps already created by SR-SLAM algorithms in order to obtain the full map (Sayed et al.,

2020). Depending on the map representation there are different algorithms to achieve this. In the following, we restrict ourselves to an approach that uses visual SLAM and a pose-graph representation of the map (Mingachev et al., 2020). The underlying SR-SLAM algorithm is based on LDSO (Gao et al., 2018) which uses a visual odometry frontend. The frontend extracts photogrammetric information to optimize the intrinsic camera parameters, estimate depth, and subsequently poses (Engel et al., 2018). Loop closures are performed by evaluating ORB features of points of interest.

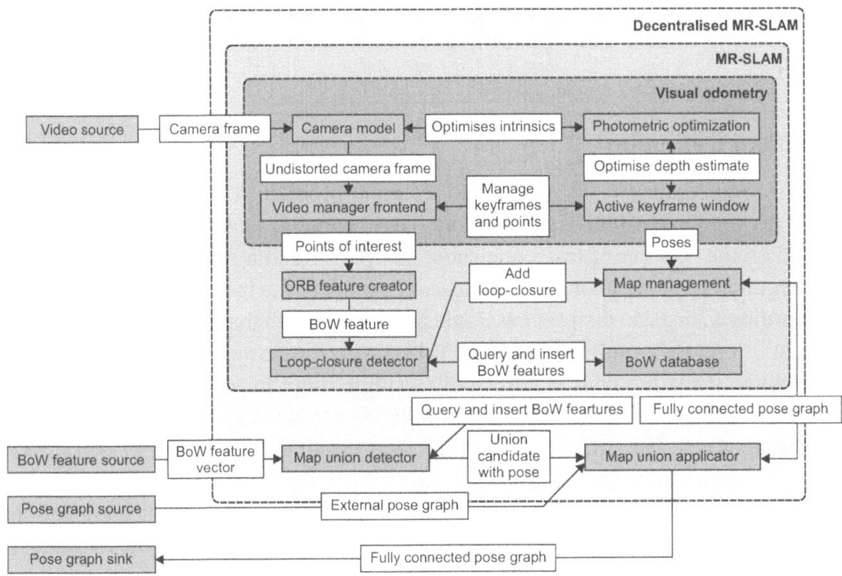

Fig. 4 Schematic structure of the chosen decentralised MR-SLAM.

Figure 4 presents a concept to extend LDSO to an MR-SLAM. The main idea is to compare ORB features obtained from different robots and check for matches that indicate loop closure so that the local maps can be stitched. Since the features of key frames are each stored in a local bag of words, they can be used independently of the actual location of the robot. This also reduces communication load since only the detected features have to be broadcast to other robots. Once a match is detected, the pose graphs of the two robots are fused, i.e., a union of the graphs is computed and a new relative pose constraint as edge connecting the graph nodes is introduced. The relative pose can be determined by epipolar geometry for 3D pose reconstruction. The connected pose graph thus generated will be shared between the participating robots. Further additions to the map will be broadcast by each robot for consistent updating. In this way the robots can continuously extend the mapped environment and thus extend their range. This also applies for dynamically morphing environments like fluid shop floors. In this application scenario, the layout is constantly changing due to AMRs reorganizing and regrouping to perform production and logistics tasks.

Task Allocation for Multi-Robot Fleets

With the introduction of more autonomy for each autonomous mobile robot, both the cost and the complexity of the implementation increases, yet this allows AMR systems to perform more advanced tasks. Building on this, a decentralised control structure can be introduced where each robot is a self-sufficient unit that acts in teams to solve tasks and shares information with other robots in the system. The division of tasks among multiple robots is called multi-robot task allocation (MRTA).

Problem Definition

Gerkey and Matarić (2004) provided a formal problem definition drawing from prior research in the field of game theory. They proposed a taxonomy of the problem based on the number of tasks one robot can perform at a time, how many robots are required to perform one task, and how dependent the tasks are onto each other. Accordingly, one can distinguish single task robots (ST) that can only perform one task at a time while multi-task robots (MT) can perform a number of tasks at a time. Moreover, if the problem requires only a single robot for completion, it is called single robot task or SR. By contrast, a multi robot task (MR) needs multiple robots for completion. The information available at the time of assignment determines whether future allocations can be considered (time-extended assignment or TA) or the task has to be assigned instantaneously without taking future allocations into account (instantaneous assignment or IA). This makes it possible to describe task allocation problems as triplets. In the decentralized industrial framework considered in this chapter, AMRs can be assumed to be both single-task and multi-task robots while the tasks themselves will typically require several robots to cooperate. The planning horizon should ideally be extended to incorporate available information about the tasks, so that ST-MR-TA as well as MT-MR-TA should be the most common problem classes for industrial applications. These cases require multiprocessor scheduling algorithms that are strongly NP-hard. Nevertheless, several solution approaches exist which can be applied in practice. Different optimization variables and objectives can be used to set them up. In an industrial application, the utility function will typically depend on any one or a combination of resource costs, quality of task execution, reward for completing the task and priority of the task. In terms of optimization, the total sum of all costs or the average cost per task can be minimized. Alternatively, criteria like maximum throughput or balancing of robots can be employed especially in an industrial context.

Most common multi-robot task allocation algorithms are implemented on a dedicated server. The employed algorithms yield near optimal assignments since they account for the global information of the robot fleet. However, their lack in performance makes them unviable for larger fleets as utilized in production. The number of robots, tasks and task constraints exponentially increases the complexity of the optimization problem directly affecting performance. Additionally, as all centralised approaches, they introduce a single point of failure. It is therefore desirable to distribute the decision making for task allocation into the robot fleet.

Distributed Task Allocation

Distributed task allocation approaches lack the global information of the robot fleet, but are able to scale better with the increasing complexity of larger robot fleets. These algorithms will in turn compute only approximate solutions and are more complex in subsequent optimisation steps. The widest set of distributed task allocation algorithms is marked-based (Braquet & Bakolas, 2021). One of the best performing amongst them is "auction with consensus" (Antidio Viguria et al., 2008; Choi et al., 2009; Otte et al., 2020). Apart from this method, there is also research in particle swarm optimization (PSO) based algorithms (Lim & Isa, 2015; Salman et al., 2002). In the remainder, auction with consensus will be further outlined. The algorithm starts with initializing an auction to assign a pending task. When the auction winner has been chosen by the auctioneer, a consensus process starts. The goal of the consensus process is to achieve a task assignment closer to the global optimum by reassigning already assigned tasks based on their marginal costs in order to use positive synergies between tasks. The marginal cost is the cost of concluding the tasks in a certain sequence potentially lowering the overall cost of the task. Using the example of two robots R1 and R2 and two tasks at waypoints W1 and W2, Figure 5 displays an example of marginal cost reassignment based on the distance measurement as cost function. Figure 5 (A) depicts a scenario in which robot R2 is closest to the initial waypoint W1 and therefore receives the corresponding task. When a second task at waypoint W2 is added, the respective auction without marginal cost bidding yields robot R1 as winner, see Figure 5 (B). Figure 5 (C) depicts the task allocation with marginal cost bidding: Now task 2 is

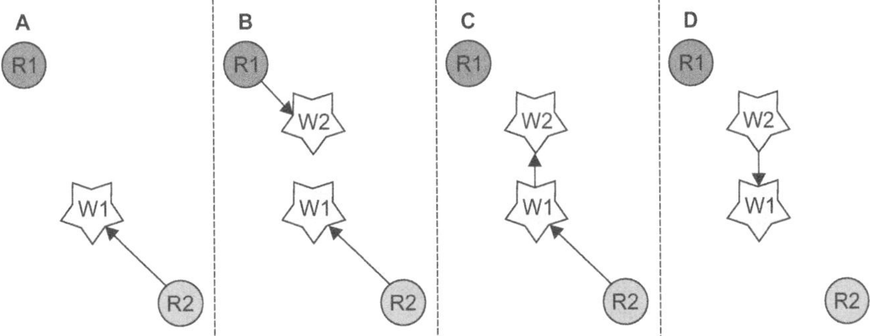

Fig. 5 An example of marginal cost using distance.

also assigned to R2 since it is closer to W2 after processing task 1 at W1. However, this clearly constitutes a nonoptimal allocation considering the overall distance would be shorter for R1 taking over both tasks. This is fixed via re-bidding on the set of both tasks as shown in Figure 5 (D) resulting in more optimal task allocation.

In terms of implementation, the auction starts with an incoming task from the task seller, which is a superordinate management system like an MES. This task will be given to an auctioneer. In a centralized architecture, the auctioneer is either a central server or a particular robot of the fleet whereas in a decentralized system the role of auctioneer is taken over by different robots using a round robin or a token procedure. Depending on the algorithm used, the literature suggests only having one auctioneer role at a time not to interfere with the consensus part of the auction. The whole process is depicted in Figure 6. Beginning in the top left the auctioneer starts an auction by broadcasting a call for bids for the task. The bidders calculate their bid and reply the result to the auctioneer (II). Next, the auctioneer assigns the task preliminarily to the lowest bidder, who supplies the auctioneer with the list of tasks it has at hand and therefore starts the search for a consensus (III). All of the tasks in the task list are offered in another auction (IV), in which the bidders calculate their bids for each task based on the best marginal cost (V). Based on this bid the tasks are reassigned (VI) shifting already assigned tasks to bidders with more synergy and therefore a solution closer to the global optimum. With the reassignment completed the task seller will be informed and the process is completed. The example implements SR-ST-IA. However, the procedure can also be adopted for all other types of task allocation problems. In and MR-ST-TA scenario, for example, the task may allow for multiple winners in the auction, based on the number of robots needed to complete the task. Furthermore, time-extended assignments are covered via the consensus step since the bidding with marginal cost will account for prior assignments.

Joint Exploration Strategies

Joint exploration requires the robot fleet to share a map, distribute movement commands for exploration purposes and make decisions on where to explore next. Using the introduced decentralised MR-SLAM and the distributed task allocation technique "auction with consensus", a strategy how to determine exploration points has to be developed for joint robot exploration. Three different levels of coordination can be distinguished (Burgard et al., 2005):

- **No information exchange** between the exploring robots.
- **Implicit coordination (uncoordinated)** via the sharing of a joint map and the communication of individual maps and poses to a central mapping system.
- **Explicit coordination** by the assignment of target points to robots, communication of individual maps and poses to coordinate and plan exploration targets.

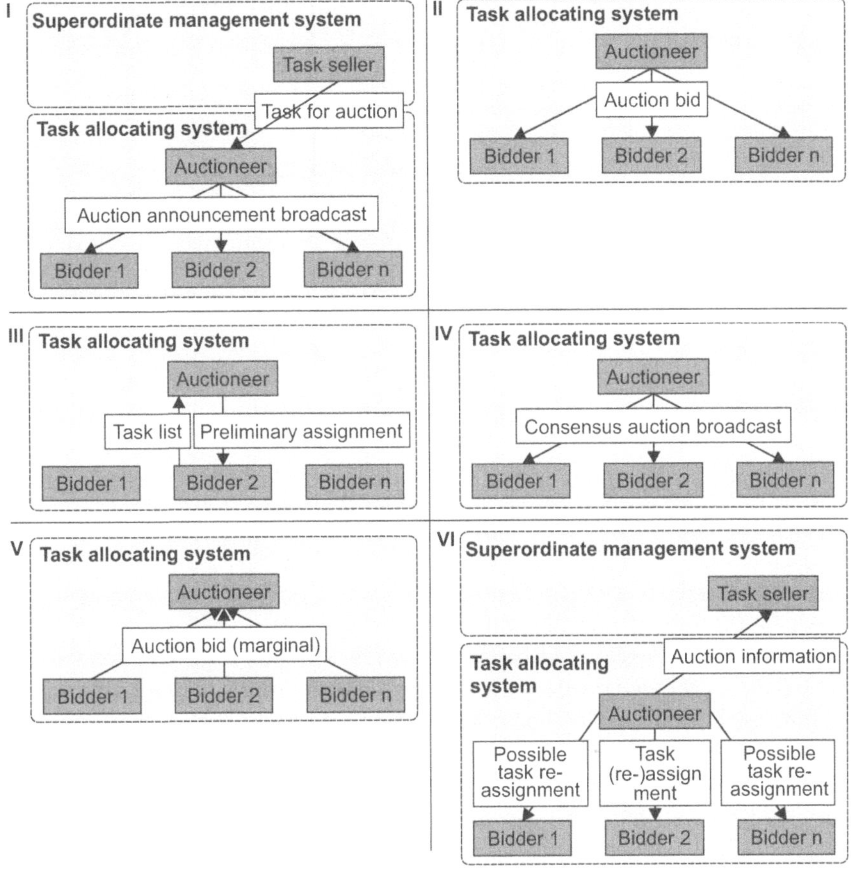

Fig. 6   Auction with consensus.

In industrial applications, exploration is used to acquire an up-to-date map of the shop floor as a prerequisite for AMR fleets to flexibly perform production and logistics processes. As a result, robots are explicitly coordinated performing MR-SLAM (Section 3) to speed up the map building process. Many approaches require the exploration mechanism to allocate exploration tasks itself (Burgard et al., 2005), but marked driven task allocation can be also used for this endeavour (Zlot et al., 2002). It is therefore reasonable to use the algorithm presented in Section 4.2 for the allocation of dedicated exploration tasks. The reward functional used during auctioning has to account for the expected information gain and the number of robots in the exploration group. The boundary between open space and unexplored space, the so-called frontier (Yamauchi, 1998, 1999), yields the highest probability of information gain g. Using an occupancy grid map, a frontier cell f_i can be defined as a cell which is probably empty (cell value close to 0, white) bordering to a cell with no knowledge (cell value equal to 0.5, grey), see Figure 7. Connected frontier

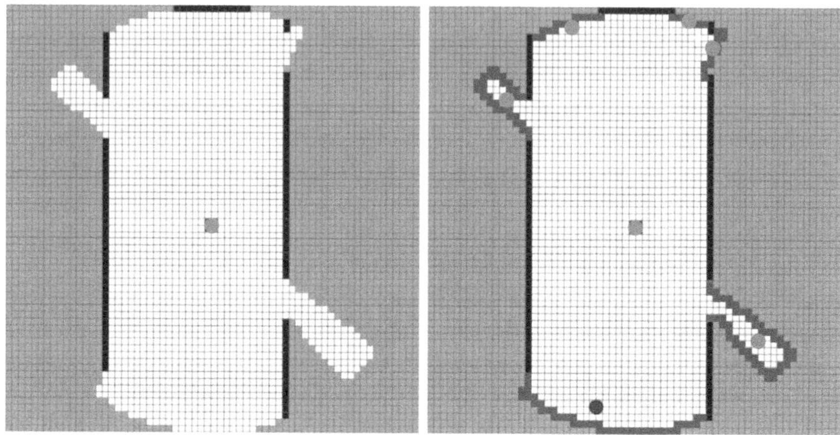

Fig. 7   Frontier detection. Left: Occupancy grid map (OGM) when initializing the robot at the starting pose p_{Robot}. Right: Frontier cells (red) with estimated centre of mass (blue) for each frontier f_m; $m = [1,6]$.

cells are grouped to frontiers $f_m = \{f_i, i = 1, 2, ..., nf_m\}$. Some algorithms use these frontiers for additional segmentation [51].

Since shop floors could be both cluttered and free, a general approach to calculate the expected information gain for each frontier is proposed in the following. First, the exploration mechanism calculates the centre of mass for each identified frontier. Choosing the centre of mass as exploration target, its reward in terms of information gain is calculated. A ray-cast with maximum sensor reach is conducted by each robot according to its sensor characteristics like field of view to determine which cells can be explored. Cells concealed by obstacles (cell value equal to 1.0, black) are excluded. For each frontier the information gain is calculated as a sum over the probability value of each ray casted cell $c_i(f_m)$ according to Equation 1.

Equation 1

$$g(f_m) = \sum_{i=0}^{n_{f_m}} 1 - \left\| prob\left(c_i(f_m)\right) - 0.5 \right\|$$

Whenever the fleet has to explore the shop floor, each robot computes its information reward for current frontiers to bid for exploration tasks. After MRTA has reached a consensus, the winning bidders start exploring. The strategy is broadcast by the auctioneer so that the individual robots can update their maps and recalculate frontiers. As described, joint exploration helps AMR fleets to quickly explore or update information about the shop floor they operate in. As with production and logistics tasks, a distributed mechanism based on auction with consensus provides a quasi-optimal strategy in terms of information gain.

Conclusion

New manufacturing paradigms like fluid production systems require technological advances in many areas. AMR systems offer the potential to flexibly arrange factory layouts replacing static installations like conveyor belts and robotic cells. However, there is a broad choice of decision-making strategies and control architectures. To ensure accurate navigation and exploration within constrained environments and efficient task allocation for fleets of AMRs a decentralized approach is proposed in this chapter.

Visual SLAM is well-suited for AMRs operating on the shop floor since they are equipped with optical sensors. Feature based loop closing can be implemented efficiently as a decentralized MR-SLAM. Once AMR fleets can jointly navigate, they can be employed in industrial production to perform manufacturing tasks. For this aim, mechanisms to optimally allocate and complete tasks have to be developed. This chapter presents an auction-based strategy for task allocation and extends it to joint exploration. Distributed decision making enhances robustness and leverages economies of scale rendering decentralized architectures the structure of choice for industrial AMR application.

As a future direction of research, the lower level control of collaborative robot tasks such as trajectory planning and force feedback remains a challenging problem not addressed in this work.

References

Abdulgalil, M.A., Nasr, M.M., Elalfy, M.H., Khamis, A. and Karray, F. (2019). Multi-robot SLAM: An Overview and Quantitative Evaluation of MRGS ROS Framework for MR-SLAM. *In*: J.-H. Kim, H. Myung, J. Kim, W. Xu, E.T. Matson, J.-W. Jung and H.-L. Choi (Eds.), *Advances in Intelligent Systems and Computing. Robot Intelligence Technology and Applications, 5 , 751*, 165–83. Springer International Publishing. https://doi.org/10.1007/978-3-319-78452-6_15.

Alatise, M.B. and Hancke, G.P. (2020). A Review on Challenges of Autonomous Mobile Robot and Sensor Fusion Methods. *IEEE Access, 8*, 39830–39846. https://doi.org/10.1109/ACCESS.2020.2975643.

Antidio Viguria, Iván Maza and Aníbal Ollero (2008). S+T: An algorithm for distributed multi-robot task allocation based on services for improving robot cooperation. *2008 IEEE International Conference on Robotics and Automation*, 3163–68. https://api.semanticscholar.org/CorpusID:9713944.

Bauernhansl, T., Fechter, M. and Dietz, T. (2020). *Entwicklung, Aufbau und Demonstration einer wandlungsfähigen (Fahrzeug-) Forschungsproduktion. ARENA2036.* Springer Vieweg. https://doi.org/10.1007/978-3-662-60491-5.

Bauernhansl, T., Hompel, M. ten and Vogel-Heuser, B. (2014). *Industrie 4.0 in Produktion, Automatisierung und Logistik: Anwendung, Technologien, Migration.* Springer Vieweg. https://doi.org/10.1007/978-3-658-04682-8.

Birk, A [A.] and Carpin, S. (2006). Merging Occupancy Grid Maps from Multiple Robots. *Proceedings of the IEEE, 94*(7), 1384–97. https://doi.org/10.1109/JPROC.2006.876965.

Braquet, M. and Bakolas, E. (2021). Greedy Decentralized Auction-based Task Allocation for Multiagent Systems. *IFAC-PapersOnLine, 54*(20), 675–80. https://doi.org/10.1016/j.ifacol.2021.11.249.

Burgard, W., Moors, M., Stachniss, C. and Schneider, F.E. (2005). Coordinated multi-robot exploration. *IEEE Transactions on Robotics, 21*(3), 376–86. https://doi.org/10.1109/TRO.2004.839232.

Carpin, S. [Stefano]. (2008). Fast and accurate map merging for multi-robot systems. *Autonomous Robots, 25*(3), 305–16. https://doi.org/10.1007/s10514-008-9097-4.

Choi, H.-L., Brunet, L., and How, J.P. (2009). Consensus-based Decentralized Auctions for Robust Task Allocation. *IEEE Transactions on Robotics*, *25*(4), 912–26. https://doi.org/10.1109/TRO.2009.2022423.

Cupek, R., Drewniak, M., Fojcik, M., Kyrkjebø, E., Lin, J.C.-W., Mrozek, D., Øvsthus, K. and Ziebinski, A. (2020). Autonomous Guided Vehicles for Smart Industries: The State-of-the-Art and Research Challenges. *In*: V.V. Krzhizhanovskaya, G. Závodszky, M.H. Lees, J.J. Dongarra, P. M.A. Sloot, S. Brissos, and J. Teixeira (Eds.), *Lecture Notes in Computer Science. Computational Science: ICCS 2020*, *12141*, 330–43. Springer International Publishing. https://doi.org/10.1007/978-3-030-50426-7_25.

Engel, J., Koltun, V. and Cremers, D. (2018). Direct Sparse Odometry. *IEEE Transactions on Pattern Analysis and Machine Intelligence*, *40*(3), 611–25. https://doi.org/10.1109/TPAMI.2017.2658577.

Fries, C., Fechter, M., Ranke, D., Trierweiler, M., Assadi, A.A., Foith-Förster, P., Wiendahl, H.-H. [Hans-Hermann] and Bauernhansl, T. (2021). Fluid Manufacturing Systems (FLMS). *In*: P. Weißgraeber, F. Heieck, and C. Ackermann (Eds.), *ARENA2036. Advances in Automotive Production Technology: Theory and Application: Stuttgart Conference on Automotive Production (SCAP2020)*; Philipp Weißgraeber, Frieder Heieck, Clemens Ackermann (Eds.), (37–44). Springer Vieweg. https://doi.org/10.1007/978-3-662-62962-8_5.

Gao, X., Wang, R., Demmel, N. and Cremers, D. (2018). Ldso: Direct Sparse Odometry with Loop Closure. *In*: *Towards a Robotic Society: 2018 IEEE/RSJ International Conference on Intelligent Robots and Systems*. October, 1–5, Madrid, Spain, Madrid Municipal Conference Centre, 2198–2204. IEEE. https://doi.org/10.1109/IROS.2018.8593376.

Gerkey, B.P. and Matarić, M.J. (2004). A Formal Analysis and Taxonomy of Task Allocation in Multi-Robot Systems. *The International Journal of Robotics Research*, *23*(9), 939–54. https://doi.org/10.1177/0278364904045564.

H. Chang, C.S. Lee, Y.C. Hu and Yung-Hsiang Lu (2007). Multi-robot SLAM with topological/metric maps. *2007 IEEE/RSJ International Conference on Intelligent Robots and Systems*. https://www.semanticscholar.org/paper/Multi-robot-SLAM-with-topological-metric-maps-Chang-Lee/3a687395f5008e551039fad80451c7380e0edab3.

Hamner, B., Koterba, S., Shi, J., Simmons, R. and Singh, S. (2010). An autonomous mobile manipulator for assembly tasks. *Autonomous Robots*, *28*(1), 131–49. https://doi.org/10.1007/s10514-009-9142-y.

Hoogeveen, R. (Ed.). (2022). *ARENA2036. Interorganisationale kollaborative Gemeinschaftsforschung: Forschungscampus für den Automobilbau der Zukunft: ARENA2036*. Springer Vieweg. https://link.springer.com/content/pdf/10.1007/978-3-662-62958-1.pdf.

Hoßfeld, M. and Ackermann, C. (2020). Der Forschungscampus ARENA2036. *Entwicklung, Aufbau und Demonstration einer wandlungsfähigen (Fahrzeug-) Forschungsproduktion*, 1–3. https://doi.org/10.1007/978-3-662-60491-5_1.

Kathmann, T., Reh, D. and Arlinghaus, J.C. (2023). Exploiting the technological capabilities of autonomous vehicles as assembly items to improve assembly performance. *Advances in Industrial and Manufacturing Engineering*, *6*, 100111. https://doi.org/10.1016/j.aime.2022.100111.

Kiefl, N., Wulle, F., Ackermann, C. and Holder, D. (2023). *Advances in Automotive Production Technology: Towards Software-Defined Manufacturing and Resilient Supply Chains*. Springer International Publishing. https://library.oapen.org/handle/20.500.12657/63563?show=full https://doi.org/10.1007/978-3-031-27933-1.

Konolige, K., Fox, D., Limketkai, B., Ko, J. and Stewart, B. (2003). Map merging for distributed robot navigation. *In*: *Proceedings / 2003 IEEE/RSJ International Conference on Intelligent Robots and Systems (IROS 2003)*. October 27 –31, Las Vegas, Nevada, 212–17. IEEE Operations Center. https://doi.org/10.1109/IROS.2003.1250630

Lazaro, M.T., Paz, L.M., Pinies, P., Castellanos, J.A. and Grisetti, G. (2013). Multi-robot SLAM using condensed measurements. *In*: *2013 IEEE/RSJ International Conference on Intelligent Robots and Systems: (IROS 2013)*. Tokyo, Japan, 3–7 November [conference digest], 1069–76. IEEE Service Center. https://doi.org/10.1109/IROS.2013.6696483.

Li, H., Tsukada, M., Nashashibi, F. and Parent, M. (2014). Multivehicle Cooperative Local Mapping: A Methodology Based on Occupancy Grid Map Merging. *IEEE Transactions on Intelligent Transportation Systems*, *15*(5), 2089–2100. https://doi.org/10.1109/TITS.2014.2309639.

Lim, W.H. and Isa, N.A.M. (2015). Particle swarm optimization with dual-level task allocation. *Engineering Applications of Artificial Intelligence*, *38*, 88–110. https://doi.org/10.1016/j. engappai.2014.10.022.

Mehami, J., Nawi, M. and Zhong, R.Y. (2018). Smart automated guided vehicles for manufacturing in the context of Industry 4.0. *Procedia Manufacturing*, *26*, 1077–86. https://doi.org/10.1016/j. promfg.2018.07.144.

Michalos, G., Kousi, N., Makris, S. and Chryssolouris, G. (2016). Performance Assessment of Production Systems with Mobile Robots. *Procedia CIRP*, *41*, 195–200. https://doi.org/10.1016/j. procir.2015.12.097.

Mingachev, E., Lavrenov, R., Tsoy, T., Matsuno, F., Svinin, M., Suthakorn, J. and Magid, E. (2020). Comparison of ROS-Based Monocular Visual SLAM Methods: Dso, LDSO, ORB-SLAM2 and DynaSLAM. *In*: A. Ronzhin, G. Rigoll, and R. Meshcheryakov (Eds.), *Springer eBook Collection, 12336, Interactive Collaborative Robotics: 5th International Conference, ICR 2020*, St. Petersburg, Russia, October 7–9, 2020, *Proceedings* (1st Edn.), 2020, 222–33). Springer International Publishing; Imprint Springer. https://doi.org/10.1007/978-3-030-60337-3_22.

Mur-Artal, R., Montiel, J.M.M. and Tardos, J.D. (2015). ORB-SLAM: A Versatile and Accurate Monocular SLAM System. *IEEE Transactions on Robotics*, *31*(5), 1147–63. https://doi. org/10.1109/TRO.2015.2463671.

Nikolay Atanasov, Jerome Le Ny, Kostas Daniilidis and George J. Pappas. (2016). *Decentralized Active Information Acquisition: Theory and Application to Multi-Robot SLAM*. http://erl.ucsd.edu/ref/ Atanasov_ActiveInformationAcquisition_ICRA15.pdf.

Otte, M., Kuhlman, M.J. and Sofge, D. (2020). Auctions for multi-robot task allocation in communication limited environments. *Autonomous Robots*, *44*(3–4), 547–84. https://doi.org/10.1007/s10514-019-09828-5.

Rooker, M.N. and Birk, A. [Andreas] (2007). Multi-robot exploration under the constraints of wireless networking. *Control Engineering Practice*, *15*(4), 435–45. https://doi.org/10.1016/j. conengprac.2006.08.007.

Roozbahani Hamid, H.H. (2017). *Developing a Mobile Assembly Robot for Hazardous Environments With Compact RIO and ROS: A case study*. http://sine.ni.com/cs/app/doc/p/id/cs-17477.

Ryck, M. de, Versteyhe, M., and Debrouwere, F. (2020). Automated guided vehicle systems, state-of-the-art control algorithms and techniques. *Journal of Manufacturing Systems*, *54*, 152–173. https:// doi.org/10.1016/j.jmsy.2019.12.002.

Salman, A., Ahmad, I. and Al-Madani, S. (2002). Particle swarm optimization for task assignment problem. *Microprocessors and Microsystems*, *26*(8), 363–371. https://doi.org/10.1016/S0141-9331(02)00053-4.

Sayed, A.S., Ammar, H.H. and Shalaby, R. (2020). Centralized Multi-agent Mobile Robots SLAM and Navigation for COVID-19 Field Hospitals. *In: The 2nd Novel Intelligent and Leading Emerging Sciences Conference: Niles 2020*, : 24–26 Oct. Conference will be held online at Nile University Premises, 444–49. IEEE. https://doi.org/10.1109/NILES50944.2020.9257919.

Schmuck, P., and Chli, M. (2019). CCM-SLAM: Robust and efficient centralized collaborative monocular simultaneous localization and mapping for robotic teams. *Journal of Field Robotics*, *36*(4), 763–81. https://doi.org/10.1002/rob.21854.

Smith, R., Self, M. and Cheeseman, P. (March 1987). Estimating uncertain spatial relationships in robotics. *In: Proceedings. 1987 IEEE International Conference on Robotics and Automation*, 850. Institute of Electrical and Electronics Engineers. https://doi.org/10.1109/ROBOT.1987.1087846.

Tyroller, Q., Bix, J., Bartscher, S. and Shah, J. A., "Mobile Robots for Moving-Floor Assembly Lines: Design, Evaluation, and Deployment," in IEEE Robotics & Automation Magazine, vol. 25, no. 2, pp. 72–81, June 2018, doi: 10.1109/MRA.2018.2815639

Ueda, K., Hatono, I., Fujii, N. and Vaario, J. (2001). Line-less Production System Using Self-Organization: A Case Study for BMS. *CIRP Annals*, *50*(1), 319–22. https://doi.org/10.1016/ S0007-8506(07)62130-1.

Ullrich, G. (2014). *Automated Guided Vehicle Systems: A Primer with Practical Applications* (1. Aufl.). Springer-Verlag. https://doi.org/10.1007/978-3-662-44814-4.

Unhelkar, V.V., Dörr, S., Bubeck, A., Lasota, P.A., Perez, J.I., Siu, H.C., Boerkoel, James C., Jr., Q., Bix, J., Bartscher, S. and Shah, J.A., obile Robots for Moving-Floor Assembly Lines: Design,

Evaluation, and Deployment, in IEEE Robotics & Automation Magazine, vol. 25, no. 2, pp. 72–81, June 2018, doi: 10.1109/MRA.2018.2815639

Vogel-Heuser, B., Bauernhansl, T. and Hompel, M. ten (Eds.). (2020). *Springer Reference. Handbuch Industrie 4.0* (3. Auflage). Springer Vieweg.

Wiendahl, H.-P., ElMaraghy, H.A., Nyhuis, P., Zäh, M.F., Wiendahl, H.-H [H.-H.], Duffie, N. and Brieke, M. (2007). Changeable Manufacturing: Classification, Design, and Operation. *CIRP Annals, 56*(2), 783–809. https://doi.org/10.1016/j.cirp.2007.10.003.

Yamauchi, B. (1998). Frontier-based exploration using multiple robots. *In*: K.P. Sycara (Ed.), *ACM Conferences, Proceedings of the Second International Conference on Autonomous Agents*, 47–53. ACM. https://doi.org/10.1145/280765.280773.

Yamauchi, B. (1999). Decentralized coordination for multirobot exploration. *Robotics and Autonomous Systems, 29*(2–3), 111–18. https://doi.org/10.1016/S0921-8890(99)00046-9.

Zlot, R., Stentz, A., Dias, M.B. and Thayer, S. (2002). Multi-robot exploration controlled by a market economy. *In: 2002 IEEE International Conference on Robotics and Automation: May 11–15, Washington, D.C.; Proceedings*, 3016–23. IEEE Service Center. https://doi.org/10.1109/ROBOT.2002.1013690.

Zobl, H.-P. (2022). Produktion 4.0: Digital, intelligent, und vernetzt. *ATZ: – Automobiltechnische Zeitschrift, 124*(1), 76–78. https://doi.org/10.1007/s35148-021-0815-y.

7

Advances in the Applications of Swarm Intelligence in Portfolio Optimization, Stock Price Prediction and Risk Management
A Comparison with Traditional Methods

Natasha Pankunni[1][*] and *Praseetha Pankunni*[2]

Introduction to Swarm Intelligence (SI)

Intelligence is a term that has varied definitions and interpretations. The most collective definition of intelligence is "the ability to acquire and apply knowledge and skills". And intelligence is grown with human interactions and deliberations. With the advent of computer and internet, intelligence moved to the next level of being the ability to find a solution for large problems quickly. Then marked the advent of the term "Artificial Intelligence", which is the simulation of the human intelligence process by machines. The developments in AI have led to the conceptualization of "Swarm Intelligence". Even though studies have been made on the behavior of insects and its natural swarm systems, it was in the late 1980s that scientists began to apply it in AI (Ahmed & Glasgow, 2012). It was in 1989 that the term "Swarm Intelligence" was first used by Gerardo Beni and Jing Wang for a study of cellular robotic systems.

Just like the term 'swarm' indicates, the intelligence is based on the group or flock behavior. The swarm intelligence refers to the collective intelligent behavior of birds, ants, and other swarm animals where they apply self-organizing processes and patterns. These birds or insects or swarm individuals that are commonly termed

[1]   University of Calicut, Kerala, India.

[2]   Govt. Polytechnic College, Palakkad, Kerala, India.
    Email : praseethapankunni@gmail.com

[*]   Corresponding author: natashapankunni@gmail.com

as 'agents' in the swarm, are with limited capabilities and naive on its own or taken individually. However, when they come in a group, their collective behavior is different and interactions among the team enable them to achieve tasks more successfully and effectively. These interactions may be either direct or indirect (Belal et al., 2006). The common example of direct interaction is that of honey bees (Figure 2), whereby they interact with sound and movements. Indirect interaction is often depicted by ants (Figure 3) where they leave behind some traces for their followers to keep track of their path.

Fig. 1   Photo of Dr. Gerardo
Source: https://www.adscientificindex.com/scientist/gerardo-beni/1770576

Fig. 2   A picture representing bee colony
Source: https://www.istockphoto.com/photo/bees-constructing-a-chain-gm468430274-60985276

Fig. 3   A picture representing ant colony
Source: https://medium.com/zerone-magazine/swarm-intelligence-676e40968473

Principles of Swarm Intelligence

The basic principles on which one can categorize the collective or swarm behavior is laid down by Mark Millonas (2004). Although not definitive, these broad categories are very helpful according to him. The five principles laid down by him are:

1. **Proximity Principle:** According to this principle, the group should be able to compute time and elementary space around them. Mark defines computation as "a direct behavioral response to environmental stimuli which, in some sense, maximizes the utility to the group as a whole of some type of activity" (Millonas, 2004).

2. **Quality Principle:** The quality principle calls for the group not only be aware of computations, but also to be aware of the quality of foodstuffs or the safety of location.

3. **Principle of Diverse Response:** As per this principle, the group should seek diverse options to keep resources rather than narrowing them, to keep up with the future environmental fluctuations.

4. **Principle of Stability:** The group should be consistent with their behavior and not shift according to environmental fluctuations soon, as it will require energy and cost which will affect the output of the investment.

5. **Principle of Adaptability:** Essentially, this principle is the other side of principle of stability. Stability does not mean rigidness, and adaptability calls for shifting when the rewards from the energy and cost incurred in shifting are worthy of the energy invested (Millonas, 2004).

The application of SI is all-pervasive and very vast. Its peculiarity is that it applies simple logic, but can solve complex difficulties. It is like what Helen Keller mentioned, "Alone we can do so little; together we can do so much." Rather than sticking upon one single opinion or solution, make it collective. That is what SI aims at, and it has a galloping effect than other strategies sometimes. It is worthy to remember what the Pearl Harbor attack mastermind, Admiral Isoroku Yamamoto, once said: "The fiercest serpent may be overcome by a swarm of ants."

The application of SI requires enough logic and algorithms. When the input is given wrong, so will be the output be incorrect. Academicians have gone through the difficulties in implementing the same as well. One work among them is of Bonabeau, Dorigo, and Theraulaz (1999), where they attempted on modeling an insect behavior. According to them,

"It is, however, fair to say that very few applications of swarm intelligence have been developed. One of the main reasons for this relative lack of success resides in the fact that swarm-intelligent systems are hard to 'program', because the paths to problem solving are not predefined but emergent in these systems and result from interactions among individuals and between individuals and their environment as much as from the behaviors of the individuals themselves. Therefore, using a swarm-intelligent system to solve a problem requires a thorough knowledge

not only of what individual behaviors must be implemented but also of what interactions are needed to produce such or such global behavior" (Bonabeau, et al., 1999).

However, scholars like James Kennedy, Russell C. Eberhart, and Yuhui Shi have contradicted the above opinion. According to them, "swarm intelligent systems are quite easy to program, and that a knowledge of individual behaviors and interactions is not needed. Rather, these behaviors and interactions emerge from very simple rules" (Kennedy et al., 2001).

One may find logic in both the arguments. However, before deciding that, it is essential to understand the applications of swarm intelligence in various fields and its repercussions. Beginning from its applications in Science fields, it has extended to Social Science researches as well. It is interesting to note that SI can be applied to the field of Finance as well. The vital areas of Finance in which the SI is applied include portfolio optimization (PO), stock price prediction, risk management, prediction of market performance, automating operations, managing frauds, personalizing services and products, cost reduction, cybersecurity and fraud detection, tracking market trends, loan and credit decisions, compliance of regulations, insurance, and so on. Through the chapter, the applications of SI in PO, stock price prediction, and risk management are discussed along with the traditional methods used for the same for comparison.

Applications of Swarm Intelligence in Portfolio Optimization

As the application of SI is a modern approach to PO, it will be enlightening to go through the traditional methods before addressing the same. The word 'traditional' is used here in the sense of conventional methods. It does not mean the methods are outdated or not in practice in the contemporary situation. As the concept of SI is a new and advanced method, other techniques are generally considered as 'traditional' procedures for the present chapter by the authors. Hence, the section is divided in to two parts, viz., PO using traditional methods and PO using SI.

Portfolio Optimization using Traditional Methods

Investment can be said as a productive economic activity leading to capital formation in an economy. It is not a mere purchase of things, which is meant for immediate use or consumption. Investment is the sacrifice of current money for future benefits. More specifically, "investment is the current commitment of dollars for a period of time in order to derive future payments that will compensate the investor for (1) the time the funds are committed, (2) the expected rate of inflation, and (3) the uncertainty of the future payments" (Reilly & Brown, 2006). It means that the fund is committed today in anticipation of future returns. There come the two significant aspects of investment: risk and return.

Return is the income expected from an investment. Risk is the other side of the coin. It is the uncertainty that the actual return can be less than the expected return (Pankunni & Kumar, 2023). One cannot completely eliminate the risk, but can minimize the risk using various methods, one of which is diversification. Diversification was proposed by Harry Markowitz (1952), as a strategy through which a combination of negatively correlated financial assets are selected for investment.

It is similar to the old proverb saying "don't put all eggs in one basket". That is, when the basket falls, all the eggs will be broken. Instead, if the eggs are put in separate baskets, even if one basket falls, the others will be saved. Thus, through diversification, those risks which can be reduced by diversification (unsystematic risk) is reduced. What that will remain in the total risk after diversification will be the systematic risk, which cannot be eliminated through diversification, as shown in Figure 4. The more the diversification, that is, the number of stock, the lesser will be the unsystematic risk and thus the total risk.

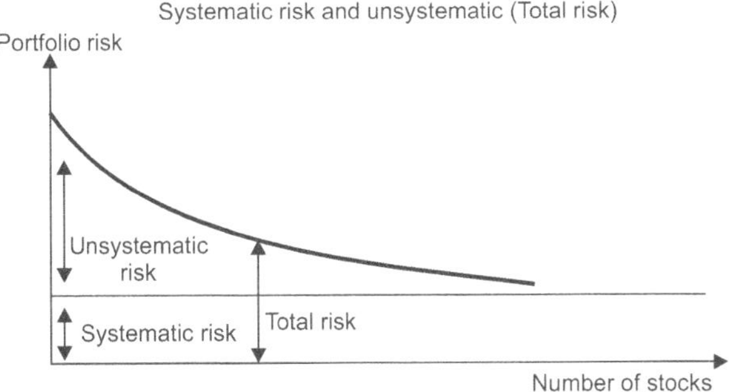

Fig. 4   An illustration of diversification process
Source: https://financialfreedomisajourney.com/reduce-unsystematic-risk-through-proper-diversification/

The relevance of diversification lies in the fact that it is implemented through **portfolios**. Portfolio is a collection of financial assets for investment. Markowitz (1952) had pointed out that, instead of investing in single securities, investing in a portfolio will give better returns with lesser risk. The justification for the same is made through **Markowitz's Portfolio Optimization**, which is considered as the pioneering optimization work in portfolio management. For that contribution of modern portfolio theory, Markowitz was given Nobel Prize in Economics and he is called as the father of modern portfolio theory. Therefore, when we discuss PO, it is imperative to go through the pioneering work in it.

In a nutshell, the optimization of portfolio as per Markowitz involves choosing the optimal portfolio among the available portfolios. He pointed out three parameters based on which a portfolio opportunity set can be constructed;

(a) *Expected return,*
(b) *Variance, and*
(c) *Co-variance.*

Expected return is the mean return expected from the portfolio. The variance measures the sum of squared deviations from the mean as a measure of volatility or variability in returns. Co-variance measures the variance of one asset with that of other assets in the portfolio. Thus, when the available portfolios are analyzed, the investor can identify the *feasible portfolios*, and among the feasible set, some will dominate over the others. Those portfolios that dominate over other portfolios are called the *efficient portfolios*. The efficient portfolios lie on the efficient frontier and the portfolio at which the efficient frontier is tangent with the indifference curve of the investor will give the *optimal portfolio*, as depicted in Figure 5. The portfolio M is the optimal portfolio in this case.

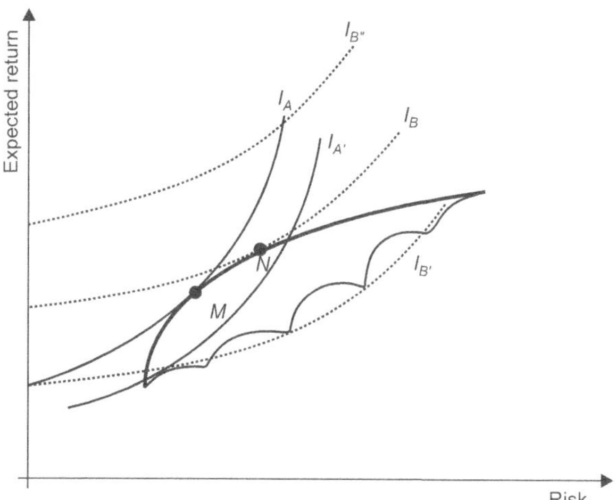

Fig. 5   An illustration of optimal portfolio
Source: https://www.researchgate.net/publication/46417913_Component_unit_pricing_theory/figures?lo=1&utm_source=google&utm_medium=organic

The PO is done by Markowitz by applying complex quadratic algorithms. Let us take an example of Markowitz optimization of a portfolio. Let ri be the random variable associated with the rate of return for asset i, for $i = 1, 2... n$, and define the random vector as given in Equation 1.

Equation 1

$$Z = \begin{pmatrix} r_1 \\ r_2 \\ \cdot \\ \cdot \\ r_n \end{pmatrix}$$

Set $\mu_i = E(r_i)$, $m = (\mu_1, \mu_2, \ldots, \mu_n)^T$, and $cov(z) = \Sigma$. If $w = (w_1, w_2... w_n)$, T is a set of weights associated with a portfolio, then the rate of return of this portfolio $r = \sum_{i=1}^{n} r_i w_i$ is also a random variable with mean $m^T w$ and variance $w^T \Sigma w$. If μ_b

is the acceptable baseline expected rate of return, then in the Markowitz theory an optimal portfolio is any portfolio solving the quadratic program given in Equation 2, where e always denotes the vector of ones (Anon., 2013). Based on that, one can reach at an optimal portfolio.

Equation 2

$$M \text{ minimize } \frac{1}{2} w^T \Sigma w$$

subject to $m^T w \geq \mu_b$ and $e^T w = 1$

Other methods through which PO can be done are M-V model (mean-variance model), semi-variance (SV) model, Variance with Skewness (VwS) model, Value-at-Risk (VaR) model, conditional VaR model, linear programming, nonlinear programming, mixed integer programming, copula based programming, stochastic programming or multistage portfolio optimization, etc.

Portfolio Optimization Using Swarm Intelligence

Markowitz's Portfolio Optimization (1954), being the fundamental and pioneering effort, is not free from limitations of complexities. Therefore, attempts had been made by theoreticians to simplify the process more intelligently. Studies reveal the application of SI tools like Grey Wolf Optimizer (GWO), a renowned meta-heuristic from evolutionary computation, for portfolio construction and unsystematic risk reduction (Mazumdar, Dongmo, & Guo, 2020). In addition to that, Particle Swarm Optimization (PSO), Firefly Optimization Algorithm, Artificial Bee Colony Optimization Algorithm, Ant Colony Optimization Algorithm, Bat Optimization Algorithm, Dragonfly Optimization Algorithm, Crow Search Optimization Algorithm, Cuckoo Search Optimization Algorithm, Krill Herd Optimization Algorithm, Fruitfly Optimization Algorithm, etc. also have found application in Portfolio Optimization.

Optimization algorithms are classified in many ways. One of the common ways is to classify it as:

(a) Deterministic algorithm, and
(b) Stochastic algorithm (Yang, 2010).

The deterministic algorithm can be said as arduous process, where the values and path are repeatable like a hill-climbing (Hassanein & Emary, 2016). On the contrary, stochastic algorithm has a randomness included in it. They are non-repeatable and the conclusion varies according to the situations applied. **PSO** is such a stochastic algorithm. As far as PO is concerned, PSO is most commonly used as the PO is also not repeatable. The portfolio set depends on the investor who selects the portfolio and her risk perception and objective regarding investment. The portfolio optimization ***includes portfolio construction, selection, and management***. There is Unconstraint portfolio optimization (UC) as well with boundary constraints

(BC) addressed. Therefore, the application of PSO for PO can be addressed under *parameter optimization, modeling, and the hybrid approaches* (Thakkar & Chaudhari, 2021).

An instance of parameter optimization by PSO is a study of Li et al. (2011), where a mutual fund is taken as a portfolio and has been tried to be optimized using PSO by incorporating time-scale features. These time-scale features were decomposed using maximum overlap discrete wavelet transform, where the weights are optimized as in the objective function as in Equation 3 (Z. et al., 2011). Thereby, $\hat{\alpha}$ and $\hat{\beta}$ indicated the ordinary least-squares regression intercept and slope; $\lambda\alpha$, $\lambda\beta \geq 0$ defined weighting values. The weighted time-scale features are then used for constructing homogeneous portfolios to select the optimal portfolio.

Equation 3

$$\text{Minimize} \quad \lambda\alpha\,|\hat{\alpha} - 0| + \lambda\beta\,|\hat{\beta} - 1|$$

When it comes to the modeling part of the portfolio, several PSO based heuristic approaches have been made through studies. Attempts have been made to extend the Markowitz optimization model with PSO. Moreover, in selection of securities, a combination of binary PSO (BiPSO) and improved PSO (IPSO) has been applied to deal with constraints and to attain the desired return (Golmakani & Fazel, 2011). Also, modeling based on multi-objective PSO (MOPSO) and dynamic search space PSO (DSSPSO) have resulted in good portfolio modeling.

The applications of PSO on PO is incomplete without addressing the hybrid approaches of PSO used by scholars in various studies. One such salient work where, to improve the PSO from being trapped into local optima, beetle antennae search (BAS) was integrated with PSO to form beetle swarm optimization (BSO) (Chen, et al., 2018). The other variants of PSO applied for PO is given in Table 1 (Thakkar & Chaudhari, 2021).

It is interesting to note that, within the variants, some may give more optimization results than the other. One such insight is from a work where the PO is done through both Ant Colony Optimization (ACO) and PSO. The study reveals that ACO performs better than PSO in the case of small-scale and large-scale portfolio, whereas PSO performance is better than the ACO technique in the case of a medium-scale portfolio (Zhu & Chen, 2010).

Applications of Swarm Intelligence in Stock Price Prediction

Of the available investment avenues, equity or stock investment is often considered as the most attractive. Despite knowing about the risk involved in investing in ownership securities, investors still continue to invest in the stock market. That leads to a conscience that the risk itself is the attraction of stock investment.

However, that does not mean stock investment is for speculation or gambling, where artificial risk is being created. A rational investor is always risk-averse. She knows that there is risk in the market and she tries to maximize return by

Table 1   Variants of Particle Swarm Optimization (PSO).

Sl. No.	Variants of PSO
1	Bare-Bones PSO
2	Binary PSO
3	Competitive Co-evolutionary PSO
4	Comprehensive Learning PSO
5	Constriction Factor-based PSO
6	Cooperative Random Learning PSO
7	Continuous Velocity PSO
8	Dimension Decreasing PSO
9	Decimal PSO
10	Drift PSO
11	Dynamic Search Space PSO
12	Evolutionary PSO
13	Fuzzy Simulation-based Multi-objective PSO
14	Fuzzy Clustering-based PSO
15	Hybrid Constraint-handling Multi-objective PSO
16	Improved PSO
17	Many Optimization Liaisons PSO
18	Multi-objective PSO
19	Multi-swarm of Improved Self-adaptive PSO
20	Adaptive PSO
21	Normalized PSO
22	Self-Regulating Multi-objective PSO
23	Sparse Velocity PSO
24	Turbulent PSO
25	Time Variant PSO

minimizing the risk. That is the uniqueness of investment. However, the question of risk still remains as to the possibility that the actual return may vary from the expected return. When investment is viewed as a coin, return and risk are the two sides of it. Risk cannot be separated from investment as the return is received in future which is uncertain and unpredictable.

As per Finance theory of Random Walk Hypothesis and Efficient Market Hypothesis (1969), stock prices are said to be random and unpredictable. That means the current market price is not dependent on its previous market price. In that case, investors will be in dilemma without knowing about the future of their hard-earned money invested. The solution to the problem is getting an idea of what will be the future stock price. It is a difficult task for investors because they do not know whether stock price can be predicted. Thus, whether the stock returns are predictable or not has always been a concern for the investors of all-time (P. & Kumar, 2019).

As stock prices are time-series data and thus a stochastic process, future stock prices can be estimated by assigning probabilities even though an exact prediction is not possible. Various models and methods have been developed for stock price predictions, including traditional methods and swarm intelligent-based methods.

Stock Price Prediction using Traditional Methods

Stock price prediction is the forecasting of future prices based on the historical prices and other market data. The traditional methods include the asset pricing models like Capital Asset Pricing Model (CAPM), Arbitrage Pricing Theory (APT), Fama's three-factor model, and Fama's five-factor model. In addition to that, various modifications to the above models have also been attempted by several authors. Such a modification is the Equity Pricing Model, where risk-adjusted efficient market return will be the expected return from a stock (Pankunni & Kumar, 2020). Another significant way of stock price prediction is through econometric models which include Auto-Regressive (AR) model, Moving Averages (MA) model, Auto-Regressive Moving Averages (ARMA) model, Auto-Regressive Integrated Moving Averages (ARIMA) model, and so on.

Stock Price Prediction using Swarm Intelligence

SI has been widely applied for predicting stock prices and returns. The prediction duration may be either short term or long term. Short-term prediction maybe for intra-day, inter-day, or daily. The weekly, bi-weekly, monthly, quarterly, and yearly are considered as long-term prediction. When the duration of prediction increases, it is likely that the accuracy decreases since it is regarding the future. Therefore, predicting with accuracy is the challenge before investors. One may predict the price using a model, but how far it stands well cannot be guaranteed. All the advancements in the technology and other fields are utilized to develop new methods for predicting stock prices. Application of Swarm Intelligence is with that hope. One of the vital swarm intelligent methods used for stock price prediction is the combination of machine learning techniques like Artificial Neural Network (ANN) and the Support Vector Machine (SVM) with SI. PSO still remains the basic model for stock prediction by applying SI.

Studies also proposed to predict stock e-exchange price using Elman network, Generalized Regression Neural Network (GRNN), and Wavelet Neural Network (WNN) with SVM NN (Xiao, et al., 2014). Attempts have also been done to optimize the NNs bias by applying Decimal PSO (DePSO) and architecture optimization by BiPSO for predicting daily stock returns. Scholars have also proposed to optimize premise and consequent parameters by applying Cooperative Random learning PSO (CRPSO) for making a better prediction model. Various advancements over PSO also became popular. One such advancement is Simplified Swarm Optimization (SSO) where, when it is applied, more stability in results have been depicted (Yeh, 2013).

Applications of Swarm Intelligence in Risk Management

Managing risk requires strong diligence and appropriate tools. Risk is inherent element in investment, that we all are aware. So, to avoid risk, selling off the investment is not the logical solution. It is like when a family member gets sick, she is treated instead of leaving her. Selling cannot be a solution to risk in investment. A rational investor should be able to protect or manage her investment by taking appropriate strategies. There lies the significance of risk management. An investor need not be an individual. An investor can be an institution, a corporate, or so on. That means the risk management is needed by not only the individual investors, but by the corporates, banks, other entities, etc. Advances have been introduced in risk management techniques as well. The section is divided into two for discussing the traditional methods and the swarm intelligent methods.

Risk Management Using Traditional Methods

The usual methods through which the risk is managed in investment is through purchase of derivatives. They derive value from the underlying assets and instead of buying and selling the assets, derivatives can be traded. The most commonly used derivatives are futures, options, and swaps. Another way of risk management technique to protect from cyclical stocks is to buy defensive stocks. Also, one should take care while selecting a portfolio is to incorporate negatively correlated securities.

Advance methods of risk management include regression analysis, Value-at-Risk (VaR), and scenario analysis. Regression analysis helps to find how one variable is influenced or affected by other variables and manage them accordingly. VaR gives a maximum loss for a given position in the time period mentioned. It will enable the investors to foresee and decide earlier. Scenario analysis takes into considerations all scenarios of the investment, say optimistic, pessimistic, and most likely conditions and the risk associated with it. This can enhance the risk management since people often decide based on the most likely or optimistic condition only. It is better to avoid high variance assets if the investor is risk-averse.

Risk Management Using Swarm Intelligence

SI has put forward a worthy leap in the risk management process. The most highlighted SI application in risk management is the GWO (Mazumdar et al., 2020). And when it comes to long term risk management, it is better to go for PSO. It is because, the PSO will enable the best parameters to be selected and thus the effective risk management in long-run (Azevedo et al., 2010). Moreover, Artificial Bee Colony Optimization Algorithm, Ant Colony Optimization Algorithm,

Bat Optimization Algorithm, Dragonfly Optimization Algorithm, Crow Search Optimization Algorithm, Cuckoo Search Optimization Algorithm, Krill Herd Optimization Algorithm, etc. are also used for mitigating and managing the risk in investment.

Conclusion

Change is the only thing that does not change. Therefore, when the advancement of technology happens, it is imperative to adapt it and apply it for the good. Swarm intelligence has paved a thoughtful way towards the optimization problems of all fields like Science, Social Sciences, and others. As far as the area of finance is considered, similar to science fields, considerable change is happening. More specifically, in the field of investment management, technology advancement has made a great impact. The applications of SI in portfolio optimization, stock price prediction, and risk management shows that clearly. Such advancements were path-breaking since only traditional methods of modeling were used till then. As the scope of SI and its applications in finance is wide, more advanced methods based on it can be applied for managing investment in future also.

References

Ahmed, H. and Glasgow, J. (2012). *Swarm Intelligence: Concepts, Models, and Applications.* Kingston, Ontario, Canada K7L3N6: School of Computing, Queen's University.

Azevedo, F., Vale, Z., Oliveira, P.M. and Khod, H. (2010). A long-term risk management tool for electricity markets using swarm intelligence. *Electric Power Systems Research, 80*(4), 380–89.

Belal, M., Gaber, J., El-Sayed, H. and Almojel, A. (2006). Swarm Intelligence. *In: Handbook of Bioinspired Algorithms and Applications.* Chapman and Hall (Eds.)

Bonabeau, E., Dorigo, M. and Theraulaz, G. (1999). *Swarm Intelligence: From Natural to Artificial Systems.* New York: Oxford University Press.

Chen, T., Zhu, Y. and Teng, J. (2018). Beetle swarm optimization for solving investment portfolio problems. *J. Eng., 16,* 1600.

Golmakani, H. and Fazel, M. (2011). Constrained portfolio selection using particle swarm optimization. *Expert Syst. Appl., 38*(7), 8327.

Hassanein, A.E. and Emary, E. (2016). *Swarm Intelligence: Principles Advances, and Applications.* Boca Raton: CRC Press.

Kennedy, J., Eberhart, R.C. and Shi, Y. (2001). *Swarm Intelligence.* San Francisco, USA: Morgan Kaufman Publishers.

Markowitz Mean-Variance Portfolio Theory. (2013). St. Louis City: University of Washington. Retrieved from https://sites.math.washington.edu//~burke/crs/408/fin-proj/mark1.pdf.

Mazumdar, K., D.Z. and Guo, Y. (2020). Portfolio selection and unsystematic risk optimization using swarm intelligence. *Journal of Banking and Financial Technology, 4,* 1–14.

Mazumdar, K., Zhang, D. and Guo, Y. (2020). Portfolio selection and unsystematic risk optimization using swarm intelligence. *Journal of Banking and Financial Technology, 4,* 1–14.

Millonas, M.M. (2004). Swarms, phase transitions, and collective intelligence. *In: Artificial Life III,* pp. 417–45. Redwood city: Addison-Wesley.

Mukhopadhyay, S. and Banerjee, S. (2010). Cooperating swarms: A paradigm for collective intelligence and its application. *International Journal of Computer Applications, 6*(10), 31–41.

P., N. and Kumar, D.S. (2019). Market Efficiency of Indian Stock Market: A Study of Selected Stocks. *Review of Research,* 1–8.

Pankunni, N. and Kumar, D.S. (2020). Risk Adjusted Efficient Market Return for Stocks. *Pacific Business Review International, 13*(2), 86–92.

Pankunni, N. and Kumar, S.R. (2023). Volatility of Returns in Stock Market Investments: A Study of BRICS Nations. *Finance: Theory and Practice, 27*(2), 87–98.

Reilly, F.K. and Brown, K.C. (2006). *Investment Analysis and Portfolio Management* (8th Edn.). New Delhi: Centage Learning.

Thakkar, A. and Chaudhari, K. (2021). A Comprehensive Survey on Portfolio Optimization, Stock Price and Trend Prediction Using Particle Swarm Optimization. *Archives of Computational Methods in Engineering, 28*, 2133–64.

Xiao, Y., Xiao, J., Lu, F. and Wang, S. (2014). Ensemble ANNs-PSO GA approach for day-ahead stock e-exchange prices forecasting. *Int. J. Comput. Intell. Syst., 7*(2), 272.

Yang, X.-S. (2010). *Nature-Inspired, Metaheuristic Algorithms.* Luniver Press.

Yeh, W. (2013). New parameter-free simplified swarm optimization for artificial neural network training and its application in the prediction of time series. *IEEE Trans. Neural Netw. Learn. Syst., 24*(4), 661.

Z, L., Y, L., S, T., B, L. and J, L. (2011). A novel time-scale feature based hybrid portfolio selection model for index fund. *Fourth International Conference on Business Intelligence and Financial Engineering.* (IEEE).

Zhu, H. and Chen, Y. (2010). Swarm Intelligence Algorithms for Portfolio Optimization. *LNCS 6145*: ICSI 2010, Part I.

Towards Next Generation Data-Driven Management
Leveraging Predictive Swarm Intelligence to Reason and Predict Market Dynamics

Dominik Brunner,[1] *Christoph Legat*[2] *and Uwe Seebacher*[3*]

Introduction

As we stand on the cusp of a new era in business management, the increasing complexity and velocity of market dynamics demand innovative approaches to data analysis and decision-making. An accelerating metamorphosis can be observed within the domain of data-driven management (U. Seebacher & Garritz, 2021), which is marked by an unprecedented proliferation of data, catalyzing a paradigm shift from traditional analytical practices to a comprehensive, predictive, and prescriptive data ethos. In antecedence, data-driven decision-making predominantly hinged upon retrospective analysis. Organizations would dissect historical datasets—financial records, consumer interactions, and erstwhile market trends—to inform future decisions. This analytical retrospection, while valuable, is rendered inadequate in the contemporary Big Data era, characterized by availability of voluminous amount of real-time data streams. The contemporary landscape necessitates not only the collection and storage of data but also its expeditious and cogent analysis. Accordingly, the crux of the contemporary challenge lies in distilling actionable insights from these vast oceans of data.

[1]   University of Applied Sciences Munich, Germany
[2]   Technical University of Applied Sciences Augsburg, Germany.
[3]   University of Applied Sciences Munich, Germany.
*   Corresponding author: uwe.seebacher@hm.edu

The current milieu advances beyond the precincts of conventional statistical modeling and business intelligence. It demands systems that transcend descriptive analytics, venturing into the realms of predictive and prescriptive intelligence through learning algorithms that discern patterns and forecast forthcoming dynamics.

Predictive Intelligence (PI) (U. Seebacher, 2021a) is predicated on the capability to anticipate future events with a high degree of accuracy. It encapsulates the transition from reactive to proactive management strategies, marking a departure from traditional data analysis that often lagged the pace of market changes. Seebacher posits that the contemporary business environment, marked by volatility and uncertainty, necessitates an anticipatory approach where organizations can dynamically recalibrate their strategies in response to emerging data trends.

In an era of high dynamics, where the timeliness of information changes rapidly, and the proper interpretation of available data is more important than the availability of the data itself, there is a growing trend toward collaborative collaboration. The utilization of collective ideas and underlying behavior patterns has the potential to shift from conventional linear and isolated analysis methods to a multidimensional and interconnected approach. Accordingly, combining PI and collective behavior patterns lead to the novel approach of Predictive Swarm Intelligence (PSI), for highly dynamic, multidimensional forecasting, and decision support improving the understanding of market and organizational phenomena. PSI draws from the natural world's exemplars of collective intelligence—such as avian flocks, ichthyic schools, and formicary colonies—where decentralized agents governed by rudimentary protocols aggregate to manifest complex, adaptive behaviors. Translated into the language of business analytics, these principles underpin a collaborative array of algorithms that function in unison to parse data and prognosticate outcomes.

This chapter originates to provide an overview of the PSI with a roadmap of the discussions to follow. It will highlight the critical aspects of PSI, from its theoretical underpinnings to practical applications, and the impact it can have on understanding and leveraging the complex networks of value and supply chains. Through this exploration, we will illustrate how PSI provides an essential competitive edge by enabling organizations to anticipate change, optimize operations, and avert disruptions with unprecedented agility and accuracy.

Predictive Swarm Intelligence Foundations

In the vanguard of computational innovation, PSI) emerges as a paradigm-shifting methodology, synthesizing the collective behavioral insights of swarm theory with the computational prowess of PI. In the following, an attempt is made to explain the conceptual foundations and practical applications of PSI and to outline its potential for the metamorphosis of data-driven decision-making paradigms in the context of the modern digital enterprise.

Swarm Theory Foundations

PSI is anchored in the precepts of swarm intelligence (SI), a domain initially illuminated by the seminal works on natural swarm behaviors by Bonabeau et al., (1999) and later expanded upon by Kennedy et al., (2001) through the conceptualization of particle swarm optimization algorithms. These foundational studies postulate that the decentralized, self-organized behaviors observed in biological entities such as ant colonies and bird flocks can be algorithmically modeled to solve complex problems (Bonabeau et al., 1999; Kennedy et al., 2001).

Expanding upon these principles, PSI integrates the predictive analytics framework, which traditionally harnesses statistical models and machine learning techniques to forecast future events based on historical data (Han et al., 2012). In PSI, however, the predictive process is not merely a linear extrapolation but a dynamic, iterative mechanism wherein the swarm's collective intelligence is continually refined through feedback loops and adaptive learning (Engelbrecht, 2005).

The theoretical framework for PSI is multifaceted, incorporating elements of computational intelligence, data analytics, and biological swarm behavior modeling. Engelbrecht (2005) provides a comprehensive overview of the algorithms that underpin swarm-based computational models, while Han et al., (2012) detail the methodologies that drive predictive analytics. The convergence of these disciplines within PSI offers a robust paradigm for tackling the complex, multidimensional challenges characteristic of big data environments.

An essential aspect of PSI's theoretical underpinnings is its alignment with the principles of complexity science, particularly the concept of emergence, where simple interactions at the micro-level lead to sophisticated patterns at the macro-level (Holland, 1995, 2006). PSI systems exemplify this phenomenon, as the collective behavior of the swarm's agents results in an emergent (PI) that cannot be ascribed to any single agent (Miller & Page, 2007).

The principle of decentralization is central to swarm theory. In natural swarms, control is distributed among individuals, with no single entity dictating the group's actions. This decentralization results in high levels of system robustness and redundancy, as the failure of individual members does not precipitate system collapse (Bonabeau et al., 1999). In computational terms, this translates to algorithms that are fault-tolerant and can function even when individual components fail (Brambilla et al., 2013).

Another fundamental principle is the use of local rules and simple interactions. In biological swarms, individual behavior is governed by local interactions with neighbors or the environment, rather than by global knowledge (Parrish & Edelstein-Keshet, 1999). This local interaction leads to the emergence of global patterns and intelligence, a concept that has been effectively harnessed in multi-agent systems and robotics to create complex, coordinated behaviors from simple rules (Şahin, 2005).

Self-organization is a defining feature of swarm theory, wherein structure and order emerge from the interactions of individuals within the swarm. This process is dynamic, adaptable, and often results in the formation of efficient structures and pathways, as seen in the construction behaviors of termites and the trail formation of ants (Ben-Jacob et al., 2000; Zhongshan Zhang et al., 2014). Computational models of self-organization have profound implications for the development of autonomous systems and have been a focus of research in distributed artificial intelligence (Casadei, 2023; Valentini et al., 2017).

The principle of stigmergy (Theraulaz & Bonabeau, 1999), or indirect communication through the environment, is another tenet of swarm theory that has been translated into computational algorithms. Stigmergy allows individuals within a swarm to coordinate indirectly by modifying and responding to the environment, as is the case in ant pheromone trails (Grassé, 1959). This form of communication has inspired the development of algorithms for collaborative robotics and distributed problem-solving (Forrest, 1991; Van Dyke Parunak, 2006).

Scalability is inherent to swarm-based systems, a principle that allows for the expansion or contraction of the system without significant loss of functionality. This scalability is evident in natural swarms, where the group can dynamically adjust its size based on environmental conditions or the task at hand (Couzin et al., 2002). In artificial systems, this scalability is critical for applications that require adaptability to varying operational scales (Dorigo & Stützle, 2004).

Thus, the principles of swarm theory provide a robust theoretical framework that spans disciplines and informs the design of systems characterized by decentralized control, local interactions, self-organization, stigmergy, and scalability. These principles have been instrumental in the advancement of algorithms and technologies that mimic the remarkable capabilities of natural swarms, leading to innovative solutions in complex system optimization and artificial intelligence. As research in this field continues to evolve, the fusion of swarm theory with emerging technologies promises to yield further insights and applications, solidifying its place at the confluence of natural and computational sciences.

Predictive Intelligence Foundations

PI stands as a paradigmatic shift in data analytics, offering profound insights by forecasting future events and trends with significant accuracy. Its foundations are deeply rooted in statistical theory, machine learning, and cognitive computing, amalgamating to form a multifaceted approach to predictive analytics. This scholarly review seeks to expound on these foundations, anchoring its discourse in the seminal contributions of leading researchers in the field, and highlighting the interplay between theoretical advancements and practical implementations.

Statistical foundations of PI are traditionally associated with the predictive models that extrapolate future data points from historical trends. Canonical works by Box et al., (2016) on time-series analysis provided the initial methodological framework for such predictive models, utilizing autoregressive integrated moving average (ARIMA) processes. Subsequent enhancements in statistical methodologies have led to the development of more sophisticated models that account for nonlinearities and high-dimensional data structures (Brockwell & Davis, 2002).

Machine learning has revolutionized the landscape of predictive analytics by introducing algorithms capable of learning from data without being explicitly programmed. The foundational work of (Cortes & Vapnik, 1995) on Support Vector Machines (SVMs) and the conceptualization of the kernel trick have been pivotal in the development of predictive models that can handle complex, high-dimensional datasets. Furthermore, the emergence of ensemble methods, such as Random Forests and Gradient Boosting Machines, as elucidated by (Breiman, 2001), has significantly improved the predictive performance by reducing variance and bias in model predictions.

The integration of cognitive computing into PI marks a significant advance, incorporating aspects of human-like reasoning and decision-making into predictive models. Researchers like Strohmeier have explored the potential of cognitive computing to enhance PI by emulating the processes of human thought, thus enabling machines to make predictions in a manner akin to human experts (Strohmeier, 2022). This approach to PI foregrounds the importance of understanding the cognitive aspects of prediction, as well as the underlying computational processes.

Recent advancements in the field have been marked by a focus on deep learning, a subset of machine learning characterized by neural networks with many layers. Pioneering work by LeCun et al., (2015) has demonstrated the profound capabilities of deep learning in predictive analytics, enabling the extraction of high-level features from raw data and improving prediction accuracy across various domains, from image recognition to natural language processing.

Contemporary scholars such as U. Seebacher (2021a) and Brunner (2023) have furthered the discourse on PI by incorporating insights from complex systems and network theory into predictive models. For enhancing predictive accuracy, understanding the interdependencies within data and the emergent properties of complex systems is essential (U. Seebacher, 2021b). This has been proven also by (Brunner, 2023; Legat et al., 2023) in their works defining a framework for analyzing the structural relationships within data and markets, thus facilitating the prediction of system dynamics.

The theoretical underpinnings of PI are further enriched by advancements in data management and processing. The development of Big Data technologies and the increasing availability of large-scale datasets have necessitated the creation of predictive models that can handle the volume, variety, and velocity of Big Data (Mayer-Schönberger & Cukier, 2013). This has led to innovative approaches to PI that leverage distributed computing and in-memory processing to perform real-time analytics (Zaharia et al., 2012).

The foundations of PI are deeply interdisciplinary, drawing from statistical theory, machine learning, cognitive computing, and can be enriched by novel emerging technologies. The seminal works of researchers in these fields have provided the methodological and theoretical frameworks that underpin PI, enabling its application across various industries. As PI continues to evolve, driven by technological advancements and theoretical innovations, its role in forecasting and decision-making processes is set to become increasingly significant, offering transformative potential for data-driven insights and actions.

Predictive Swarm Intelligence

The integration of PSI within the domain of data-driven management exemplifies a transformative convergence of complex systems theory, artificial intelligence (AI), and swarm intelligence (SI). This synthesis promises to reorient strategic and operational decision-making, harnessing the nuanced dynamics of collective behavior to forecast market trends and organizational needs with enhanced precision.

The theoretical underpinnings of PSI can be traced to swarm theory (as detailed before) and its key principles of decentralization, self-organization, and local interaction which dictate the collective actions of simple agents to engender complex, emergent behavior.

At the juncture where PSI meets predictive analytics, a robust statistical heritage is evident. Grounded in the methodologies developed by pioneers such as (Box et al., 2016), the discipline has evolved from linear statistical models to encompass the intricate capabilities of machine learning. This evolution, marked by advancements in ensemble and deep learning techniques, has expanded the predictive scope, allowing for the extraction of subtle patterns from expansive and intricate datasets.

AI, especially its adaptive learning mechanisms, infuses predictive models within PSI with the agility for continuous evolution and real-time adaptability. Cognitive computing further augments this framework, bringing to bear models that mirror human cognitive processes in decision-making, thus offering nuanced and sophisticated solutions to complex market predictions.

Furthermore, complexity science enriches the theoretical framework of PSI by offering insights into emergent behaviors and system dynamics. The understanding of emergence and the application of network theory provide critical perspectives for modeling the interconnectedness prevalent within complex market systems. It's within this interplay of networked agents that the power of PSI is truly realized, as the cooperative and competitive interactions within the swarm culminate in a simulation of market behaviors that yield predictive insights.

The integration of neurocomputing with swarm principles introduce a neural basis for decision-making into the PSI framework. Bio-inspired algorithms that draw from neural mechanisms offer the potential for PSI to emulate natural decision processes, enhancing the capability of predictive models to reason and simulate complex market dynamics accurately more closely.

Nevertheless, this theoretical construct does not come without challenges. The computational complexity of scaling PSI models to accommodate the real-world intricacy and volume of data remains a significant hurdle. Additionally, there are ethical considerations that must be at the forefront, especially concerning the transparency and interpretability of PSI models, which are imperative for their application in sensitive areas such as market prediction and strategic decision-making.

In a nutshell, the PSI framework beckons a commitment to interdisciplinary research, integrating insights from biology, computer science, and cognitive studies to refine its predictive capabilities. The horizon of PSI applications spans across diverse domains, each standing to benefit from the sophisticated analysis and forecasting that PSI offers. From finance to logistics, healthcare, and beyond, the promise of PSI is one of a future where the cascades of data intrinsic to modern business are not merely managed but harnessed, providing a strategic vantage point from which organizations can anticipate and navigate the ever-evolving market landscapes.

Modeling Market Dynamics in Complex Industrial Value Chains

In the intricate tapestry of modern industry, understanding market dynamics requires a panoramic view that extends beyond isolated analysis. This section delves into the transformative shift towards a holistic examination of market dynamics within complex industrial value chains, examining the interdependencies and systemic influences that pervade today's business ecosystems. In this context, the following aspects are being critically evaluated and discussed:

- **Shifting from Isolated Market Modeling to Industry and Value Chain-Based Analysis**
  Traditionally, market modeling has focused on the performance of individual entities, but the emergence of globalized economies and intricate supply networks necessitates a broader perspective. We explore the transition from this siloed approach to one that encompasses the full spectrum of industry and value chain interactions, revealing the intricate web of influences that define market success in the contemporary landscape.

- **Ripple Effects of Market Changes on Ecosystem Partners**
  Market fluctuations can send shockwaves through the entire value chain, impacting suppliers, distributors, and consumers alike. This section investigates the cascade of effects that market changes can precipitate, emphasizing the critical nature of understanding these dynamics to maintain equilibrium and capitalize on emergent opportunities within the industrial ecosystem.

- **Importance of Integrated Approach in Decision-Making**
  An integrated approach to market analysis and decision-making is no longer a luxury but a necessity. Here, we underscore the importance of considering the multifaceted relationships and feedback loops that characterize modern industrial value chains, advocating for decision-making frameworks that are both inclusive and system-oriented.

- **Leveraging Predictive Swarm Intelligence for Chain-Based Analysis**
  Finally, we turn to the potential of PSI as a tool for analyzing and forecasting market dynamics within these complex value chains. PSI's ability to synthesize vast quantities of data and model collective behaviors offers a promising avenue for businesses to gain a comprehensive understanding of their market position and navigate the future with confidence.

Shifting from Isolated Market Modeling to Industry and Value Chain-Based Analysis

The scholarly landscape of industrial economics and strategic management has witnessed a fundamental paradigm shift from traditional market modeling methodologies, characterized by a focus on isolated industry variables, towards a more integrated and systemic view encapsulating the entire value chain. This evolution in analytical perspective is pivotal in an epoch where intricate global supply chains and the interconnectivity of market entities are omnipresent. This discourse seeks to illuminate this transition, unpacking its theoretical implications for predictive analytics and strategic business practice.

Historically, market modeling was predominantly predicated upon classical economic theories, with an emphasis on isolated market variables or single business entities, typically situated within a static analytical framework. This approach, deeply rooted in the economic doctrines of perfect competition (Marshall, 2013) and monopolistic competition (Chamberlin, 1969), has served as a cornerstone in the edification of foundational business strategies. However, the burgeoning complexity of global marketplaces, propelled by technological innovations and multifaceted supply networks, has precipitated the obsolescence of such methodologies.

The seminal conception of the value chain by (Porter, 1985) signified an initial shift towards a broader interpretive perspective, accentuating the necessity of a comprehensive understanding of how various business activities synergistically contribute to competitive advantage. This was a precursor to a broader realization within the scholarly and practitioner communities that the value chain does not exist in isolation; rather, it is part of a more extensive industry value system that includes a myriad of stakeholders including suppliers, distributors, customers, and even competitors (Kaplan & Norton, 2004).

The acknowledgement that market dynamics are not confined to singular firms culminated in the formulation of industry-wide models that take into account the complex interactions between a multitude of market participants. The conceptual expansion of this perspective by (Dyer & Singh, 1998) into relational rents, arising from network-specific assets, knowledge-sharing routines, complementary resources, and efficacious governance, underscored the limitations of isolated modeling. Relational rents are contingent upon the broader network or ecosystem of value creation, demanding an analysis that transcends traditional models. (Moore, 1996) exposition of business ecosystems further underscored the interconnected nature of modern industries, proposing that companies co-evolve capabilities around innovations, engaging in both cooperative and competitive behaviors to support new products and satisfy customer needs. This ecosystemic perspective necessitates an analytical approach that is sensitive to the dynamic interrelations between various actors.

The integration of complexity science into market analysis offers profound insights into the behavior of complex adaptive systems. Markets, when viewed through the lens of complexity theory, exhibit emergent properties that are the byproduct of the interactions of numerous agents, leading to market patterns that cannot be anticipated through the examination of any individual agent in isolation. Arthur's (2015) application of complexity theory to market analysis has emphasized the critical nature of a systemic approach, one that accounts for feedback loops, nonlinear interactions, and the influence of stochastic events in shaping market dynamics.

The concept of the extended enterprise was introduced to acknowledge that a company's performance is influenced by its suppliers' and customers' operations. This requires advanced modeling techniques capable of capturing the nuanced interactions within the entire value network (Davis & Spekman, 2004). Predictive analytics has responded to this call, providing methodologies that can analyze large datasets to identify patterns indicative of future market behavior. The incorporation of machine learning algorithms, such as neural networks (G. Zhang et al., 1998) and support vector machines (Cortes & Vapnik, 1995), into predictive models has enabled the accommodation of the nonlinear and complex nature of market dynamics within industrial value chains.

The advent of big data analytics has further facilitated this paradigm shift, providing the technological means to collect and process the voluminous amounts of information generated across value chains (Manyika et al., 2011). Such analytical prowess allows for the examination of not only structured data but also unstructured data from a plethora of sources, proffering a more comprehensive overview of market dynamics. Latest advancements in Industry and Value Chain-based Analysis have been realized by Seebacher, Legat, and Brunner (Legat et al., 2023) in their works as part of the Predictores.ai[1] research project and spin-off startup. As part of the PI studies, different industrial and value chain-based analysis in the fields of automotive, printing, and photo as well as video streaming B2C businesses were deployed attempting to showcase the predictability of industries as well as value chains. These attempts were realized by the combinatory usage of specially selected trustworthy AI systems by also integrating social listening and community-building solutions.

Yet, the implementation of PSI in value chain analysis is confronted with considerable challenges. The quality and granularity of the input data, the computational power required to process and model this data, and the development of sophisticated algorithms that can precisely represent the dynamics of the value chain are formidable hurdles (Engelbrecht, 2005). Furthermore, ensuring the

---

[1] http://predictores.ai

interpretability of models generated by PSI remains paramount for stakeholders who must comprehend the rationale behind the predictions to make informed strategic decisions.

In conclusion, the potential of PSI for industry and value chain analysis is substantial. As firms increasingly operate within interconnected ecosystems, the capacity to predict how changes in one segment of the value chain will affect the entire system becomes a valuable strategic asset. PSI models can inform decisions on inventory management, production planning, and market entry strategies, offering firms a competitive edge in the fast-paced global market.

Ripple Effects of Market Changes on Ecosystem Partners

In the contemporary milieu of industrial economics, the intricacies of market dynamics are not merely confined to the boundaries of individual firms but extend across a complex network of ecosystem partners. This interconnectivity implies that perturbations in the market can propagate through this network, instigating a series of ripple effects that can significantly alter the competitive landscape. In the following discourse, we will explore the ripple effects of market changes on ecosystem partners, elucidating the underpinnings of this phenomenon and its implications within the context of PSI).

The concept of an industrial ecosystem operates under the premise that businesses are interlinked in a symbiotic relationship, akin to organisms within a biological ecosystem (Moore, 1996). Each entity, from suppliers to consumers, operates within an intricate web of dependencies. A perturbation in one node, such as a sudden shift in consumer demand or a disruption in supply, can cascade through the entire system, akin to the butterfly effect described in chaos theory (Shen et al., 2022).

The mechanisms through which market changes affect ecosystem partners are manifold. The inter-firm linkages act as conduits for the transmission of these effects, which can manifest as shifts in supply chain demands, alterations in strategic alliances, or changes in competitive strategies. For instance, a technological innovation by one firm can necessitate adaptations in the operations of its suppliers and partners (Teece, 1986, 2006).

The quantification of ripple effects requires sophisticated modeling techniques that can account for the nonlinear and dynamic nature of these interactions. Recent advancements in network theory have provided the tools to analyze these complex interdependencies, offering insights into how changes propagate through networks (Barabási & Albert, 1999). These models have been instrumental in the development of PSI, which can simulate and predict the outcomes of such intricate dynamics (Kennedy et al., 2001).

Empirical studies have provided evidence of the ripple effects of market changes on ecosystem partners. For instance, the automotive industry, characterized by a dense network of suppliers and manufacturers, has exhibited significant susceptibility to market fluctuations (Helper & Sako, 2010; Helper & Henderson,

2014). A single manufacturer's shift in design can necessitate changes across its supply network, impacting numerous partners and potentially altering the competitive dynamics within the industry.

PSI stands at the forefront of analytical tools that can capture and predict the ripple effects of market changes on ecosystem partners. By leveraging the collective intelligence and adaptive capabilities of swarm-based algorithms, PSI can model the reactions of individual agents within an ecosystem to various market stimuli, providing valuable foresight into the potential impacts on the network (Kennedy et al., 2001).

Understanding the ripple effects of market changes is vital for strategic management, as it enables firms to anticipate and mitigate potential risks. The ability to predict how these effects will traverse the ecosystem can inform decision-making processes, from risk management to strategic planning (Kaplan & Norton, 2004). Firms that can adeptly navigate these ripple effects can gain a competitive advantage by preemptively adapting to market changes.

Despite the advances in PSI and network theory, challenges remain in accurately modeling the ripple effects of market changes. The inherent unpredictability of market behavior, coupled with the complexity of inter-firm relationships, makes precise prediction an arduous task (Engelbrecht, 2005). Furthermore, the need for high-quality, real-time data to feed into PSI models remains a significant hurdle for many firms.

Future research in this domain is likely to focus on refining PSI models to better account for the complexity of industrial ecosystems. This may involve the integration of big data analytics and AI to enhance the predictive accuracy of these models (Manyika et al., 2011). Additionally, interdisciplinary approaches that combine insights from economics, network theory, and complexity science are expected to yield new frameworks for understanding and managing the ripple effects of market changes.

The ripple effects of market changes on ecosystem partners constitute a critical aspect of industrial dynamics that firms must navigate. The integration of PSI into strategic management practices offers a promising avenue for understanding and leveraging these effects to maintain competitive advantage. As the field evolves, the continued development of sophisticated models and analytical tools will be imperative in enabling firms to proactively respond to the complex interplay of forces that shape the industrial landscape.

Importance of Integrated Approach in Decision-making

In the ever-expanding complexity of today's marketplaces, decision-making has evolved from a linear, reductionist process to a multifaceted, integrative endeavor. The essence of this transition lies in the recognition that the myriad elements comprising modern industrial systems are interconnected, where actions taken in one domain can have profound, often unpredictable consequences in another. This interconnectedness necessitates an integrated approach to decision-making—a synthesis of data, models, and strategies across multiple domains and time scales (Porter, 1996; Allen & Helms, 2006).

Integrated decision-making is predicated on systems thinking, a concept that views a system as a coherent whole, characterized by complex interactions among its parts (Checkland, 1981). The shift towards systems thinking in business strategy emerged from the need to address the multifarious challenges that accompany globalization, rapid technological advancement, and increased market volatility (Senge & Sterman, 1992).

In the realm of PSI, the integrated approach is particularly salient. PSI embodies the principles of emergent intelligence, derived from the collective behavior of decentralized agents (Kennedy et al., 2001). The decision-making process within the PSI framework is inherently integrative, amalgamating insights from numerous agents—each representing a discrete data source or model—into a cohesive predictive force. This collective intelligence is greater than the sum of its parts and capable of navigating the complexities of market dynamics with a sophistication unattainable by isolated models (Millonas, 1993).

The integrated approach in PSI breaks down disciplinary silos, merging insights from economics, cognitive science, computer science, and organizational theory. It aligns with the transdisciplinary ethos advocated by scholars like Nicolescu (2002), who posits that addressing complex problems necessitates a convergence of disciplines. In the context of PSI, this implies combining market analysis with computational models that draw from natural swarm behaviors and cognitive processes.

The advent of big data analytics has significantly enhanced the capacity for integrated decision-making. By harnessing large volumes of data from diverse sources, organizations can gain a holistic view of the market and its potential trajectories (Mayer-Schönberger & Cukier, 2013). This comprehensive data integration is fundamental in PSI, where the predictive models are only as robust as the data they synthesize.

However, the path towards effective integrated decision-making is fraught with challenges. The complexity of integrating data and models necessitates advanced computational capabilities and sophisticated algorithms that can handle the intricacies of data fusion and model aggregation (Lycett, 2013). Moreover, the integrated approach requires a cultural shift within organizations, moving away from compartmentalized decision-making towards a more collaborative, systemic paradigm (Kotter, 1996).

An integrated decision-making approach imbues organizations with strategic flexibility and adaptability, allowing them to pivot in response to emerging trends and disruptions (Eisenhardt & Martin, 2000). This agility is crucial in a business landscape where the ability to anticipate and respond to change can confer a significant competitive advantage.

Adopting an integrated approach has significant implications for organizational structure and leadership. It demands a more networked organizational form, eschewing rigid hierarchies for more fluid, adaptive structures that facilitate cross-functional collaboration (Ashkenas et al., 2002). Leadership in such organizations must foster a culture that values systemic thinking and collective intelligence, aligning closely with the principles of PSI (Heifetz et al., 2009).

In conclusion, the integrated approach to decision-making is indispensable in the realm of PSI and data-driven management. It requires the orchestration of diverse elements—data, models, strategies, and human insights—into a unified decision-making framework that is greater than the sum of its parts. As organizations strive to navigate the complexities of modern industrial systems, the integrated approach will become increasingly central to their strategic and operational success. As we advance, the continuous refinement of integrative methodologies, supported by technological advancements and organizational evolution, will remain a pivotal endeavor for scholars and practitioners alike.

Leveraging Predictive Swarm Intelligence for Chain-based Analysis

As the global economy continues to evolve into an intricately connected web of value chains, the utilization of PSI for chain-based analysis has become an imperative for achieving strategic insights and operational excellence. This scholarly treatise examines the fusion of PSI within the domain of value chain analysis, providing a scientific overview of its methodologies, advantages, and the latest advancements that underpin its growing significance.

PSI emerges as a confluence of the decentralized decision-making principles found in natural swarms and the advanced predictive capabilities of artificial intelligence (Kennedy et al., 2001). Within the context of industrial value chains, PSI serves as a dynamic analytical tool capable of synthesizing complex data from various chain links to anticipate market trends and optimize decision-making processes (Engelbrecht, 2005).

Chain-based analysis using PSI involves the application of swarm-based algorithms to decipher the multifaceted interactions and dependencies within value chains. Such analysis transcends the capabilities of traditional linear models by accommodating the nonlinear and adaptive behaviors characteristic of complex industrial ecosystems (Bonabeau et al., 1999). By simulating the collective behaviors of agents within a value chain, PSI facilitates a granular understanding of how changes in one segment can propagate and impact the entire chain.

The methodological framework of PSI in chain-based analysis integrates principles from various disciplines. It utilizes algorithms inspired by biological swarms, such as particle swarm optimization (PSO) and ant colony optimization (ACO), to process and analyze data through a distributed network of agents (Dorigo & Stützle, 2004). Each agent, acting as a proxy for a component within the value chain, contributes to the generation of a comprehensive predictive model.

Advancements in computational techniques and data analytics have significantly bolstered the effectiveness of PSI in value chain optimization. Machine learning algorithms, particularly those rooted in deep learning, have been integrated into PSI models to enhance their predictive accuracy (LeCun et al., 2015). These algorithms enable PSI systems to identify intricate patterns and correlations within vast datasets, accounting for the complexities of modern supply chains.

Real-world applications of PSI in industry demonstrate its transformative impact on value chain analysis. In the automotive sector, PSI has been leveraged to optimize supply chain resilience and efficiency, providing manufacturers with the foresight to mitigate disruptions and maintain production continuity (J.-H. Zhang et al., 2017). In retail, PSI models have been employed to forecast consumer behavior, enabling retailers to align their inventory and distribution strategies with market demand (Ngai et al., 2009).

Despite its potential, the implementation of PSI for chain-based analysis is not without challenges. The complexity of designing and training PSI models that accurately reflect the dynamics of value chains requires significant computational resources and expertise (Engelbrecht, 2005). Moreover, the need for comprehensive and high-quality data to feed into these models remains a critical challenge for many organizations.

Future research in PSI is poised to explore the integration of more advanced AI techniques, such as reinforcement learning and generative adversarial networks, to further refine the predictive capabilities of swarm-based models (Goodfellow et al., 2014). Additionally, there is a growing interest in the development of hybrid models that combine PSI with other analytical frameworks, such as game theory and network analysis, to provide a more nuanced understanding of value chain dynamics.

In summation, leveraging PSI for chain-based analysis represents a frontier in data-driven management, offering a sophisticated approach to understanding and forecasting the complexities of global value chains. As organizations continue to grapple with the intricacies of interconnected markets, the strategic deployment of PSI models stands as a testament to the potential of artificial swarm-based systems in driving business intelligence. The ongoing evolution of computational technologies and data analytics methodologies will undoubtedly amplify the role of PSI in shaping the decision-making landscapes of the future.

Swarm Theory in Capturing Market Dynamics

In the rapidly advancing domain of market analysis, the introduction of swarm theory into the exploration of market dynamics marks a significant innovation. Therefore, we delve into the utilization of swarm principles to decode and predict the complex behaviors of markets. By examining the mechanisms of swarm intelligence within economic contexts, we uncover how these systems can not only replicate but also illuminate the intricacies of market movements and industry trends.

Firstly, we explore fundamental constructs of swarm systems and the parameters that govern their operation. Next, the imperative of continuous data feeding into swarm systems to garner real-time insights is considered because the velocity and volume of economic data generation necessitate systems that are not only responsive but also adaptive to the constant influx of information. Subsequently, the incorporation of swarm-inspired algorithms for forecasting market trends is detailed. Finally, the dual roles of cooperative and competitive agents within swarm systems as they apply to market and industry life cycle simulations are discussed for reflecting the various stakeholders in a market and their (potentially competing) interests and intentions.

Swarm Systems and Their Parameters

In the sphere of economic analysis, the application of swarm systems has garnered substantial interest for their potential to elucidate and forecast market dynamics. Swarm systems leverage the principles of swarm intelligence to model complex systems. This exposition delineates the intricacies of swarm systems and their parameters within the context of capturing market dynamics, integrating the latest advancements in the field.

Swarm systems operate on the premise that collective behavior in nature can inform the design of algorithms and systems to tackle complex, dynamic problems. The functionality of a swarm system is underpinned by a set of parameters that guide the behavior of individual agents and their interactions within the collective. These parameters include agent velocity, neighborhood size, quality of position, and alignment rules, which collectively determine the system's ability to explore and exploit the solution space (Kennedy et al., 2001). In the context of economic modeling, these parameters are calibrated to reflect the intricacies of market behaviors and trends.

Agent-based Models and Economic Forecasting

Agent-based models (ABMs) within swarm systems serve as microcosms of larger market mechanisms, where each agent represents an entity such as a consumer, firm, or financial instrument. The parameters of these agents—such as their decision rules, interaction protocols, and adaptation strategies—are crucial in simulating the market's response to various stimuli (Tesfatsion & Judd, 2006). The calibration of these parameters is informed by empirical data and theoretical constructs, allowing for simulations that mirror real-world economic phenomena.

Adaptation and Learning in Swarm Systems

A pivotal feature of swarm systems is their capacity for adaptation and learning. By employing algorithms that adjust parameters based on feedback—akin to the biological processes of evolution and learning—swarm systems can evolve over time, enhancing their predictive accuracy (Engelbrecht, 2005). This adaptive nature is particularly pertinent in volatile markets where conditions and trends can shift rapidly.

Data-driven Calibration of Swarm Parameters

The calibration of swarm system parameters is increasingly data-driven, harnessing the vast amounts of information generated by market activities. Big data analytics facilitate the processing of this data, enabling the refinement of agent parameters to reflect emerging trends and patterns (Manyika et al., 2011). The integration of machine learning techniques, such as reinforcement learning, allows for the automated tuning of these parameters, ensuring that the system remains attuned to the latest market dynamics (Sutton & Barto, 2018).

Swarm Systems and Market Complexity

The complexity of markets, characterized by non-linear interactions and emergent phenomena, aligns with the capabilities of swarm systems. By modeling markets as well as industries as deployed by (Legat et al., 2023) as part of their studies on industry and value chain predictions as complex adaptive systems, swarm systems can account for the multifaceted nature of economic interactions and the feedback loops that drive market evolution (Arthur, 2015). This approach stands in contrast to traditional models that often assume linear interactions and equilibrium states.

Real-time Market Simulation with Swarm Systems

One of the foremost advancements in the application of swarm systems to economics is their ability to perform real-time simulation and forecasting. By continuously integrating market data into the swarm model, the system can provide up-to-date insights into market movements, enabling decision-makers to respond proactively to changes (Tett et al., 2017; Frank & Sanati, 2018). In this context also the works of (U.G. Seebacher, 2021) the growth model towards PI he brings into discourse Template-based Intelligence (TBI) for predictive and interactive scenario planning and simulations. His works are based on the advancements and findings from the field of the game theory (Owen, 2013).

Challenges in Swarm System Implementation

Despite their potential, the implementation of swarm systems in economic analysis is not devoid of challenges. The complexity of accurately modeling market dynamics necessitates advanced computational infrastructure and expertise in both economics and systems engineering (Macal & North, 2010). Moreover, ensuring the validity and reliability of the data that informs swarm system parameters remains a critical concern.

Continuous Data Feeding for Real-time Insight

The ever-accelerating pace of the global marketplace necessitates a paradigm of continuous data analysis for businesses to remain competitive and adaptive. The emergence of technologies enabling continuous data feeding has revolutionized the landscape of real-time market insight, offering a transformative approach to operational and strategic decision-making. This scholarly exposition explores the mechanisms, advancements, and implications of continuous data feeding for real-time insight, particularly within the realm of PSI.

Continuous data feeding refers to the ongoing collection, processing, and analysis of data streams, which enables organizations to garner insights into market dynamics as they unfold. The concept of real-time data analysis is not novel; however, its application and scope have expanded dramatically with the advent of advanced computational technologies and the Internet of Things (Gubbi et al., 2013). The continuous flow of data—from consumer transactions, social media, sensor outputs, and more—provides a rich tapestry of information that, when analyzed effectively, can inform a nuanced understanding of market conditions.

The technological infrastructure that supports continuous data feeding is multifaceted. It encompasses advancements in data storage, such as cloud computing, which offers scalable solutions for the vast amounts of generated data (Mell & Grance, 2011). In conjunction with this, edge computing brings data processing closer to the source of data generation, reducing latency and enhancing the speed of insight derivation (W. Shi et al., 2016).

A pivotal development in the processing of continuous data streams is the evolution of real-time analytics platforms. These platforms employ advanced algorithms and computational models capable of handling high-velocity data, providing businesses with instantaneous analytical outputs (Kambatla et al., 2014). The integration of machine learning techniques, particularly online learning algorithms, enables these platforms to adapt to new data, learning and evolving without the need for system downtime (Domingos, 2012).

PSI represents an innovative confluence of collective behavior modeling and predictive analytics. In PSI, the continuous feeding of data is critical as it ensures that the swarm algorithms are informed by the most current market information, enabling accurate and timely predictions. The swarm's distributed nature allows for parallel processing of data, enhancing the speed and scalability of real-time analysis (Kennedy et al., 2001).

Swarm algorithms are uniquely suited to dynamic data streams due to their inherent adaptability and flexibility. PSO, for instance, has been adapted for real-time data analysis, with each particle adjusting its trajectory based on new information, akin to traders adjusting their strategies in response to market news (Poli et al., 2007). ACO algorithms have also been employed to optimize routing and logistics based on continuous traffic flow data (Dorigo & Stützle, 2004).

The implementation of continuous data feeding systems poses several challenges. The sheer volume of data can overwhelm traditional data processing systems, necessitating the development of more efficient data management and analysis (Marjani et al., 2017). Additionally, ensuring data quality and integrity is paramount, as the insights derived are only as reliable as the data on which they are based (Sadiq, 2013).

The constant collection and analysis of data also raise significant ethical and privacy concerns. Organizations must navigate the delicate balance between insight and intrusion, ensuring that the data is used responsibly and in compliance with increasingly stringent data protection regulations (Taylor et al., 2017).

Looking to the future, the field of continuous data feeding for real-time insight is poised for further growth. The proliferation of 5G technology promises to enhance the speed and reliability of data transmission, expanding the potential for real-time analytics (Andrews et al., 2014). Furthermore, advancements in AI are expected to yield more sophisticated analytical models that can extract deeper insights from continuous data streams (Jordan & Mitchell, 2015).

Continuous data feeding constitutes a cornerstone of modern market analysis, providing businesses with the real-time insights necessary to navigate the complexities of the global economy. The integration of PSI with continuous data analysis offers a powerful approach to understanding and predicting market dynamics. As computational technologies continue to evolve, the potential for real-time analytics to inform decision-making processes will likely become an indispensable aspect of business operations.

Incorporating Swarm-inspired Algorithms for Forecasting

In the current era of data-driven decision-making, the integration of swarm-inspired algorithms for forecasting represents a significant leap forward in predictive analytics. These algorithms, inspired by the collective behavior of biological entities, have been increasingly recognized for their potential to model complex systems and provide accurate forecasts. This essay explores the incorporation of swarm-inspired algorithms in forecasting, touching upon their theoretical foundations, methodological applications, and the latest advancements in the field.

Swarm-inspired algorithms are grounded in the principles of SI, a field that examines the collective behaviors of decentralized, self-organized systems, typically found in nature (Bonabeau et al., 1999). These behaviors are characterized by robustness, flexibility, and the ability to find optimal solutions to problems through individuals' simple interactions. SI has been mathematically and computationally modeled through various algorithms, most notably PSO and ACO, which simulate the foraging behavior of birds and ants, respectively (Dorigo & Stützle, 2004; Kennedy et al., 2001).

In the domain of predictive analytics, swarm-inspired algorithms have been employed to forecast complex phenomena where traditional predictive models may fall short due to the non-linearity and high-dimensionality of the data involved. PSO, for example, has been applied to financial market forecasting, utilizing the collective searching capability of the swarm to identify patterns and trends that inform future market behavior (Garro & Vázquez, 2015). Similarly, ACO has been used in logistics to forecast optimal routing paths based on dynamic traffic conditions (Bell & McMullen, 2004).

Recent methodological advancements have expanded the application of swarm-inspired algorithms beyond their initial scope. Hybrid models that combine SI with other forms of AI, such as neural networks and fuzzy logic, have demonstrated enhanced predictive capabilities (Blum & Li, 2008). These hybrid models capitalize on the strengths of each approach, using the adaptability and exploration features of SI to refine the learning and generalization abilities of neural networks.

The practical applications of swarm-inspired algorithms for forecasting are manifold. In supply chain management, these algorithms have been used to predict demand fluctuations, allowing for more efficient inventory control (He et al., 2023; J.-H. Zhang et al., 2017). In energy markets, PSO has been utilized to forecast electricity prices, a critical component for strategy formulation in the energy sector (Azadeh et al., 2008).

While the benefits of incorporating swarm-inspired algorithms in forecasting are clear, several challenges remain. One of the primary issues is the calibration of algorithm parameters, which can significantly impact the accuracy of the forecast. The selection of parameters such as population size, inertia weight, and learning factors in PSO requires careful tuning and is often problem-specific (Y. Shi & Eberhart, 1998). Furthermore, the stochastic nature of these algorithms means that they can converge on local rather than global optima, necessitating multiple runs to ensure reliability (Engelbrecht, 2005).

As with all data-centric approaches, ethical and privacy implications must be considered. The use of swarm-inspired algorithms for forecasting involves the processing of large amounts of data, which may include sensitive personal information. Ensuring that these algorithms adhere to ethical guidelines and privacy regulations is essential to maintain public trust and compliance with legal standards (Taylor et al., 2017).

Cooperative and Competitive Agents in Market and Industry Lifetime Simulation

In the expansive field of economic modeling and market simulation, the incorporation of both cooperative and competitive agents stands as a testament to the complexity and veracity of market behaviors. Drawing from the rich tapestry of swarm intelligence and game theory, this integration facilitates simulations that mirror the multifaceted interactions within markets and industries over their lifetime. This exposition elucidates the significance of incorporating such agents into simulations, referencing seminal work and latest advancements in this domain.

Market and industry simulations have traditionally hinged on the principle that agents operate within a framework of competition, vying for market share and resources. However, recent advancements posit that incorporating both cooperative and competitive behaviors among agents can lead to more robust and accurate simulations of market dynamics (Axelrod, 1997). These simulations mirror real-life scenarios where firms collaborate through alliances and partnerships while simultaneously competing in other arenas (Brandenburger et al., 1996).

The theoretical underpinning of this approach is rooted in the interdisciplinary study of complex systems and game theory. Complex systems theory suggests that the aggregate behavior of individual components within a system gives rise to the system's overall behavior (Holland, 2006, 1995). Game theory, on the other hand, provides a mathematical framework to model and analyze strategic interactions among rational decision-makers (Neumann et al., 1944). The amalgamation of these two theories in economic simulations allows for a nuanced understanding of market and industry dynamics.

Agent-based modeling (ABM) has become a cornerstone in simulating economic ecosystems, wherein each agent is imbued with a set of strategies and rules governing their behavior. SI, particularly the concepts of stigmergy and self-organization, further enriches ABM by introducing agents that adapt based on the collective behavior of the group (Kennedy et al., 2001). These adaptive agents can represent various market entities, from individual consumers to multinational corporations, each with distinct objectives and behaviors.

Cooperative strategies in ABM are encapsulated by agents that share information, resources, or strategies for mutual benefit. This cooperation can manifest in various forms, such as joint ventures, strategic alliances, or collective bargaining (Axelrod & Hamilton, 1981). The modeling of such cooperative behaviors requires careful consideration of the agents' utility functions and payoff matrices, ensuring that the cooperative behavior aligns with each agent's goals.

Conversely, competitive strategies are characterized by agents acting in their self-interest, often at the expense of others. The concept of Nash equilibrium, where each agent's strategy is optimal given the strategies of other agents, is particularly relevant in competitive ABM (Nash, 1950). Competitive ABM can simulate market scenarios, such as price wars or product launches, providing insights into how these events may disrupt market equilibrium and industry stability.

The simulation of an industry's life cycle, from inception to maturity and potential decline, requires a dynamic modeling approach that incorporates both cooperative and competitive behaviors. The life cycle theory suggests that industries go through stages of introduction, growth, maturity, and decline (Klepper, 1997). ABM can simulate these stages by adjusting the prevalence and intensity of cooperative and competitive interactions among agents over time. These results receive further corroboration through the research conducted by (Legat et al., 2023), who focused on the industry life cycle within the realm of predictive intelligence. Their research sought to forecast impending industrial evolutions by analyzing data and information from three to five years prior to their significant influence. A notable case study within their work highlighted the Blockbuster corporation, utilizing predictive intelligence to chart the industry life cycle shift from traditional video rental business models to the emergence of video streaming platforms. In a scholarly examination of industry transformation done by the same group, another case study was conducted on a preeminent German newspaper to analyze the sector's evolution. Utilizing data from the late 1990s, the study aimed to forecast significant shifts within the newspaper industry. The predictive model accurately anticipated the strategic alterations undertaken by prominent entities such as *The New York Times* at the turn of the century. Specifically, it foresaw their transition from a reliance on traditional print advertising to a diversified revenue model increasingly focused on online channels (Dewenter & Rösch, 2015; Chyi, 2005). This shift underscores the broader industry trend of digitalization and the reconfiguration of revenue strategies in response to the changing media consumption landscape (Picard, 2011).

Advancements in simulation algorithms have focused on enhancing the scalability and complexity of ABM. High-performance computing (HPC) and parallel processing techniques have allowed for the simulation of large-scale models with numerous agents (Axtell, 2000). Furthermore, the integration of machine learning algorithms within ABM can enable agents to learn and evolve their strategies based on outcomes, further enriching the simulation (Tesauro & Kephart, 2002).

Despite these advancements, challenges persist in the accurate representation of cooperative and competitive behaviors. One such challenge is ensuring that the modeled behaviors accurately reflect the strategic decision-making processes of real market participants. Additionally, ethical considerations must be addressed, particularly when simulations influence policy decisions that could have socioeconomic impacts (Yang & Gilbert, 2008; N. Gilbert & Stoneman, 2015).

In conclusion, the incorporation of cooperative and competitive agents in market and industry life cycle simulations offers a profound opportunity to deepen our understanding of economic phenomena. By leveraging the frameworks of complex systems and game theory, and harnessing the latest computational advancements, these simulations can yield critical insights into the strategic behaviors that drive market evolution. As computational techniques continue to advance, the potential for these simulations to inform economic policy and business strategy is vast, promising to unlock new frontiers in the predictive analysis of market and industry dynamics.

Designing a Predictive Swarm Intelligence System

For running a PSI system successfully, a methodical sound approach is evident. Therefore, this section serves as a blueprint for practitioners and researchers interested in building sophisticated predictive systems that capitalize on the collective wisdom of SI, providing a step-by-step approach to the design and implementation of a PSI system. By exploring selected building blocks derived from dissecting the process into its fundamental components, the basic steps are derived.

The foundation of any robust PSI system is predicated on the systematic collection and preprocessing of data. Therefore, we delve into the methodologies for gathering relevant data, the strategies for handling the vast influx of information, and the crucial preprocessing steps necessary to ensure data integrity and suitability for input into the swarm model.

Subsequently, the agent-based simulation which is a key component of a PSI system since it allows for the dynamic representation of individual agents and their interactions within the swarm. Finally, the integration of PSI with existing PI tools systems for increasing efficacy is discussed.

Data Collection and Preprocessing

In the domain of PSI, the acquisition and preprocessing of data constitute critical preliminary steps that underpin the system's predictive accuracy and reliability. This detailed discourse explicates the methodologies and considerations that govern the collection and preprocessing of data within PSI systems, integrating academic insights and the latest technological advancements in the field.

Data Collection: The Bedrock of Predictive Swarm Intelligence

Data collection for PSI is an intricate process that involves gathering vast quantities of diverse data types. This stage is paramount, as the integrity of the entire PSI system hinges on the quality and relevance of the data collected. Data sources for PSI are manifold, ranging from structured databases and transaction logs to unstructured social media feeds and sensor data (Han et al., 2012).

Technological Advancements in Data Acquisition

Technological advancements have dramatically expanded the capabilities and efficiency of data collection. The advent of the Internet of Things and edge computing has facilitated the real-time capture of data from a multitude of distributed sources (W. Shi et al., 2016). Additionally, cloud technologies have enabled scalable storage solutions that can accommodate the exponential growth of data volumes (Hashem et al., 2015).

Preprocessing: Conditioning Data for PSI Systems

Preprocessing is an essential phase where raw data is transformed into a clean, structured format suitable for analysis. This phase encompasses several critical processes:

1. **Data Cleaning:**   The removal of inaccuracies and inconsistencies in the data set is a foundational preprocessing step. Techniques such as error detection algorithms and outlier removal are employed to enhance data quality (Heer et al., 2018; Kandel et al., 2011).
2. **Data Normalization:**   This process involves scaling numerical data to fall within a specific range, facilitating uniformity and preventing bias towards certain features within algorithms (García et al., 2015; Fernández et al., 2018).
3. **Feature Selection:**   The selection of pertinent features is crucial for reducing dimensionality and focusing the PSI system on the most informative data aspects. This process can be accomplished through techniques such as principal component analysis (PCA) or genetic algorithms (Guyon & Elisseeff, 2003).

4. **Data Transformation:**   Data transformation methods, such as discretization and the construction of derived attributes, aid in enhancing the predictive power of the subsequent modeling stages (Pyle, 1999).

Dealing with High-Dimensional Data

PSI systems often deal with high-dimensional data, which poses significant challenges for processing and analysis. Dimensionality reduction techniques are employed to distill the most significant information from the data while minimizing the loss of critical information (Lee & Verleysen, 2007).

Real-Time Data Streams and Online Preprocessing

The real-time nature of many PSI applications necessitates the ability to preprocess data on the fly. Online preprocessing techniques have been developed to address this need, allowing for incremental updates to the system without the requirement for batch processing (Aggarwal, 2007).

Integration of Machine Learning in Preprocessing

Machine learning techniques have become integral to the preprocessing phase, particularly in the context of automated feature extraction and selection. Deep learning models, such as autoencoders, have demonstrated effectiveness in identifying complex patterns and reducing dimensionality in large datasets (Bengio et al., 2012).

Remarks and Conclusion on Data Collection

The data collection and preprocessing stages are not solely technical endeavors; they also encompass ethical considerations, particularly concerning data privacy. With regulations such as the General Data Protection Regulation (GDPR) in Europe setting stringent guidelines for data handling, PSI systems must ensure compliance and protect individual privacy rights (Bussche & Voigt, 2017). Accordingly, it is evident to identify required regulatory requirements with respect to the economic market, the PSI system should operate.

Despite advancements, the field continues to face challenges, notably in the areas of data veracity and the processing of heterogeneous data types. Future research directions are likely to focus on the development of more sophisticated preprocessing algorithms that can handle the growing complexity and variability of data sources.

In sum, data collection and preprocessing are fundamental to the effective deployment of PSI systems. The processes involved in transforming raw data into a structured and meaningful format are as complex as they are vital. With the continuous evolution of data acquisition and preprocessing technologies, the capabilities of PSI systems are expected to advance correspondingly, leading to increasingly accurate and insightful predictive models. As such, the field stands on the precipice of significant growth, driven by both technological innovation and a deepening understanding of the intricacies of data science.

Swarm Model Development

Swarm model development draws on the intricate interactions observed in natural systems, such as ant colonies or bird flocks, where simple rules at the individual level give rise to complex collective behaviors (Bonabeau et al., 1999). In the economic and industrial spheres, these models have gained traction for their potential to simulate market dynamics and forecast industry trends with unprecedented accuracy.

The biological foundations of swarm behavior are rooted in ethology and the study of social insects. These foundations provide insights into how individual behaviors, governed by local rules and interactions, can lead to the emergence of global patterns and intelligent decision-making capabilities (Camazine, 2003). By understanding these biological principles, researchers can derive computational analogs that retain the adaptability and efficiency characteristic of natural swarms.

The computational models of swarming behavior encompass a variety of algorithms, each tailored to capture specific aspects of collective intelligence. Algorithms such as PSO and ACO are predicated on mimicking the foraging behavior and path-finding abilities of birds and ants, respectively (Dorigo & Stützle, 2004; Kennedy et al., 2001). These algorithms are designed to solve complex optimization problems by iteratively improving a candidate solution with regard to a given measure of quality.

A critical aspect of swarm model development is parameterization, which involves setting the rules and variables that govern agent behavior and interactions within the model. Parameters such as the number of agents, the radius of interaction, and the weighting of individual versus collective knowledge must be finely tuned to reflect the system being modeled (Kennedy et al., 2001). The calibration of these parameters is both an art and a science, often requiring iterative experimentation and validation against real-world data.

Recent advancements in swarm algorithm design have seen the incorporation of machine learning techniques to enhance the adaptability and predictive power of swarm models. Deep reinforcement learning, for instance, has been utilized to enable agents within a swarm to learn and evolve their strategies based on environmental feedback (Mnih et al., 2015). These developments have broadened the scope of swarm models, allowing them to tackle more complex, dynamic systems such as those found in financial markets or supply chain logistics.

The integration of swarm models with other predictive techniques has been a focus of recent research. Hybrid models that combine the exploratory power of SI with the predictive accuracy of time-series analysis or regression models offer a more comprehensive approach to forecasting (Wen-Jun Zhang & Xiao-Feng Xie, 2003; Petropoulos et al., 2022). These integrated models can leverage the strengths of each method, yielding superior predictive performance.

Despite the progress made in swarm model development, challenges remain. The stochastic nature of swarm algorithms can lead to variability in outcomes, necessitating robust validation methods to ensure the reliability of predictions (Engelbrecht, 2005). Additionally, the computational complexity of these models can pose significant demands on processing power, particularly for large-scale applications.

The development of swarm models represents a confluence of natural observation and computational innovation, offering profound implications for the field of predictive analytics. As technological advancements continue to propel the capabilities of these models, their application across various domains promises to yield rich insights and foresights. The interdisciplinary nature of swarm model development ensures that it remains a vibrant area of research, with ongoing contributions from the fields of biology, computer science, and economics.

Agent-based Simulation Design

The advent of ABM has engendered a paradigmatic shift in the epistemological frameworks that underpin economic and industrial simulations. This intricate essay delineates the vanguard methodologies in the development of ABM systems, drawing upon contemporary academic discourse and the latest technological breakthroughs that facilitate a granular understanding of market dynamics through PSI.

The genesis of ABM lies in the granular codification of individual actors within a system, a methodology that affords a unique lens through which to examine the emergent phenomena inherent in complex economic landscapes. The traditional paradigm of ABM, while robust, often presupposes a level of rationality and information parity that diverges from the empirical reality of market environments (Epstein & Axtell, 1996).

An innovative trajectory in ABM is the explicit codification of complexity and heterogeneity among agents. This acknowledgment—of variations in agents' access to information, cognitive capacities, and decision-making heuristics—augments the verisimilitude of simulations (Thompson & Stappenbeck, 1999). Such heterogeneity allows for the exploration of the competitive and cooperative interplay of behavioral strategies within the market, offering profound insights into the mechanisms of innovation and market perturbation.

The integration of adaptive learning mechanisms into agent behaviors constitutes a significant departure from static rule-based paradigms. The application of machine learning algorithms, particularly those predicated on neural computation and reinforcement paradigms, enables agents within a simulation to modulate their strategies in response to environmental feedback, engendering a simulation environment where strategies are evolutionary rather than static (Sutton & Barto, 2018). This endows ABM systems with a dynamism that parallels the adaptive learning processes observed among real economic actors.

Advancing the concept of agent-based simulation design is the explicit modeling of network effects and agent interactivity. Tools derived from social network analysis proffer the means to map and quantify the interconnective relationships between agents, thus elucidating how these linkages influence macroeconomic outcomes (Jackson, 2008). This is particularly salient in simulating markets where network externalities are pronounced, such as in digital marketplaces or platform-centric industries.

Interdisciplinary Convergence for Enhanced Simulations

The design of agent-based simulations is inherently benefited by interdisciplinary convergence, amalgamating insights from diverse fields such as economics, sociology, psychology, and computer science. Such collaboration begets the development of comprehensive models that simulate not only the economic outcomes but also consider the societal and psychological factors that shape those outcomes (N. Gilbert & Stoneman, 2015; P. Gilbert et al., 2008). Models that incorporate such breadth are better equipped to inform policymaking and strategic planning.

Challenges and Ethical Imperatives

Despite the advancements in simulation design, challenges persist. The complexity of representing agent behaviors that accurately mirror human decision-making, the computational demands of executing large-scale simulations, and the requisite for interdisciplinary expertise represent considerable obstacles. Additionally, the ethical stewardship of data and the potential repercussions of simulation-based decisions on real-world economies necessitate careful deliberation (Moss & Edmonds, 2005).

In summation, agent-based simulation design stands poised for transformative development, with emerging methodologies extending from the integration of advanced computing to novel applications of big data. These simulations promise to offer an unprecedented depth of insight into the operational mechanics of economic systems, guiding decisions that shape the contours of markets and industries. As the discipline advances, it is imperative to navigate the challenges and ethical considerations with rigor and foresight, ensuring that the proliferation of ABM contributes constructively to our collective understanding of economic phenomena.

Integration with Predictive Intelligence Tools

The fusion of SI principles with predictive analytical tools signifies an evolutionary leap in data-centric strategic management, catalyzing the progression toward a new paradigm of market analysis. This scholarly exposition posits an array of innovative methodologies and conceptual advancements, underpinned by seminal scholarly work and recent technological breakthroughs, aimed at refining the capacity of PSI systems to dissect and anticipate market phenomena.

Multidisciplinary Convergence for Augmented Forecasting

An intricate PSI system's scaffolding is constructed from a multidisciplinary tapestry that harnesses the collective insights from data science, computational learning, and econometrics. This harmonization transcends the confines of traditional market analysis, equipping PSI systems with a multifaceted acumen for elucidating economic behaviors (Bishop, 2006; Shmueli, 2010). Such systems move beyond mere predictive output, delivering a composite understanding of the forces at play within markets.

A novel conceptual avenue involves embedding sophisticated learning architectures into swarm-based models. Algorithms rooted in reinforcement learning and neural networks can be harnessed to extract nuanced patterns from data, which in turn inform the strategic interactions of agents within a PSI framework (LeCun et al., 2015). This symbiosis between predictive technology and swarm adaptation engenders a dynamic system where the collective intelligence of the swarm is continuously honed.

Cognitive Informatics within PSI

Furthering this trajectory is the amalgamation of cognitive informatics with PSI constructs. Cognitive informatics endeavors to replicate the intricacies of human reasoning within computational frameworks, thus providing an additional interpretive layer to analytical outputs (Modha et al., 2011). With the incorporation of such informatics, PSI systems can articulate explanations for emerging market trends, rendering complex forecasts accessible to decision-makers.

Behavioral Economic Insights for Predictive Nudging

An intriguing concept is the potential for PSI systems to not only predict but also to nudge market dynamics. Informed by the probabilistic forecasts of market behavior, these systems could empower strategists to fine-tune economic parameters, such as market positioning or resource allocation, to steer market trajectories (Thaler & Sunstein, 2008). This active modulation of market direction represents a departure from reactive strategies, embodying a more deliberate and strategic manipulation of economic outcomes.

Enriching PSI systems further is the integration of behavioral economic insights, which reveal the non-rational drivers of market actor behaviors (Kahneman & Tversky, 1979). By weaving these principles into PSI, the systems gain a nuanced lens through which to view and predict market movements, accounting for the often-irrational nature of economic decision-making.

Distributed Learning for Enhanced Privacy and Predictive Scope

The implementation of distributed learning frameworks, such as federated learning, presents an innovative strategy for enhancing the predictive scope of PSI systems. These frameworks offer a decentralized model-training environment that upholds data privacy while synergizing the intelligence garnered from disparate data silos (Konečný et al., 2016). Federated learning's adoption within PSI could bolster predictive accuracy while maintaining compliance with stringent privacy regulations.

The journey toward fully realized PSI systems is laden with complexities, from the technical rigor required in their construction to the ethical quandaries they present. The stewardship of vast data sets and the responsible application of predictive insights stand as key challenges in this endeavor (Bussche & Voigt, 2017).

The integration of advanced predictive tools within PSI systems ushers in an era of sophisticated market analysis. The exploration of these systems—replete with advanced algorithms, cognitive informatics and insights from behavioral economics—heralds a future where economic forecasting and strategic market manipulation are profoundly enhanced. The ongoing refinement and responsible application of these integrated systems will indubitably serve as a cornerstone in the advancement of data-driven strategic management, shaping the analytical acumen of future market ecosystems.

Case Studies in Predictive Swarm Intelligence with Asian, South American, and European Vignettes

The exploration of PSI through practical applications provides a rich tapestry for understanding its efficacy across global markets. This chapter delves into empirical instances where PSI has been employed to tackle real-world industrial challenges. Each case study, drawn from distinct geographical regions, showcases the adaptability of PSI in addressing complex, domain-specific problems.

Collectively, these vignettes not only highlight the global resonance of PSI but also illustrate its practical deployment across various industry verticals. They exemplify how predictive models grounded in swarm theory can be leveraged to deliver actionable insights, thereby informing strategic decision-making in complex economic landscapes. Through these diverse applications, the chapter contributes a valuable perspective on the operationalization of PSI in contemporary business environments.

Case Study 1: Predictive Supply Chain Monitoring and Optimization for Indian MNC

In this case study, we will examine an Indian multinational corporation's (MNC) journey toward refining its supply chain operations using PSI. The objective is to present a comprehensive narrative that captures the company's transition to next-generation data-driven management.

Company, Market, and Industry Description

Our subject is an Indian MNC with a presence across multiple continents, employing over 50,000 individuals. As of the latest fiscal year, the company reported sales figures surpassing $5 billion, marking a year-over-year growth rate of 12%. Operating in the highly competitive consumer electronics sector, the MNC faces stiff competition from both established global giants and nimble local startups. While the company holds a 15% market share in India, global competitors command upwards of 20% to 30%, with sales figures reaching $20 billion.

The 'As-Is' Problem Situation

The impetus for the PSI project was a confluence of escalating costs, logistical inefficiencies, and burgeoning market volatility. The MNC's supply chain was plagued by opaque vendor relationships and an inability to respond swiftly to market changes. The trigger for the project materialized when the company witnessed a critical stockout during a peak sales period, directly attributable to its inflexible and slow-reacting supply chain structure.

The Process

"We were navigating in the dark, reacting to market changes with a lag that was unacceptable in today's dynamic environment," recounted the Chief Operations Officer. The company embarked on an ambitious project to overhaul its supply chain management by leveraging PSI. The project team, consisting of data scientists, market analysts, and logistics experts, initiated by collating vast datasets encompassing vendor performance, inventory levels, demand forecasts, and market trends.

The team adopted a swarm-based approach, simulating various market scenarios using a swarm of intelligent agents representing different supply chain components. "Each agent was programmed with simple rules, but when they worked together, the insights we gained were far from simple," explained a lead data scientist. The PSI model was continuously refined, incorporating real-time market data to make predictive judgments, thus transitioning the company towards a more agile, data-driven decision-making paradigm.

The Results

The results of the PSI initiative were striking. Post-implementation, the company observed a 25% reduction in stockouts and a 15% decrease in excess inventory levels. Forecast accuracy improved by 30%, and the company could respond to market changes twice as fast as before. "The PSI system became our eyes and ears in the market, giving us the foresight, we desperately needed," a project manager shared. A senior manager added, "This isn't just about technology; it's about transforming our business culture to one that's proactive rather than reactive."

This narrative illustrates the potential of PSI in revolutionizing supply chain operations and catalyzing a shift towards data-driven management. The Indian MNC's case provides a blueprint for how companies can harness the power of swarm intelligence to adapt to and pre-empt market dynamics effectively. The insights gleaned from this project could serve as a benchmark for similar enterprises looking to optimize their supply chain and market prediction capabilities.

Case Study 2: Market Demand Prediction for South American High-tech Startup

In the dynamic and rapidly evolving tech sector of South America, a high-tech start-up (HTS), stands as a beacon of innovation and growth. This case study delves into HTS's strategic embrace of PSI to enhance its market demand forecasting abilities. Facing the dual challenges of intense competition and market volatility, the journey reveals how advanced data analytics and the collective behavior modeling of PSI can transform a company's understanding of market dynamics. Through this exploration, we uncover the challenges and triumphs of integrating next-generation data-driven management techniques, showcasing the pivotal role of PSI in forecasting and strategic planning within the vibrant South American technological landscape.

Company, Market, and Industry Description

Based in São Paulo, Brazil, HTS has emerged as a force within the competitive landscape of South American technology innovators. Specializing in AI-driven analytics solutions, HTS has grown from a team of five to a workforce of 200 within three years. With a growth rate hovering around 40% year-on-year and annual sales figures reaching $50 million, the company has begun to challenge established players with sales figures ranging from $200 to $300 million and market shares approximating 25%. Yet, despite its rapid growth, the South American market's volatility has required the company to adopt sophisticated demand prediction models to maintain its upward trajectory.

The 'As-Is' Problem Situation

"Our success was double-edged," reflects the CEO, "We were growing, but so was the complexity of our market. Traditional demand forecasting methods were no longer cutting it." The trigger for the project was a pivotal quarter where the flagship product experienced a severe forecast miss, resulting in overproduction and sunk costs. This incident highlighted the need for a more nuanced demand prediction model that could navigate the rapidly shifting tech landscape of South America.

The Process

The journey towards advanced market demand prediction was marked by the integration of PSI into their strategic management arsenal. "We embarked on a path less traveled, where each market signal became a data point in a vast swarm of information," shares a data scientist from the project team. The company pooled cross-disciplinary talent to develop a PSI model that simulated the behavior of market agents—consumers, competitors, and regulators—to predict demand more accurately. "Each agent, a cog in the predictive machine, brought us closer to understanding the market's pulse," a project analyst notes. Through iterative design and continuous data refinement, HTS crafted a model that learned from market behavior and adjusted its forecasts in real-time.

The Results

The implementation of the PSI model marked a turning point. Demand forecast accuracy improved by 35%, directly leading to a 20% reduction in inventory waste and a 15% increase in customer satisfaction due to better stock availability. "The PSI model didn't just predict demand; it gave us a window into the future of the market," the CEO remarked. A product manager added, "Our meetings transformed from post-mortems on our forecasting failures to strategic sessions on our next innovation." The results were a testament to the power of PSI in decoding the complex market dynamics of the South American tech industry.

This case study illustrates the transformative impact that PSI can have on a high-tech company's approach to market demand prediction. By harnessing the collective behavior of market data through advanced modeling, the HTS not only enhanced its predictive accuracy but also gained deeper strategic insights, securing its place as a rising star in the South American tech scene.

Case Study 3: Early Predictive Detection of Industry Disruption for Hidden European Champion

In the heart of Europe's industrial fabric lies Electronix, a company colloquially termed as the "Hidden Champion" due to its discreet yet dominant position within the niche market of high-precision electronics. This case study chronicles Electronix's foray into PSI as a means to foresee and navigate impending industry disruptions, securing its competitive edge in a landscape prone to rapid change.

Company, Market, and Industry Description

Electronix, headquartered in Germany, boasts a rich heritage in engineering, with over 10,000 employees and an annual turnover eclipsing €3 billion. With a steady growth rate of 8% per annum, Electronix has outpaced its closest competitors, holding a market share of 35% within Europe. Its clientele spans various industries, from automotive to aerospace, each demanding unparalleled precision and reliability. Despite its success, Electronix faces the looming challenge of disruptive technologies and new market entrants, with competitors investing heavily in areas like AI and IoT, pushing for a shift in the industry's trajectory.

The 'As-Is' Problem Situation

"The velocity of technological innovation is relentless, and we found ourselves at a crossroads," the CEO reflects. Electronix's trigger point was the realization that traditional market analysis methods were inadequate in predicting the waves of disruption brought about by emerging tech startups and shifting consumer preferences. The need for a more advanced, proactive approach to market intelligence became evident following a downturn in market share due to an unforeseen technological leap by a competitor.

The Process

Determined to revamp their approach, Electronix embarked on a project titled "Towards Next Generation Data-Driven Management". "We were in search of a crystal ball, and predictive swarm intelligence became our closest answer," quips the head of strategy. The project assembled a diverse team of data scientists, market analysts, and product developers, each bringing a unique perspective to the PSI system. "Our task was to harmonize the cacophony of market signals into a symphony of actionable insights," a data scientist muses. Through continuous iterations and leveraging real-time market data, the team developed a PSI model that encapsulated both the cooperative and competitive signals of market dynamics.

The Results

The fruits of their labor were not long in coming. Electronix experienced a 25% improvement in the accuracy of detecting market shifts, equipping them to pivot their strategies effectively. "We transformed from being reactive to proactive, often outpacing our competitors in adapting to new market trends," a senior manager shares. The finance director highlights, "Our investments became more targeted, our R&D more prescient, leading to a 15% improvement in ROI within the first fiscal year of implementation." The project team's efforts solidified Electronix's position as a leader adept at not just weathering disruption but capitalizing on it.

This case study showcases Electronix's strategic application of PSI to pre-emptively detect and adapt to industry disruption. It illustrates the power of next-generation data-driven management in fostering a proactive business culture that thrives on foresight and innovation. Through Electronix's story, we gain insights into the transformative impact of PSI on maintaining market leadership in the ever-evolving European tech industry.

Results and Insights

In this section, we meticulously scrutinizes the efficacy of PSI across various industry applications, evaluating its precision, influence on decision-making, and versatility. Through this analytical lens, we extract pivotal lessons from PSI's deployment and gauge its transformative potential within the corporate sphere.

Analyzing the Accuracy of Predictive Swarm Intelligence

In the intricate realm of market analytics, the deployment of PSI has heralded a new epoch wherein computational heuristics are leveraged to forecast market trajectories with notable precision. This detailed examination of PSI's accuracy ventures beyond mere empirical assessment; it endeavors to dissect the intricate layers of PSI models, delineating their capacity to forecast with verisimilitude the mercurial patterns of market trends and industry shifts.

Deconstructing PSI Model Precision

The precision of PSI models is not solely a function of their algorithmic design but also a reflection of their synergistic interaction with diverse datasets. The fidelity of these models in mirroring market realities is assayed through a systematic comparison with subsequent market events. This juxtaposition, a meticulous calibration of PSI-generated forecasts against tangible market evolutions, offers a quantitative measure of PSI's acumen in market prognostication. The granularity of this comparison extends to evaluating the models' temporal accuracy—how the timing of predicted market shifts aligns with real-world occurrences.

Forecasting Fidelity in Fluid Markets

To further understand the forecasting fidelity of PSI, one must consider the fluidity of markets—constantly shaped and reshaped by a confluence of economic, sociopolitical, and technological forces. PSI models, constructed on principles of swarm theory, emulate the collective intelligence emergent in nature, which, when applied to market data, must navigate the stochastic nature of human economic interactions. An exploration into how these models adapt to the entropy inherent in economic systems reveals the intricacies of their predictive capabilities.

Nuanced Assessment of Predictive Outcomes

The nuanced assessment of PSI outcomes demands more than a binary appraisal of accuracy; it necessitates a multi-dimensional analysis that considers the probabilistic nature of economic forecasting. It is here that the creative insights of PSI models are brought to the fore—how they process and interpret market signals, and how the interplay of agents within these models correlates with the multifaceted dynamics of market economies. The section illuminates how PSI models discern subtle market signals that may elude traditional analytical frameworks, thereby providing an advanced predictive lens.

PSI's Analytical Prowess and Market Dynamics

In delving deeper into PSI's analytical prowess, one must acknowledge the convergence of simplicity in agent rules with the complexity of market dynamics.

This section showcases the adaptability of swarm intelligence in decoding the layered nuances of market behavior, where the sum of simple agent interactions culminates in the sophisticated prediction of market movements. The emergent properties of these agent-based interactions are dissected to comprehend how collective intelligence forecasts not only the direction but also the potential magnitude of market shifts.

Conclusively, this section offers a profound exploration of the PSI models' predictive accuracy, elucidating their nuanced capacity to distill the essence of market dynamics into actionable foresight. By engaging in a deep dive into the comparative analysis of PSI forecasts and real-world market developments, we uncover the layers of complexity that underpin PSI's analytical prowess. The insights garnered from this investigation not only enhance our understanding of PSI's potential within the domain of predictive analytics but also catalyze the ongoing refinement of these models to meet the evolving demands of market forecasting.

Impact on Strategic and Operational Decision-making

The domain of strategic and operational decision-making within business management has been and will be profoundly impacted by the advent of PSI. This section rigorously investigates the extent to which PSI-derived insights have permeated the strategic echelons of corporations, effectuating paradigmatic shifts in strategic planning, risk mitigation, and the enhancement of operational workflows.

Strategic Formulation and PSI Integration

The strategic formulation within the contemporary corporate milieu has been traditionally guided by market analysis and forecast models that are often linear and retrospective in nature. The incorporation of PSI, however, has introduced a forward-looking dynamism into this process. Executives and strategists are now armed with anticipatory insights that inform long-term planning, allowing for a proactive recalibration of corporate trajectories. The creative integration of swarm-based forecasting models has enabled firms to construct strategic roadmaps that are not only reactive to current market conditions but are also pre-emptive of future market evolutions.

Risk Management through Predictive Insights

In the sphere of risk management, the application of PSI has engendered a more nuanced approach to identifying and mitigating potential threats. The collective intelligence gleaned from PSI models provides a multifaceted view of potential risk factors, offering a spectrum of probabilistic outcomes rather than a singular path of future events. This approach allows risk managers to craft contingency strategies that are robust, diverse, and informed by a comprehensive understanding of the possible vicissitudes in market conditions.

Operational Efficiency Gains via PSI Applications

On an operational level, the deployment of PSI has translated into significant efficiency gains. The insights offered by PSI have facilitated more informed decision-making in areas such as inventory management, logistics, and supply chain coordination. The real-time nature of PSI analytics ensures that operational decisions are reflective of the latest market data, leading to a reduction in waste, optimization of resources, and an overall increase in productivity. This transition towards a data-driven operational framework has not only enhanced efficiency but has also fostered a culture of continuous improvement and agility within organizations.

Integrative Insights and Organizational Transformation

The integration of PSI insights has had a transformative effect on organizational structures and processes. Decision-making hierarchies have evolved to accommodate more collaborative and data-informed discussions, moving away from intuition-based or hierarchical decree. This shift has not only democratized decision-making but has also improved the strategic alignment across various organizational tiers, ensuring that decisions at the operational level are congruent with overarching strategic objectives.

In conclusion, the infusion of PSI into the strategic and operational decision-making landscape has catalyzed a shift towards a more agile, anticipatory, and data-centric business ethos. By employing PSI, corporations' benefit from a heightened sensitivity to market dynamics, an improved capability to preemptively navigate through potential risks, and a marked elevation in operational efficiency. The insights derived from PSI have indeed redefined the parameters of corporate decision-making, underscoring the innovative potential of swarm intelligence in sculpting the future of business strategy and operations.

Scalability and Adaptability in Different Industry Contexts

In the contemporary corporate theater, the application of Predictive Swarm Intelligence (PSI) has emerged as a lynchpin for companies aiming to harness the dynamism of the Fourth Industrial Revolution. This section meticulously articulates the scalability and adaptability of PSI as a transformative tool across various industry sectors. Through an incisive examination of case studies, we distill the essence of how PSI has been customized to meet the discrete challenges of different industrial milieus, ultimately reinforcing organizational resilience and fostering a culture of innovation.

PSI's Versatility across Diverse Economic Sectors

PSI's versatility is most strikingly observed in its application across a gamut of economic sectors, each with its unique set of market conditions and operational complexities. The adaptive algorithms at the heart of PSI facilitate a level of customization that allows companies to model scenarios germane to their specific industry—from the volatility of commodities trading to the rapid innovation cycles in the tech sector.

Scalable Swarm Intelligence in Manufacturing

In the manufacturing industry, PSI has demonstrated remarkable scalability. Large-scale manufacturing entities have utilized SI to optimize production lines and supply chain networks. This scalability is critical in an industry where a single decision can impact hundreds of subsequent processes. Through PSI, manufacturers have been able to model production scenarios at scale, accounting for variables such as seasonal demand fluctuations and supplier reliability.

Tailoring PSI for the Service Industry

Conversely, in the service industry, the adaptability of PSI is leveraged to model consumer behavior and service delivery outcomes. Businesses have tailored PSI algorithms to predict customer service needs, enhancing response times and personalizing service offerings. By modeling the 'swarm' behavior of consumer data points, service companies can anticipate market trends and align their service portfolios accordingly.

Adaptability in High-Velocity Tech Markets

The tech industry, characterized by its high-velocity and innovation-driven market dynamics, presents a different set of challenges for PSI application. Tech companies have adapted PSI models to forecast market receptiveness to new product launches and to predict the direction of technological advancements. The dynamic nature of PSI allows these companies to remain agile, adapting their strategic initiatives in concert with the rapid pace of industry change.

Resilience Building in Financial Markets

Financial markets, with their inherent complexity and susceptibility to external shocks, have utilized PSI to build resilience against market turbulence. Financial institutions employ swarm-based models to simulate various economic scenarios, thereby stress-testing their portfolios against potential market disruptions. The predictive capacity of PSI in this context is not merely about foresight but about preparing organizations to withstand and adapt to unexpected market shifts.

Customization of PSI in Healthcare

In healthcare, customization of PSI models takes on life-impacting significance. Healthcare providers use PSI to predict patient influx, optimize resource allocation, and improve patient care outcomes. By simulating various healthcare delivery scenarios, these models help in devising strategies for efficient patient flow management and resource utilization.

Scalability in Supply Chain Optimization

Based on the case study in this article our examination begins with an Indian multinational corporation (MNC) that harnessed PSI to surmount the challenges of supply chain optimization. Here, PSI's scalability was put to the test in a complex network spanning numerous international borders and regulatory environments. The MNC's PSI deployment allowed for real-time responsiveness to logistical disruptions, showcasing the system's scalability in synchronizing operations across a global supply chain fraught with variability.

Adaptable Forecasting for Market Demand

Transitioning to the South American tech scene, we explore the adaptability of PSI in a high-tech startup's quest to predict market demand. In an industry where the product life cycle is often as volatile as the market's whims, the startup's PSI model adeptly navigated through proliferating datasets to provide predictive insights. This case exemplifies PSI's adaptability in drawing from a confluence of market signals to render precise demand forecasts, allowing the startup to pivot its strategies with agility.

Early Detection of Industry Disruption

The narrative then shifts to Europe, where a "Hidden Champion" within the manufacturing sector utilized PSI for the early detection of industry disruption. The PSI model was calibrated to the unique tempo and innovation cadence of the European market, demonstrating the system's adaptability in predicting and preparing for industry-wide changes. This case study illustrates PSI's role in bolstering the company's strategic foresight, ensuring that it remained at the vanguard of industry trends.

Synthesizing Insights across Industries

These case studies collectively affirm the hypothesis that PSI's real-world applications are as diverse as the challenges they address. From supply chain optimization to demand forecasting and preemptive disruption analysis, PSI's deployment is characterized by a high degree of scalability and adaptability. These qualities enable organizations to calibrate PSI models to their specific operational contexts and industry landscapes, leading to enhanced decision-making capabilities.

The "Results and Insights" section, with its deep dive into PSI's application across different industries, concludes that the utility of PSI extends beyond theoretical models. Its real-life applicability in the corporate realm has been validated through tangible success stories across continents. The insights gleaned from this exploration into PSI's scalability and adaptability offer a promising outlook for businesses seeking to leverage the power of SI to navigate the complexities of today's market dynamics.

To sum it up, this section elucidates the inherent scalability and adaptability of PSI, accentuating its potency as a decision-making tool across disparate industries. Through the lens of case studies, we discern the capacity of businesses to harness PSI's flexibility to address industry-specific challenges, bolster resilience, and stimulate innovation. The insights gleaned from this analysis not only reinforce the utility of PSI in diverse industry contexts but also underscore its role as a catalyst for organizational evolution in an era marked by rapid economic and technological transformations.

Societal and Regulatory Considerations

As PSI continues to redefine the boundaries of what's possible within the realms of data analytics and market prediction, it becomes imperative to address the societal and regulatory landscape that surrounds its application. Therefore, we provide a critical analysis of the moral, privacy, and legal dimensions associated with the deployment of PSI technologies.

Societal Implications of Predictive Swarm Intelligence

The integration of PSI into decision-making processes represents a revolutionary shift in the analytic capabilities of organizations. However, this innovation does not come without its societal and ethical consequences.

At its core, PSI harnesses the collective behavior of numerous agents—each agent representing a data point or model—working together to predict future scenarios. The societal implications of PSI emerge from its potential impact on individuals, societies, and markets. The decisions informed by PSI could have profound effects, from personal consumer experiences to broader economic shifts.

Societal and ethical stewardship in the deployment of PSI is paramount. As organizations predict and influence market trends, questions arise about the fairness of exploiting analytical advantages that PSI provides. There are societal and ethical obligations to consider whether such predictions could inadvertently lead to market manipulation or contribute to the widening of existing economic disparities. Human-centered PSI deployment requires the consideration of societal and ethical aspects, providing transparency in the algorithms used, the data sourced, and the intentions behind their use.

A pressing concern is the potential for algorithmic bias within PSI models. Algorithms are only as unbiased as the data they are fed and the designers who create them. In the case of PSI, where numerous agents interact to generate predictions, ensuring that these interactions are free of bias is crucial. The case studies of PSI applications in various industries—such as the Indian MNC optimizing its supply chain or the European manufacturer detecting industry disruptions—must critically evaluate the data inputs for biases that could skew outcomes and lead to unfair competitive advantages or market distortions.

Another consideration is the impact of PSI on individual autonomy. For instance, in market demand prediction for a South American high-tech startup, PSI's ability to predict consumer behavior raises questions about consumer privacy and the potential for influencing consumer choices without explicit consent. There is a thin line between predicting market demand and manipulating it, and organizations must navigate this with ethical foresight.

The broader societal impact of PSI predictions cannot be understated. Predictions that influence market dynamics have the power to shape economic landscapes, potentially affecting employment, GDP, and the financial health of regions. For example, PSI-driven decisions that lead to the automation of jobs must be weighed against the social costs of potential unemployment.

In sum, as PSI continues to evolve and integrate into various facets of corporate strategy and operations, the societal and ethical implications grow increasingly complex. Organizations must commit to practices that ensure the responsible use of PSI, considering the long-term effects on markets, societies, and individuals. The ethical deployment of PSI is not merely a legal obligation but a moral imperative to foster trust, equity, and sustainability in the age of advanced analytics. As we continue to explore the frontiers of PSI's capabilities, let us also fortify the societal and ethical frameworks that will underpin its application now and in the future.

Data Privacy and Security Concerns

Data privacy and security are pivotal concerns in the realm of PSI, especially when considering the scale and scope of data necessary for its operations. As PSI systems assimilate and process vast quantities of data to predict market trends and facilitate strategic decision-making, they inevitably intersect with sensitive and proprietary information. The stewardship of such data is not merely a technical challenge but also a foundational element of corporate integrity and social responsibility.

Effective data governance is the cornerstone of ensuring privacy and security within PSI systems. Organizations employing PSI, such as the Indian MNC optimizing supply chains or the European manufacturing giant detecting industry disruptions, must establish stringent protocols for data handling. This involves categorizing data based on sensitivity, regulating access through robust authentication mechanisms, and continuously monitoring data usage to prevent leaks or unauthorized exploitation.

The regulatory landscape for data privacy, encompassing directives like the GDPR in Europe or the CCPA in California, provides a legal framework that mandates the protection of individual data rights. These regulations compel organizations to not only safeguard consumer data but also to provide transparency around data collection and use. For instance, the South American tech startup predicting market demand must reconcile its data analytics practices with these legal requirements to maintain consumer trust and avoid punitive repercussions.

The security concerns in PSI are amplified due to the need for integration of disparate data sources to feed the predictive models. The aggregation of data from diverse origins introduces multiple vectors for potential breaches. Ensuring the security of data as it flows through the PSI system, from collection to processing, demands advanced encryption protocols, vigilant network security practices, and rigorous penetration testing to fortify the system against cyber threats.

Anonymization techniques serve as a crucial component in addressing privacy concerns within PSI. By stripping away identifiable markers from datasets, organizations can utilize valuable insights without compromising individual privacy. Furthermore, advanced encryption techniques ensure that data, even if intercepted, remains unintelligible to unauthorized parties. These methods must be applied judiciously, balancing the need for data utility against the imperative of privacy preservation.

To fortify PSI systems against malicious exploits, organizations are increasingly resorting to ethical hacking—employing cybersecurity experts to probe for vulnerabilities. This proactive approach to security enables the preemptive identification and rectification of weaknesses within the PSI infrastructure, thereby safeguarding the data that drives predictive analytics.

In conclusion, the integration of robust data privacy and security measures is an indispensable aspect of deploying PSI systems. Organizations must navigate the twin imperatives of leveraging data for strategic advantage and upholding the sanctity of privacy and security. As PSI systems continue to evolve and become more deeply embedded in operational and strategic frameworks, the emphasis on privacy and security will only intensify. The onus is on businesses to establish an environment where data is not only a strategic asset but also a well-guarded treasure, ensuring that the march towards data-driven decision-making is both ethical and secure.

Compliance with Regulations and Standards

The deployment of PSI within the corporate sector engages a complex network of regulatory requirements and standards. As PSI systems increasingly inform critical business decisions, from supply chain logistics to market demand forecasting, adherence to the myriad of international, regional, and sector-specific regulations becomes a crucial operational imperative.

The landscape of regulatory compliance for PSI is as diverse as it is intricate. Organizations leveraging PSI, such as global manufacturing firms and agile tech enterprises, must align their data analytic practices with stringent legal mandates like the European Union's GDPR or the varying data protection statutes across the Americas, including Brazil's LGPD and the United States' CCPA. These statutes impose rigorous constraints on the collection, processing, and movement of data, compelling organizations to approach PSI with a compliance-first mindset.

Legal mandates dictate a series of steps organizations must take when employing PSI. At the data acquisition stage, the legal requirement for explicit consent and the articulation of the data's intended use are paramount. This is particularly relevant for entities like an Indian MNC operating on a global scale, which must navigate the stipulations of India's proposed Personal Data Protection Bill with its emphasis on data localization and user consent.

The transnational functionality of PSI raises pertinent questions about the legality of international data flows. Compliance with the EU's stringent measures on data transfer, for instance, requires adherence to a set of approved mechanisms, including adherence to recognized frameworks or the execution of Standard Contractual Clauses (SCCs). For PSI applications that transcend geographical boundaries, ensuring that data flows comply with such mechanisms is essential for legal operation. Recently, increasing emerging digital technologies are regulated reflecting societal protection goals formerly used for physical products, e.g., European regulations of Artificial Intelligence (AI Act), Data (Data Act), and data-driven services (Digital Service Act). Some might also provide novel opportunities for data-driven applications. Nevertheless, the resulting requirements from legislation and established standards must be taken into account in order to operate a PSI system successfully.

PSI systems must also conform to recognized cybersecurity standards, ensuring the integrity and confidentiality of the data they process. Implementing standardized security measures such as those recommended by the ISO/IEC 27001 series or the NIST framework is not only a regulatory requirement but also critical to maintaining the security of predictive analytics systems against a backdrop of evolving cyber threats.

Incorporating regulatory compliance into the architecture of PSI systems often involves the use of specialized GRC tools. These tools facilitate the management of the intricate array of regulations affecting PSI, particularly for businesses like the South American tech startup, which relies on PSI for accurate market prediction while operating within a complex regulatory environment.

The advance toward ethical AI extends beyond mere legal compliance. Organizations are progressively instituting ethical codes that address the equitable use of AI and analytics. These guidelines aim to ensure that PSI applications are not just legally compliant but are also equitable, accountable, and transparent.

Far from being a mere legal formality, regulatory compliance can be leveraged as a strategic asset. Organizations that excel in meeting regulatory requirements can enhance their market position, portraying themselves as trustworthy and conscientious entities in an era where data privacy is of paramount concern to consumers.

To conclude, compliance with regulations and standards forms an integral component of PSI system deployment. As these predictive tools become more integral to business strategy in diverse industries, their successful adoption will hinge on the ability to adeptly navigate and fulfill the expectations of a complex regulatory landscape. In doing so, organizations will not only mitigate risk but also establish themselves as paragons of corporate integrity and accountability in the utilization of advanced data analytics.

Future Directions and Challenges

As we cast our gaze forward, the "Future Directions and Challenges" section contemplates the trajectory and hurdles of PSI. This area of exploration is crucial for understanding the potential evolution of PSI, its expansion to harness the exponential growth of big data, and the proactive management of inherent limitations and risks. Within this landscape, we will dissect the innovative technologies and techniques on the horizon, consider how PSI can scale to meet the demands of burgeoning data repositories, and deliberate on the anticipated challenges that may arise in its broader application.

Evolving Technologies and Techniques

The future of PSI is intertwined with an ever-evolving technological landscape, where innovation continues to redefine the boundaries of what PSI can achieve. As we chart the course ahead, several exciting technological advancements and novel techniques are poised to shape the future of PSI.

Fusion of AI and Swarm

The convergence of AI and SI represents a promising avenue for PSI. AI algorithms, particularly deep learning models, can be integrated with swarming techniques to enhance the adaptability and intelligence of swarm agents. This fusion enables PSI systems to autonomously learn and evolve their predictive capabilities over time, mimicking the dynamic nature of real-world ecosystems.

Explainable AI for Enhanced Transparency

To address ethical concerns and ensure trust in PSI systems, the development of Explainable AI (XAI) techniques is paramount. Future PSI models will prioritize transparency and interpretability, allowing stakeholders to understand the rationale behind predictions. This advancement will not only enhance the ethical standing of PSI but also facilitate its widespread adoption in sensitive domains like healthcare and finance.

Hybrid Models for Robust Predictions

The future of PSI may witness the emergence of hybrid models that combine the strengths of diverse predictive techniques. Hybridization could involve integrating PSI with traditional machine learning, statistical methods, and domain-specific expertise. Such models could provide more robust and reliable predictions, particularly in scenarios where swarm-based approaches face challenges.

Edge Computing for Real-time Swarm Intelligence

Edge computing, characterized by decentralized data processing at the device level, offers the potential to bring real-time SI to the field. Future PSI systems may leverage edge computing to enable autonomous decision-making at the edge, reducing latency and enhancing responsiveness. This is particularly relevant in applications such as autonomous vehicles and IoT-driven smart cities.

In this era of technological innovation, the evolution of PSI will hinge on the strategic incorporation of these emerging technologies and techniques. These advancements promise to propel PSI into new domains, enhance its predictive capabilities, and amplify its impact on decision-making across industries.

As we contemplate the future of PSI, we must also acknowledge the challenges that lie ahead. These include ethical dilemmas surrounding the use of advanced predictive technologies, the need for robust cybersecurity measures to protect sensitive data, and the imperative of addressing biases in AI and swarm models. Additionally, the scalability of PSI to handle the deluge of big data and its adaptability to diverse industry contexts will remain ongoing challenges.

The path forward for PSI is not without obstacles, but with a visionary approach, innovative thinking, and a commitment to ethical and responsible deployment, we stand at the threshold of a transformative era in PI, where PSI reshapes the way, we perceive and harness collective intelligence.

Scaling Predictive Swarm Intelligence for Big Data

As PSI gains momentum, its potential to revolutionize decision-making across industries becomes increasingly evident. However, as PSI charts a course into the future, one of the most pressing challenges it faces is scaling to accommodate the ever-expanding universe of big data.

Decentralized Swarm Networks

To meet the demands of big data, PSI may transition towards decentralized swarm networks. Traditional centralized swarm models may struggle to handle massive datasets efficiently. Decentralization allows swarm agents to operate autonomously on distributed nodes, processing data in parallel. This approach not only enhances scalability but also ensures redundancy and fault tolerance.

Edge Computing and Swarm Edge Devices

Incorporating edge computing into PSI can be a game-changer for scaling. By deploying SI directly at the edge, data can be processed and analyzed in real-time, reducing the burden on central processing units. Swarm edge devices, such as IoT sensors and edge servers, can contribute to localized decision-making and data reduction before transmitting crucial insights to the core PSI system.

Federated Learning for Privacy and Scalability

Federated learning, a collaborative machine learning approach, holds promise for PSI's scalability while preserving data privacy. In this setup, individual data sources retain their data locally, and only model updates are shared centrally. PSI systems can leverage federated learning to aggregate knowledge from distributed sources, enabling scalable predictions without compromising data security.

Hybrid Swarm Models

Combining SI with traditional machine learning methods can enhance scalability. Hybrid models allow PSI to adapt its approach based on data volume and complexity. When faced with big data, PSI can switch to more data-efficient machine learning techniques while still benefiting from SI in contexts where it excels.

Furthermore, ethical considerations regarding data ownership and consent must be carefully navigated. As PSI scales to process more extensive and potentially sensitive datasets, robust data governance frameworks will be indispensable in preserving privacy and compliance.

The future of PSI hinges on its ability to confront these challenges head-on and harness the power of big data without compromising the principles of ethical and responsible predictive intelligence. By embracing innovative approaches and staying attuned to emerging technologies, PSI is poised to transcend the boundaries of scalability and shape a data-driven future that is both efficient and ethical.

Addressing Potential Limitations and Risks

As we contemplate the future trajectory of PSI, it is imperative to confront the potential limitations and risks that may impede its progress. While PSI holds immense promise, addressing these challenges will be pivotal in harnessing its full potential.

Bias Mitigation and Fairness in PSI

One of the foremost challenges is mitigating biases in PSI models. As PSI relies on historical data, it can inherit and perpetuate biases present in the data. Addressing this requires rigorous evaluation, transparency, and the development of ethical guidelines to ensure fairness in predictions. Emerging techniques like adversarial debiasing and fair swarm optimization will play a crucial role in this endeavor.

Data Privacy and Ethical Concerns

As PSI scales to process vast datasets, concerns regarding data privacy and ethical use of data become paramount. Future PSI systems must prioritize privacy-preserving techniques such as federated learning, differential privacy, and homomorphic encryption. Additionally, robust ethical frameworks and regulations are needed to govern the collection, use, and sharing of data for predictive purposes.

Explainability and Trustworthiness

Ensuring the transparency and interpretability of PSI models is vital. As PSI systems become more complex, stakeholders require insights into how decisions are made. Advancements in Explainable AI (XAI) and interpretable swarm algorithms will be critical in fostering trust and accountability in PSI.

Adapting to Evolving Threats

The landscape of cybersecurity threats is constantly evolving, posing a risk to PSI systems. PSI must evolve in tandem with emerging threats, incorporating advanced security measures such as robust encryption, anomaly detection, and real-time threat intelligence to safeguard against cyberattacks and data breaches.

Interdisciplinary Collaboration

The multifaceted nature of PSI necessitates interdisciplinary collaboration. Future PSI projects should foster partnerships between data scientists, domain experts, ethicists, and policymakers. This collaboration will enable a holistic approach to addressing limitations and risks, ensuring that PSI is developed and deployed responsibly.

Scalability Challenges

As PSI scales to handle big data, scalability challenges may emerge. Ensuring that PSI systems remain efficient and responsive even as data volumes grow exponentially will require innovative solutions such as decentralized swarm networks and edge computing integration.

Ethical Decision-making Frameworks

Developing robust ethical decision-making frameworks specific to PSI will be crucial. These frameworks should guide organizations in ethical data collection, model development, and deployment. Incorporating ethical considerations into the very fabric of PSI development is essential to mitigate risks.

Regulatory Compliance

Navigating the evolving landscape of data protection regulations and standards is an ongoing challenge. PSI systems must remain compliant with a myriad of regulatory frameworks, which vary across regions and industries. Staying abreast of regulatory changes and adapting PSI systems accordingly will be imperative.

In embracing these challenges, PSI has the potential to evolve into a transformative force for predictive intelligence. By proactively addressing limitations and mitigating risks, PSI can continue to drive innovation, enhance decision-making, and usher in an era where collective intelligence is harnessed responsibly and ethically.

As we venture into the future, PSI's ability to navigate these frontiers will determine its role in shaping the data-driven landscape of tomorrow.

Conclusion

In this comprehensive exploration of PSI, we have delved into its intricacies, applications, challenges, and ethical considerations. As we draw this journey to a close, it is imperative to reflect on the key findings and contributions that have emerged from our discourse on PSI.

Throughout our exploration, we have unearthed pivotal insights into the world of PSI. We have witnessed how PSI, an innovative amalgamation of swarm theory, predictive analytics, and AI, has the potential to revolutionize strategic and operational decision-making. It has demonstrated its prowess in forecasting market dynamics, optimizing supply chains, and even predicting industry disruptions.

We have highlighted the importance of ethical considerations, data privacy, and regulatory compliance in the deployment of PSI. These facets underscore the need for responsible and transparent development and application of PI systems.

Looking ahead, PSI holds great promise in the realm of management. Its ability to harness collective intelligence, adapt to evolving data landscapes, and provide actionable insights positions it as a game-changer. PSI empowers organizations to anticipate business opportunities, optimize operations, and navigate market dynamics with unprecedented precision.

In a nutshell, PSI has illuminated the path to next-generation data-driven management. It is a path paved with opportunities and challenges, where responsible innovation will shape the future of decision-making. The promise of PSI beckons, and it is in our collective hands to harness its potential for the betterment of organizations, industries, and society.

References

Aggarwal, C.C. (Ed.). (2007). *Data Streams: Models and Algorithms* (Vol. 31). Springer US. https://doi.org/10.1007/978-0-387-47534-9.

Allen, R.S. and Helms, M.M. (2006). Linking strategic practices and organizational performance to Porter's generic strategies. *Business Process Management Journal, 12*(4), 433–54. https://doi.org/10.1108/14637150610678069.

Andrews, J.G., Buzzi, S., Choi, W., Hanly, S.V., Lozano, A., Soong, A.C.K. and Zhang, J.C. (2014). What Will 5G Be? *IEEE Journal on Selected Areas in Communications, 32*(6), 1065–82. https://doi.org/10.1109/JSAC.2014.2328098.

Arthur, W.B. (2015). *Complexity and the Economy*. Oxford University Press.

Ashkenas, R.N., Ulrich, D., Jick, T. and Kerr, S. (Eds.). (2002). *The Boundaryless Organization: Breaking the Chains of Organizational Structure* (2nd Edn.). Jossey-Bass.

Axelrod, R. (1997). *The Complexity of Cooperation: Agent-Based Models of Competition and Collaboration*. Princeton University Press. http://www.jstor.org/stable/j.ctt7s951.

Axelrod, R. and Hamilton, W.D. (1981). The Evolution of Cooperation. *Science, 211*(4489), 1390–96. https://doi.org/10.1126/science.7466396.

Axtell, R.L. (2000). Why Agents? On the Varied Motivations for Agent Computing in the Social Sciences. *Working Paper No. 17, Center on Social and Economic Dynamics, The Brookings Institution.* https://www.brookings.edu/articles/why-agents-on-the-varied-motivations-for-agent-computing-in-the-social-sciences/

Azadeh, A., Ghaderi, S.F. and Sohrabkhani, S. (2008). A simulated-based neural network algorithm for forecasting electrical energy consumption in Iran. *Energy Policy, 36*(7), 2637–44. https://doi.org/10.1016/j.enpol.2008.02.035.

Barabási, A.-L. and Albert, R. (1999). Emergence of Scaling in Random Networks. *Science, 286*(5439), 509–12. https://doi.org/10.1126/science.286.5439.509.

Bell, J.E. and McMullen, P.R. (2004). Ant colony optimization techniques for the vehicle routing problem. *Advanced Engineering Informatics, 18*(1), 41–48. https://doi.org/10.1016/j.aei.2004.07.001.

Bengio, Y., Courville, A. and Vincent, P. (2012). *Representation Learning: A Review and New Perspectives.* https://doi.org/10.48550/ARXIV.1206.5538.

Ben-Jacob, E., Cohen, I. and Levine, H. (2000). Cooperative self-organization of microorganisms. *Advances in Physics, 49*(4), 395–554. https://doi.org/10.1080/000187300405228.

Bishop, C.M. (2006). *Pattern Recognition and Machine Learning.* Springer.

Blum, C. and Li, X. (2008). Swarm Intelligence in Optimization. *In*: C. Blum and D. Merkle (Eds.), *Swarm Intelligence*, 43–85. Springer Berlin Heidelberg. https://doi.org/10.1007/978-3-540-74089-6_2.

Bonabeau, E., Dorigo, M. and Theraulaz, G. (1999). *Swarm Intelligence: From Natural to Artificial Systems.* Oxford University Press. https://doi.org/10.1093/oso/9780195131581.001.0001.

Box, G.E.P., Jenkins, G.M., Reinsel, G.C. and Ljung, G.M. (2016). *Time Series Analysis: Forecasting and Control* (5th Edn.). John Wiley and Sons, Inc.

Brambilla, M., Ferrante, E., Birattari, M. and Dorigo, M. (2013). Swarm robotics: A review from the swarm engineering perspective. *Swarm Intelligence, 7*(1), 1–41. https://doi.org/10.1007/s11721-012-0075-2.

Brandenburger, A., Brandenburger, A.M., Nalebuff, B.J. and Nalebuff, B. (1996). *Co-opetition: A Revolutionary Mindset that Combines Competition and Co-operation; 2. The Game Theory Strategy that's Changing the Game of Business* (1st Edn.). Doubleday.

Breiman, L. (2001). Random Forests. *Machine Learning, 45*(1), 5–32. https://doi.org/10.1023/A:1010933404324.

Brockwell, P.J. and Davis, R.A. (Eds.). (2002). *Introduction to Time Series and Forecasting.* Springer New York. https://doi.org/10.1007/b97391.

Brunner, D. (2023). Fallstudie Datengetriebenes Management mit Predictive Intelligence zur Früherkennung von Exportchancen. *In*: U. Seebacher (Ed.), *Praxishandbuch B2B-Marketing*, 1111–1155. Springer Fachmedien Wiesbaden. https://doi.org/10.1007/978-3-658-40037-8_41.

Bussche, A. von dem and Voigt, P. (2017). *Data Protection in Germany: Including EU General Data Protection Regulation 2018* (2nd Edn.). C.H. Beck.

Camazine, S. (Ed.). (2003). *Self-organization in Biological Systems* (2nd print, and 1st Paperback print). Princeton Univ. Press.

Casadei, R. (2023). Artificial Collective Intelligence Engineering: A Survey of Concepts and Perspectives. *Artificial Life, 29*(4), 433–67. https://doi.org/10.1162/artl_a_00408.

Chamberlin, E.H. (1969). *The Theory of Monopolistic Competition: A Re-orientation of the Theory of Value* (8th Edn.). Harvard Univ. Press.

Checkland, P. (1981). *Systems Thinking, Systems Practice.* J. Wiley.

Chyi, H.I. (2005). Willingness to Pay for Online News: An Empirical Study on the Viability of the Subscription Model. *Journal of Media Economics, 18*(2), 131–42. https://doi.org/10.1207/s15327736me1802_4.

Cortes, C. and Vapnik, V. (1995). Support-vector networks. *Machine Learning, 20*(3), 273–97. https://doi.org/10.1007/BF00994018.

Couzin, I.D., Krause, J., James, R., Ruxton, G.D. and Franks, N.R. (2002). Collective Memory and Spatial Sorting in Animal Groups. *Journal of Theoretical Biology, 218*(1), 1–11. https://doi.org/10.1006/jtbi.2002.3065.

Davis, E.W. and Spekman, R.E. (2004). *The Extended Enterprise: Gaining Competitive Advantage through Collaborative Supply Chains*. Financial Times/Prentice Hall.

Dewenter, R. and Rösch, J. (2015). Das Konzept der zweiseitigen Märkte. *In*: R. Dewenter and J. Rösch, *Einführung in die neue Ökonomie der Medienmärkte*, 115–138. Springer Fachmedien Wiesbaden. https://doi.org/10.1007/978-3-658-04736-8_5.

Domingos, P. (2012). A few useful things to know about machine learning. *Communications of the ACM, 55*(10), 78–87. https://doi.org/10.1145/2347736.2347755.

Dorigo, M. and Stützle, T. (2004). *Ant Colony Optimization*. MIT Press.

Dyer, J.H. and Singh, H. (1998). The Relational View: Cooperative Strategy and Sources of Interorganizational Competitive Advantage. *The Academy of Management Review, 23*(4), 660. https://doi.org/10.2307/259056.

Eisenhardt, K.M. and Martin, J.A. (2000). Dynamic Capabilities: What Are They? *Strategic Management Journal, 21*(10/11), 1105–21.

Engelbrecht, A.P. (2005). *Fundamentals of Computational Swarm Intelligence*. Wiley.

Epstein, J.M. and Axtell, R. (1996). *Growing Artificial Societies: Social Science from the Bottom Up*, xv, 208. The MIT Press.

Fernández, A., García, S., Galar, M., Prati, R.C., Krawczyk, B. and Herrera, F. (2018). *Learning from Imbalanced Data Sets*. Springer International Publishing. https://doi.org/10.1007/978-3-319-98074-4.

Forrest, S. (Ed.). (1991). *Emergent Computation: Self-organizing, Collective, and Ccooperative Phenomena in Natural and Artificial Computing Networks* (1st MIT Press Edn.). MIT Press.

Frank, M.Z. and Sanati, A. (2018). How does the stock market absorb shocks? *Journal of Financial Economics, 129*(1), 136–53. https://doi.org/10.1016/j.jfineco.2018.04.002.

García, S., Luengo, J. and Herrera, F. (2015). *Data Preprocessing in Data Mining* (Vol. 72). Springer International Publishing. https://doi.org/10.1007/978-3-319-10247-4.

Garro, B.A. and Vázquez, R.A. (2015). Designing Artificial Neural Networks Using Particle Swarm Optimization Algorithms. *Computational Intelligence and Neuroscience, 2015*, 1–20. https://doi.org/10.1155/2015/369298.

Gilbert, N. and Stoneman, P. (2015). *Researching Social Life*. SAGE Publications Ltd. https://uk.sagepub.com/en-gb/eur/researching-social-life/book242913.

Gilbert, P., McEwan, K., Mitra, R., Franks, L., Richter, A. and Rockliff, H. (2008). Feeling safe and content: A specific affect regulation system? Relationship to depression, anxiety, stress, and self-criticism. *The Journal of Positive Psychology, 3*(3), 182–91. https://doi.org/10.1080/17439760801999461.

Goodfellow, I.J., Pouget-Abadie, J., Mirza, M., Xu, B., Warde-Farley, D., Ozair, S., Courville, A. and Bengio, Y. (2014). *Generative Adversarial Networks*. https://doi.org/10.48550/ARXIV.1406.2661.

Grassé, P.-P. (1959). La reconstruction du nid et les coordinations interindividuelles chezBellicositermes natalensis etCubitermes sp. la théorie de la stigmergie: Essai d'interprétation du comportement des termites constructeurs. *Insectes Sociaux, 6*(1), 41–80. https://doi.org/10.1007/BF02223791.

Gubbi, J., Buyya, R., Marusic, S. and Palaniswami, M. (2013). Internet of Things (IoT): A vision, architectural elements, and future directions. *Future Generation Computer Systems, 29*(7), 1645–1660. https://doi.org/10.1016/j.future.2013.01.010.

Guyon, I. and Elisseeff, A. (2003). An Introduction to Variable and Feature Selection. *Journal of Machine Learning Research, 3*, 1157–82.

Han, J., Kamber, M. and Pei, J. (2012). *Data mining: Concepts and Techniques* (3rd Edn.). Elsevier/Morgan Kaufmann.

Hashem, I.A.T., Yaqoob, I., Anuar, N.B., Mokhtar, S., Gani, A. and Ullah Khan, S. (2015). The rise of "big data" on cloud computing: Review and open research issues. *Information Systems, 47*, 98–115. https://doi.org/10.1016/j.is.2014.07.006.

He, Z., Ni, S., Jiang, X. and Feng, C. (2023). The Influence of Demand Fluctuation and Competition Intensity on Advantages of Supply Chain Dominance. *Mathematics, 11*(24), 4931. https://doi.org/10.3390/math11244931.

Heer, J., Hellerstein, J.M. and Kandel, S. (2018). Data Wrangling. *In*: S. Sakr and A. Zomaya (Eds.), *Encyclopedia of Big Data Technologies*, 1–8). Springer International Publishing. https://doi.org/10.1007/978-3-319-63962-8_9-1.

Heifetz, R.A., Grashow, A., Linsky, M. and Linsky, M. (2009). *The Practice of Adaptive Leadership: Tools and Tactics for Changing your Organization and the World*. Harvard Business Press.

Helper, S. and Henderson, R. (2014). Management Practices, Relational Contracts, and the Decline of General Motors. *Journal of Economic Perspectives, 28*(1), 49–72. https://doi.org/10.1257/jep.28.1.49.

Helper, S. and Sako, M. (2010). Management innovation in supply chain: Appreciating Chandler in the twenty-first century. *Industrial and Corporate Change, 19*(2), 399–29. https://doi.org/10.1093/icc/dtq012.

Holland, J.H. (1995). *Hidden Order: How Adaptation Builds Complexity*. Addison-Wesley.

Holland, J.H. (2006). Studying Complex Adaptive Systems. *Journal of Systems Science and Complexity, 19*(1), 1–8. https://doi.org/10.1007/s11424-006-0001-z.

Jackson, M.O. (2008). *Social and Economic Networks*. Princeton University Press.

Jordan, M.I. and Mitchell, T.M. (2015). Machine learning: Trends, perspectives, and prospects. *Science, 349*(6245), 255–60. https://doi.org/10.1126/science.aaa8415.

Kahneman, D. and Tversky, A. (1979). Prospect Theory: An Analysis of Decision under Risk. *Econometrica, 47*(2), 263. https://doi.org/10.2307/1914185.

Kambatla, K., Kollias, G., Kumar, V. and Grama, A. (2014). Trends in big data analytics. *Journal of Parallel and Distributed Computing, 74*(7), 2561–73. https://doi.org/10.1016/j.jpdc.2014.01.003.

Kandel, S., Heer, J., Plaisant, C., Kennedy, J., Van Ham, F., Riche, N.H., Weaver, C., Lee, B., Brodbeck, D. and Buono, P. (2011). Research directions in data wrangling: Visualizations and transformations for usable and credible data. *Information Visualization, 10*(4), 271–88. https://doi.org/10.1177/1473871611415994.

Kaplan, R.S. and Norton, D.P. (2004). The strategy map: Guide to aligning intangible assets. *Strategy & Leadership, 32*(5), 10–17. https://doi.org/10.1108/10878570410699825.

Kennedy, J.F., Eberhart, R.C. and Shi, Y. (2001). *Swarm Intelligence*. Morgan Kaufmann Publishers.

Klepper, S. (1997). Industry Life Cycles. *Industrial and Corporate Change, 6*(1), 145–82. https://doi.org/10.1093/icc/6.1.145.

Konečný, J., McMahan, H.B., Yu, F.X., Richtárik, P., Suresh, A.T. and Bacon, D. (2016). *Federated Learning: Strategies for Improving Communication Efficiency*. https://doi.org/10.48550/ARXIV.1610.05492.

Kotter, J.P. (1996). *Leading Change* (Nachdr.). Harvard Business School Press.

LeCun, Y., Bengio, Y. and Hinton, G. (2015). Deep learning. *Nature, 521*(7553), 436–44. https://doi.org/10.1038/nature14539.

Lee, J.A. and Verleysen, M. (2007). High-Dimensional Data. *In*: J.A. Lee and M. Verleysen (Eds.), *Nonlinear Dimensionality Reduction*, 1–16. Springer New York. https://doi.org/10.1007/978-0-387-39351-3_1.

Legat, C., Seebacher, U. and Brunner, D. (2023). *Reasoning about Causal Effects of Regulation and Legislation on Interconnected Markets*. https://doi.org/10.18420/INF2023_201.

Lycett, M. (2013). 'Datafication': Making sense of (big) data in a complex world. *European Journal of Information Systems, 22*(4), 381–86. https://doi.org/10.1057/ejis.2013.10.

Macal, C.M. and North, M.J. (2010). Tutorial on agent-based modelling and simulation. *Journal of Simulation, 4*(3), 151–162. https://doi.org/10.1057/jos.2010.3.

Manyika, J., Chui, M., Brown, B., Bughin, J. and Dobbs, R. (2011). *Big Data: The Next Frontier for Innovation, Competition, and Productivity* (McKinsey Global Institute).

Marjani, M., Nasaruddin, F., Gani, A., Karim, A., Hashem, I.A.T., Siddiqa, A. and Yaqoob, I. (2017). Big IoT Data Analytics: Architecture, Opportunities, and Open Research Challenges. *IEEE Access, 5*, 5247–61. https://doi.org/10.1109/ACCESS.2017.2689040.

Marshall, A. (2013). *Principles of Economics*. Palgrave Macmillan UK. https://doi.org/10.1057/9781137375261.

Mayer-Schönberger, V. and Cukier, K. (2013). *Big Data: A Revolution That Will Transform How We Live, Work, and Think* (1st. publ.). Murray.

Mell, P.M. and Grance, T. (2011). *The NIST Definition of Cloud Computing* (NIST SP 800-145; 0 ed., p. NIST SP 800-145). National Institute of Standards and Technology. https://doi.org/10.6028/NIST.SP.800-145.

Miller, J.H. and Page, S.E. (2007). *Complex Adaptive Systems: An Introduction to Computational Models of Social Life*. Princeton University Press.

Millonas, M.M. (1993). *Swarms, Phase Transitions, and Collective Intelligence*. https://doi.org/10.48550/ARXIV.ADAP-ORG/9306002.

Mnih, V., Kavukcuoglu, K., Silver, D., Rusu, A.A., Veness, J., Bellemare, M.G., Graves, A., Riedmiller, M., Fidjeland, A.K., Ostrovski, G., Petersen, S., Beattie, C., Sadik, A., Antonoglou, I., King, H., Kumaran, D., Wierstra, D., Legg, S. and Hassabis, D. (2015). Human-level control through deep reinforcement learning. *Nature, 518*(7540), 529–33. https://doi.org/10.1038/nature14236.

Modha, D.S., Ananthanarayanan, R., Esser, S.K., Ndirango, A., Sherbondy, A.J. and Singh, R. (2011). Cognitive computing. *Communications of the ACM, 54*(8), 62–71. https://doi.org/10.1145/1978542.1978559.

Moore, J.F. (1996). *The Death of Competition: Leadership and Strategy in the Age of Business Ecosystems*. Wiley.

Moss, S. and Edmonds, B. (2005). Sociology and Simulation: Statistical and Qualitative Cross-Validation. *American Journal of Sociology, 110*(4), 1095–1131. https://doi.org/10.1086/427320.

Nash, J.F. (1950). Equilibrium points in n-person games. *Proceedings of the National Academy of Sciences, 36*(1), 48–49. https://doi.org/10.1073/pnas.36.1.48.

Neumann, J. von, Morgenstern, O. and Rubinstein, A. (1944). *Theory of Games and Economic Behavior (60th Anniversary Commemorative Edition)*. Princeton University Press. http://www.jstor.org/stable/j.ctt1r2gkx.

Ngai, E. W.T., Xiu, L. and Chau, D.C.K. (2009). Application of data mining techniques in customer relationship management: A literature review and classification. *Expert Systems with Applications, 36*(2), 2592–2602. https://doi.org/10.1016/j.eswa.2008.02.021.

Nicolescu, B. (2002). *Manifesto of Transdisciplinarity*. State University of New York Press.

Owen, G. (2013). *Game theory* (4th Edn.). Emerald.

Parrish, J.K. and Edelstein-Keshet, L. (1999). Complexity, Pattern, and Evolutionary Trade-Offs in Animal Aggregation. *Science, 284*(5411), 99–101. https://doi.org/10.1126/science.284.5411.99.

Petropoulos, F., Apiletti, D., Assimakopoulos, V., Babai, M.Z., Barrow, D.K., Ben Taieb, S., Bergmeir, C., Bessa, R.J., Bijak, J., Boylan, J.E., Browell, J., Carnevale, C., Castle, J. L., Cirillo, P., Clements, M.P., Cordeiro, C., Cyrino Oliveira, F.L., De Baets, S., Dokumentov, A., Joanne Ellison, Piotr Fiszeder, Philip Hans Franses, David T. Frazier, Michael Gilliland, M. Sinan Gönül, Paul Goodwin, Luigi Grossi, Yael Grushka-Cockayne, Mariangela Guidolin, Massimo Guidolin, Ulrich Gunter, Xiaojia Guo, Renato Guseo, Nigel Harvey, David F. Hendry, Ross Hollyman, Tim Januschowski, Jooyoung Jeon, Victor Richmond R. Jose, Yanfei Kang, Anne B. Koehler, Stephan Kolassa, Nikolaos Kourentzes, Sonia Leva, Feng Li, Konstantia Litsiou, Spyros Makridakis, Gael M. Martin, Andrew B. Martinez, Sheik Meeran, Theodore Modis, Konstantinos Nikolopoulos, Dilek Önkal, Alessia Paccagnini, Anastasios Panagiotelis, Ioannis Panapakidis, Jose M. Pavía, Manuela Pedio, Diego J. Pedregal, Pierre Pinson, Patrícia Ramos, David E. Rapach, J. James Reade, Bahman Rostami-Tabar, Michał Rubaszek, Georgios Sermpinis, Han Lin Shang, Evangelos Spiliotis, Aris A. Syntetos, Priyanga Dilini Talagala, Thiyanga S. Talagala, Len Tashman, Dimitrios Thomakos, Thordis Thorarinsdottir, Ezio Todini, Juan Ramón Trapero Arenas, Xiaoqian Wang, Robert L. Winkler, Alisa Yusupova and Ziel, F. (2022). Forecasting: Theory and practice. *International Journal of Forecasting, 38*(3), 705–871. https://doi.org/10.1016/j.ijforecast.2021.11.001.

Picard, R.G. (2011). Mapping Digital Media: Digitization and Media Business Models. *Open Society Foundations*. https://www.opensocietyfoundations.org/publications/mapping-digital-media-digitization-and-media-business-models.

Poli, R., Kennedy, J. and Blackwell, T. (2007). Particle swarm optimization: An overview. *Swarm Intelligence, 1*(1), 33–57. https://doi.org/10.1007/s11721-007-0002-0.

Porter, M.E. (1985). *Competitive Advantage: Creating and Sustaining Superior Performance*. Free Press; Collier Macmillan.

Porter, M.E. (1996). What is strategy? *Harvard Business Review, 74*(6).

Pyle, D. (1999). *Data Preparation for Data Mining* (1st Edn.). Morgan Kaufmann Publishers Inc.

Sadiq, S. (Ed.). (2013). *Handbook of Data Quality: Research and Practice*. Springer Berlin Heidelberg. https://doi.org/10.1007/978-3-642-36257-6.

Şahin, E. (2005). Swarm Robotics: From Sources of Inspiration to Domains of Application. *In*: E. Şahin and W.M. Spears (Eds.), *Swarm Robotics, 3342*, 10–20. Springer Berlin Heidelberg. https://doi.org/10.1007/978-3-540-30552-1_2.

Seebacher, U. (2021a). *Predictive Intelligence for Data-Driven Managers: Process Model, Assessment-Tool, IT-Blueprint, Competence Model and Case Studies.* Springer International Publishing. https://doi.org/10.1007/978-3-030-69403-6.

Seebacher, U. (2021b). The Predictive Intelligence Case Studies. *In*: U. Seebacher, *Predictive Intelligence for Data-Driven Managers*, 245–60. Springer International Publishing. https://doi.org/10.1007/978-3-030-69403-6.

Seebacher, U.G. (2021). *Template-based Management: A Guide for an Efficient and Impactful Professional Practice.* Springer International Publishing.

Seebacher, U. and Garritz, J. (2021). *Data-driven Management: A Primer for Modern Corporate Decision-making.* AQPS Inc.

Senge, P.M. and Sterman, J.D. (1992). Systems thinking and organizational learning: Acting locally and thinking globally in the organization of the future. *European Journal of Operational Research, 59*(1), 137–50. https://doi.org/10.1016/0377-2217(92)90011-W.

Shen, B.-W., Pielke, R.A., Zeng, X., Cui, J., Faghih-Naini, S., Paxson, W. and Atlas, R. (2022). Three Kinds of Butterfly Effects within Lorenz Models. *Encyclopedia, 2*(3), 1250–59. https://doi.org/10.3390/encyclopedia2030084.

Shi, W., Cao, J., Zhang, Q., Li, Y. and Xu, L. (2016). Edge Computing: Vision and Challenges. *IEEE Internet of Things Journal, 3*(5), 637–46. https://doi.org/10.1109/JIOT.2016.2579198.

Shi, Y. and Eberhart, R. (1998). A modified particle swarm optimizer. *1998 IEEE International Conference on Evolutionary Computation Proceedings. IEEE World Congress on Computational Intelligence (Cat. No.98TH8360)*, 69–73. https://doi.org/10.1109/ICEC.1998.699146.

Shmueli, G. (2010). To Explain or to Predict? *Statistical Science, 25*(3). https://doi.org/10.1214/10-STS330.

Strohmeier, S. (Ed.). (2022). *Handbook of Research on Artificial Intelligence in Human Resource Management.* Edward Elgar Publishing.

Sutton, R.S. and Barto, A. (2018). *Reinforcement Learning: An Introduction* (2nd Edn. The MIT Press.

Taylor, L., Floridi, L. and Van Der Sloot, B. (Eds.). (2017). *Group Privacy: New Challenges of Data Technologies.* Springer International Publishing. https://doi.org/10.1007/978-3-319-46608-8.

Teece, D.J. (1986). Profiting from technological innovation: Implications for integration, collaboration, licensing, and public policy. *Research Policy, 15*(6), 285–305. https://doi.org/10.1016/0048-7333(86)90027-2.

Teece, D.J. (2006). Reflections on "Profiting from Innovation". *Research Policy, 35*(8), 1131–46. https://doi.org/10.1016/j.respol.2006.09.009.

Tesauro, G. and Kephart, J.O. (2002). Pricing in Agent Economies Using Multi-agent Q-Learning. *Autonomous Agents and Multi-Agent Systems, 5*(3), 289–304. https://doi.org/10.1023/A:1015504423309.

Tesfatsion, L. and Judd, K.L. (2006). *Handbook of Computational Economics* (Volume 2): *Agent-Based Computational Economics.* Elsevier.

Tett, L., Cree, V.E. and Christie, H. (2017). From further to higher education: Transition as an on-going process. *Higher Education, 73*(3), 389–406. https://doi.org/10.1007/s10734-016-0101-1.

Thaler, R.H. and Sunstein, C.R. (2008). *Nudge: Improving Decisions about Health, Wealth, and Happiness.* Yale University Press.

Theraulaz, G. and Bonabeau, E. (1999). A Brief History of Stigmergy. *Artificial Life, 5*(2), 97–116. https://doi.org/10.1162/106454699568700.

Thompson, A.A. and Stappenbeck, G.J. (1999). *The Business Strategy Game: A Global Industry Simulation.* McGraw-Hill Publishing Co.

Valentini, G., Ferrante, E. and Dorigo, M. (2017). The Best-of-n Problem in Robot Swarms: Formalization, State of the Art, and Novel Perspectives. *Frontiers in Robotics and AI, 4.* https://doi.org/10.3389/frobt.2017.00009.

Van Dyke Parunak, H. (2006). A Survey of Environments and Mechanisms for Human-Human Stigmergy. *In*: D. Weyns, H. Van Dyke Parunak, and F. Michel (Eds.), *Environments for Multi-Agent Systems II, 3830*, 163–86). Springer Berlin Heidelberg. https://doi.org/10.1007/11678809_10.

Wen-Jun Zhang and Xiao-Feng Xie. (2003). DEPSO: Hybrid particle swarm with differential evolution operator. *SMC'03 Conference Proceedings. 2003 IEEE International Conference on Systems, Man, and Cybernetics. Conference Theme: System Security and Assurance (Cat. No.03CH37483)*, *4*, 3816–3821. https://doi.org/10.1109/ICSMC.2003.1244483.

Yang, L. and Gilbert, N. (2008). Getting away from numbers: Using Qualitative Observation for Agent-Based Modelling. *Advances in Complex Systems*, *11*(02), 175–85. https://doi.org/10.1142/S0219525908001556.

Zaharia, M., Chowdhury, M., Das, T., Dave, A., Ma, J., McCauley, M., Franklin, M.J., Shenker, S. and Stoica, I. (2012). Resilient distributed datasets: A fault-tolerant abstraction for in-memory cluster computing. *Proceedings of the 9th USENIX Conference on Networked Systems Design and Implementation*, *2*.

Zhang, G., Eddy Patuwo, B. and Y. Hu, M. (1998). Forecasting with artificial neural networks the state of the art. *International Journal of Forecasting*, *14*(1), 35–62. https://doi.org/10.1016/S0169-2070(97)00044-7.

Zhang, J.-H., Yang, B. and Chen, M. (2017). Challenges of the development for automotive parts remanufacturing in China. *Journal of Cleaner Production*, *140*, 1087–94. https://doi.org/10.1016/j.jclepro.2016.10.061.

Zhongshan Zhang, Keping Long, Jianping Wang and Dressler, F. (2014). On Swarm Intelligence Inspired Self-Organized Networking: Its Bionic Mechanisms, Designing Principles, and Optimization Approaches. *IEEE Communications Surveys & Tutorials*, *16*(1), 513–37. https://doi.org/10.1109/SURV.2013.062613.00014.

Parkinson's Disease Detection with Deep Long Short-term Memory Networks Optimized by Modified Metaheuristic Algorithm

Nebojsa Bacanin,[1,*] *Luka Jovanovic,*[2] *Miodrag Zivkovic,*[3] *Tamara Zivkovic,*[4] *Petar Bisevac,*[5] *Milos Dobrojevic,*[5] *Marko Sarac*[6] *and Milos Antonijevic*[7]

Introduction

Diseases that lead to progressive degeneration and death of neurons are called neurodegenerative. With time, changes in the patients cognitive, motor, emotional, and memory functioning are more visible and impede on the quality of life (Dugger & Dickson, 2017). Diseases such as Parkinson's, Huntington's, and Alzheimer's belong to neurodegenerative diseases (Checkoway et al., 2011). The causes of these diseases are well researched, but the results are often inconclusive. The general consensus between medical experts and researches is that the causation is multifactorial, including genetics, environmental impacts, and lifestyle. Unfortunately, while there are preventative measures and treatment that slows the progression of the disease, a cure for neurodegenerative conditions is not yet available.

[1,2,3,5,6,7] Singidunum University, Serbia.

[4] School of Electrical Engineering, Belgrade, Serbia.

Email: luka.jovanovic.191@singimail.rs; mzivkovic@singidunum.ac.rs; zt125040p@student.etf.bg.ac.rs; pbisevac@singidunum.ac.rs; msarac@singidunum.ac.rs; mantonijevic@singidunum.ac.rs

* Corresponding author: nbacanin@singidunum.ac.rs

Still, diagnosing them early and adapting the lifestyle and receiving treatment significantly betters the patient's quality of life and life expectancy (Paulsen et al., 2013; Chekani et al., 2016; Godkin et al., 2022). Thus, many researches and medical practitioners focus exactly on this crucial aspect of battling this group of illnesses.

The early diagnosis of Parkinson's disease has significant implications for both the patients and the healthcare system itself (Hess et al., 2022; Senturk, 2020). It may allow early treatment and slowing down the progress of the disease, which will improve the patient's quality of life (Pagan, 2012; Kansara et al., 2013). It may also allow better care planning, as well as social and emotional support. Finally, it can enhance resource allocation and healthcare planning, especially in the situation of the increased demands related to chronic neurodegenerative diseases (Boina, 2022; Yang et al., 2020). This would allow reducing the long-term healthcare costs.

However, early diagnosis can pose a problem, as many symptoms in their early stages resemble symptoms of other less threatening or simply different conditions. To diagnose a patient, a group of neuropsychiatrists, neurologists, and neuropsychologists work together with many other medical experts, and use tools such as anamnesis, evaluation, neuropsychological testing, and many more (Shusharina et al., 2023; Baldacci et al., 2019). Additionally, magnetic resonance imaging (MRI) and positron emission tomography (PET) can aid in diagnosis as noninvasive scanning techniques (Koikkalainen et al., 2016; Barthel et al., 2015; Mueller et al., 2006).

As the data needed for this kind of diagnostics is immense, machine learning (ML), and artificial intelligence (AI) may be of use to process it as well as find patterns that the human eye, no matter how expert, may miss (Myszczynska et al., 2020; Shusharina et al., 2023). With the use of Shapely Additive Explanations (SHAP) and their ability to analyze features, AI may help both diagnose and understand diagnosis better (Lundberg et al., 2020).

The field of ML faces challenges when it comes to determining the best hyperparameters to implement, as no single approach is best for every problem. This idea is known as the "no free lunch theorem" (Wolpert et al., 1997). Fine-tuning a model is a complicated task known as an "NP hard" problem. Therefore, metaheuristic algorithms will be used to aid the hyperparameter tuning, as they are capable of tackling even "NP hard" problems.

This paper centers on the investigation of walking patterns, which holds a pivotal role in the diagnosis of Parkinson's disease (Jankovic, 2015; Von Coelln et al., 2021; Mirelman et al., 2019). Recent studies have emphasized the importance of gait analysis, as it can unveil abnormalities in the disease's early phases (Pistacchi et al., 2017; Di Biase et al., 2020; Ghislieri et al., 2021). Gait abnormalities in Parkinson's disease are commonly marked by shuffling steps, limited arm motion, episodes of freezing during walking, and challenges in posture maintenance (Perumal et al., 2016; Morris et al., 2001). These gait variations are intricately connected to the reduction of dopamine and other neurological modifications that affect motor control and coordination within the brain.

Because changes in walking patterns frequently emerge as one of the earliest signs of Parkinson's disease, the creation of an efficient model for classifying these gait variations offers significant potential for assisting healthcare providers in diagnosing the condition. In this research, the Long Short-Term Memory model (LSTM) (Staudemeyer et al., 2019) was selected due to the time-series nature of the data. Moreover, a more advanced iteration of the relatively recent crayfish optimization algorithm (COA) (Jia et al., 2023) was utilized to optimize the hyperparameters of the LSTM specifically for this purpose. Consequently, the main achievements of this paper can be outlined as follows:

- A modified variant of the COA metaheuristics has been proposed, aiming to overcome the known drawbacks of the basic version of the algorithm.
- This novel algorithm was then used as part of ML framework with a goal to find the optimal LSTM model hyperparameters' values for the gait prediction task.
- The model has been evaluated on the benchmark gait in Parkinson's disease dataset, and the simulation outcomes were later compared to LSTM models tuned by other cutting-edge metaheuristics algorithms, followed by the statistical analysis of the experimental findings.
- Shapley Additive Explanations approach was used for interpretation of the results and better understanding of the model and the importance of the features.

The remainder of this work is as follows: Section 2 describes preceding approaches for Parkinson's detection, presents the available body of literature, and reviews preceding methods used for detection. In Section 3, the original optimizer, observed shortcomings, and proposed modifications are presented in detail. Section 4 describes the simulation and comparative analysis setup, and Section 5 presents set simulation outcomes. A conclusion to the work is provided in Section 6 with future research proposals presented at the end.

Related Works

The application of AI and data-driven methods offer several advantages in the medical domain and the field of diagnostics (Dai et al., 2019; Anikwe et al., 2022; Basile et al., 2023). Modern diagnostics methods in Healthcare 4.0, such as Internet of Things (IoT) devices, generate a substantial volume of data, and this volume is constantly increasing (Krishnamoorthy et al., 2023; Kishor et al., 2022). As AI models are capable of analyzing vast amounts of data quickly and accurately, often outperforming human experts in diagnosing diseases and conditions, they can detect subtle patterns and variations that may not be apparent to human observers. AI models are very efficient and they can help with early diagnostics, as they are able to identify disease markers at an earlier stage, potentially allowing for earlier intervention and treatment (Hunter et al., 2022; Paul et al., 2022; Rashid et al., 2022). Early detection can lead to better patient outcomes and reduced healthcare costs

(Johnson et al., 2022; Schanbel et al., 2023; Rajpurkar et al., 2022). Utilization of AI allows the medical stuff to make informed decisions easier, increasing the speed of diagnosis and reducing the waiting times for patients, making healthcare more efficient (Tang et al., 2021; Byrne et al., 2023). Sufficiently advanced AI systems also provide consistent results regardless of the time of day or the experience level of the healthcare provider. This consistency can help reduce errors caused by human factors (Yeasmin, 2019; Haleem et al., 2020; Gaba, 2018). Finally, AI-driven diagnostics can lead to cost savings by optimizing resource allocation, reducing unnecessary tests, and preventing misdiagnoses or delayed diagnoses (Blasiak et al., 2020; Munavalli et al., 2021).

Time-series forecasting is a common challenge in different areas of medicine. ML models have been used recently to predict heart rate (Staffini et al., 2021), electroencephalogram (EEG) (Kose et al., 2018), epileptic seizures (Subasi, 2021; Chisci et al., 2010), respiratory rate (Ahmed et al., 2023; Recio-Garcia et al., 2023), cardiac arrest (Kim et al., 2019), blood glucose level (Nemat et al., 2022), etc. Light-weight ML structures are capable of processing data in real-time and perform long-term monitoring, which can be crucial in intensive care units (ICU) (Gandin et al., 2021; Portela et al., 2010), predictive diagnostics and remote patient monitoring (Shaik et al., 2023; Crowell et al., 2022). Doctors frequently do not have enough time to identify certain illnesses within the limited time, and these models can assist in making correct decisions, especially in cases of neurodegenerative diseases such as Parkinson's, Alzheimer's, and ALS (Khaliq et al., 2023; Ayaz et al., 2023; Rana et al., 2022; Das et al., 2022; Kamini et al., 2023).

Last but not least, traditional diagnostic methods are typically conducted on spot, often with invasive tests, while novel methods yield excellent results. To the best of authors' knowledge, time-series classification has not been examined thoroughly for Parkinson's disease diagnostics. On the other hand, wearable sensors (such as gyroscopes and accelerometers) implanted in the patient's shoes are cheap and a noninvasive method for diagnostics (Nahavandi et al., 2022; Zhang et al., 2023; Subhan et al., 2023). AI models can process data generated by these wearable sensors and help in early diagnostics, which is crucial to start the proper treatment as soon as possible. In this way, the patient's quality of life can be improved significantly.

Long Short-term Memory

The LSTM structure has been envisioned as an enhanced iteration of the recurrent neural network (RNN) (Salehinejad et al., 2017), with the specific aim of efficiently handling sequential time-series inputs through recognition and preservation of long-term dependencies (Hochreiter, 1998). In contrast to traditional RNNs, LSTM effectively tackles the problem of vanishing or exploding gradients (Hochreiter & Schmidhuber, 1997) and exhibits more significant adaptability to time-series data. Moreover, it represents a significant departure from conventional RNNs, where the fidelity of initial signals fades away as they traverse the propagation path, hindering the neural network's power to effectively capture past events. This method utilizes a sophisticated four-layer gating procedure, which enables dynamic and accurate control, tuning, and elimination of cell inputs to support the retention of long-term memory. Within the LSTM network, the cell state operates as a storage module engineered to preserve crucial data. The forget gate, characterized by a sigmoid function (as shown in Equation 1), plays a central role in determining what should be discarded, effectively removing irrelevant information from the cell state.

Equation 1

$$f_t = \sigma\,(W_f x_t + U_f h_{t-1} + b_f)$$

Here, f_t symbolizes the forget gate, x_t correlates to its input data during time step t, h_{t-1} stands for the preceding hidden state, and W_f and U_f refer to the weight coefficients linked to the inputs, with bf representing the bias vector. The task of determining which incoming data should be retained in the cell state is facilitated by a distinct sigmoid function referred to as the input gate (i_t), as described in Equation 2).

Equation 2

$$i_t = \sigma\,(W_i x_t + U_i h_{t-1} + b_i)$$

Here, W_i and U_i indicate the respective weight factors, and b_i signifies the bias. A supplementary set of candidate solutions is generated using a hyperbolic tangent (tanh) layer, in accordance with Equation 3.

Equation 3

$$\tilde{C}_t = \tanh(W_c x_t + U_c h_{t-1} + b_c)$$

During this procedure, it is crucial to consider the weight coefficients, specifically, W_c, U_c, and the bias term represented as b_c. By employing an element-wise product symbolized as $\odot$ in conjunction with the forget gate f_t, the prior cell state C_{t-1} is cleared to accommodate the new cell state C_t which is then populated with the actual data. The freshly calculated candidate values $\tilde{C}_t$ are multiplied by the input gate, and subsequently merged with this outcome, as described in Equation 4.

Equation 4

$$\tilde{C}_t = f_t \odot C_{t-1} + i_t \odot \tilde{C}_t$$

The original sigmoid output o_t is subjected to additional processing, as depicted in Equation 5, which is dependent of the cell state. Calculated result is subsequently transmitted through a tanh layer, as outlined in Equation 6, ultimately leading to the creation of the revised hidden state.

Equation 5

$$o_t = \sigma\,(W_o x_t + U_o\,h_{t-1} + b_o)$$

Equation 6

$$h_t = o_t \odot \tanh\,(C_t)$$

In this context, W_o and U_o denote the weight coefficients, while b_o signifies the bias value.

Metaheuristics for Optimization

In recent years, there has been a growing fascination in the realm of computer science with enhancing models. The increase in model complexity and the abundance of hyperparameters in modern algorithms have emphasized the necessity for the creation of automated approaches. Traditionally, model optimization relied on empirical methods, but now there is a pressing need to tackle this systematically. However, this poses a substantial challenge because the pursuit of ideal parameters often involves navigating an intricate terrain of both discrete and continuous values, resulting in a mixed NP-hard problem that has a significant impact on model performance.

Metaheuristic optimization algorithms constitute a robust class of algorithms that excel at addressing NP-hard problems within feasible time constraints and with acceptable computing resources. These algorithms excel at improving performance by treating the selection of parameters as an optimization problem. Among the notable branches of metaheuristics, swarm intelligence stands out. It draws inspiration from the cooperative behaviors observed in natural groups and harnesses these behaviors to accomplish optimization tasks with impressive efficiency.

Numerous algorithms have gained recognition among researchers for their effectiveness in addressing optimization tasks. These algorithms encompass Harris Hawks optimization (HHO) (Heidari et al., 2019), genetic algorithm (GA) (Mirjalili & Mirjalili, 2019), particle swarm optimizer (PSO) (Kennedy & Eberhart, 1995), artificial bee colony (ABC) (Karaboga, 2010) algorithm, and firefly algorithm (FA) (Yang & Slowik, 2020). In addition to these, more recent methods such as the LSHADE for Constrained Optimization with Levy Flights (COLSHADE) algorithm (Gurrola-Ramos et al., 2020) and the Self-Adapting Spherical Search (SASS) (Zhao et al., 2022) algorithm are recognized as potent optimization techniques.

These methods, along with algorithms implemented upon their principles, have found application in a wide range of domains and have demonstrated promising outcomes. Remarkable examples of metaheuristics utilization in optimization tasks encompass their application in predicting crude oil prices (Jovanovic et al., 2022a; Jovanovic et al., 2022b; Al-Qaness, 2022), gold prices (Jovanovic et al., 2023a; Stankovic et al., 2023), energy load balancing (Bacanin et al., 2023a; Bacanin et al., 2023b; Stoean et al., 2023; Salb et al., 2023), forecasting trends in

Ethereum and Bitcoin (Stankovic et al., 2022; Milicevic et al., 2023; Petrovic et al., 2023; Gupta et al., 2023), applications in Industry 4.0 (Jovanovic et al., 2023c; Dobrojevic et al., 2023; Para et al., 2022), healthcare 4.0 (Zivkovic et al., 2022c; Bezdan et al., 2022; Budimirovic et al., 2022; Stankovic, Gavrilovic et al., 2022; Zivkovic et al., 2021a), computer networks and intrusion detection (Zivkovic et al., 2022b; Savanović et al., 2023; Jovanovic et al., 2023d; Zivkovic et al., 2022a), cloud and fog computing (Thakur et al., 2022; Mirmohseni et al., 2022; Bacanin et al., 2022b; Zivkovic et al., 2021b), environmental sciences (Jovanovic et al., 2023b; Bacanin et al., 2022a; Kiani et al., 2022).

Methods

The following section provide a detailed description of the original algorithm. This is followed by the observed deficiencies and provisioned alterations. As the original COA is relatively novel, there is still a lot of space for experimentation using various modifications and hybridization.

Original COA

The COA mathematically models the behavior of crayfish, small freshwater crustaceans resembling shrimp. Major observed behavioral patterns of these crustaceans serve as a basis for the algorithms optimization. Feigning vacation and competitive actions all form a fundamental basis for the optimization. As a population-based algorithm, the initial stage involves generating a population of agents X. Each agent is assigned with a position that, within a search space, randomly indicates the position within a given dimension.

Equation 7

$$X = [X_1, X_2, \ldots X_N] = \begin{bmatrix} X_{1,1}, & \ldots, & X_{1,j}, & \ldots & X_{1,k}, \\ \vdots, & \ldots, & \vdots, & \ldots, & \vdots \\ X_{i,1}, & \ldots, & X_{i,j}, & \ldots & X_{1,k}, \\ \vdots, & \ldots, & \vdots, & \ldots, & \vdots \\ X_{N,1}, & \ldots, & X_{N,j}, & \ldots & X_{N,k} \end{bmatrix}$$

In Equation 7, X denotes the population, k the dimensionality of said problem and N the population limit, $X_{i,j}$ is the position of an agent in the i and j coordinate. Agents are randomly dispersed across the search space according to Equation 8 in which ll represents the lower limit, ul the upper limit, and *rnd* introduces randomness in the range [0, 1].

Equation 8

$$X_{i,j} = ll_j + (ul_j - ll_j) \times rnd$$

A major influence of agent behavior is simulated temperature defined as given in Equation 9 where *tmp* represents the temperature in the environment and *rnd* is a random value within [0, 1].

Equation 9

$$tmp = rnd \cdot 15 + 20$$

Once temperatures exceed 30, agents chose to locate a cooler region to vacation and resume foraging at a more appropriate temperature. Feeding is therefore influenced by temperature, with an optimal feeding range being at 25; however, feeding will occur at any temperature between 15 and 30. Agent intake can be approximately assumed to be normally distributed and can be determined in accordance with Equation 10 where *tmp* is the environmental temperature, μ denotes the optimal agent temperature, and ω and C1 define control parameters for the given algorithm.

Equation 10

$$p = C_1 \times \left(\frac{1}{\sqrt{2 \cdot \pi \cdot \omega}} \times \exp\left(\frac{(temp - \mu)^2}{2\omega^2} \right) \right)$$

During the vacation, strange and once triggered simulated temperatures exceed 30, agents follow Equation 11 with X_G and X_L denoting the global and current best agent positions.

Equation 11

$$X_{shadow} = \frac{(X_G + X_I)}{2}$$

Occasionally, crayfish will fight for cave space. This is simulated by the algorithm as a random event with a 0.5 probability of occurring once *tmp* exceeds 30. Once a competition stage is initiated, agents are relocated in accordance with Equation 12 with *z* denoting a random agent. Positions are, therefore, adjusted in accordance with other competing individual agents.

Equation 12

$$X_{i,j}^{t+1} = X_{i,j}^t - X_{z,j}^t + X_{shadow}$$

During foraging, the agents explore the search space moving towards the food simulated by high objective function values. The quality Q of a located solution is determined by the formula given in Equation 13 where C_3 is the food control factor.

Equation 13

$$Q = C_3 \cdot rnd \cdot \left(\frac{fitness_{agent}}{fitness_{food}} \right)$$

The assessment is compared to the optimal solution via Equation 14.

Equation 14

$$X_{food} = exp \left(-\frac{1}{Q} \times X_{food} \right)$$

Food is processed as an intensification approach, and the sine and cosine functions are used in an alternating pattern to locate promising solutions as per Equation 15.

Equation 15

$$X_{i,j}^{t+1} = X_{i,j}^{t} + X_{food} \times p \times (\cos(2 \times \pi \times rnd) - \sin(2 \times \pi \times rnd))$$

Once $Q \leq \dfrac{(C_3 + 1)}{2}$, agents move towards samples more directly as described in Equation 16.

Equation 16

$$X_{i,j}^{t+1} = (X_{i,j}^{t} - X_{food}) \times p + p \times rnd \times X_{i,j}^{t}$$

Through the foraging stage, COA will approach the optimal solution, enhancing the exploitation ability of the algorithm and making it have good convergence ability.

Modified COA Approach

While the COA demonstrated good performance during CEC evaluations, as a relatively novel algorithm its application is not fully explored. Furthermore, modifications could potentially yield further improvements. This work explores the potential of low level hybridization with the GA (Mirjalili & Mirjalili, 2019). The modified version of the algorithm has been named the COA with generic operators (COAGO).

The newly introduced mechanism is activated, and it operates as follows: It selects a random agent and combines it with the best solution achieved up to that point. The blending of their parameters is done in a uniform manner, with the degree of blending controlled by a parameter denoted as pc. In practical terms, it has been empirically determined that the optimal value for pc is 0.1. Additionally, there is another adjustment that involves modifying parameters. When this adjustment is initiated, it selects a random value within a specified parameter range. Half of this selected value is either added to or subtracted from the parameter, depending on the value of another parameter known as md. Once again, the value for md has been established through empirical research and is set at 0.1.

After creating a new solution, the weakest performing solution in the group is replaced with the newly generated agent. The evaluation of this fresh solution is deferred until the next iteration, maintaining the computational complexity of the original algorithm. In order to ensure a thorough understanding of the algorithm, we present its pseudocode in Figure1.

Initialize agent group P
while t is less than T **do**
    Rank agents in P using objectively
    for Each solution X in P do
        Applying the COA search to perform optimization
        Generate a new solution, referred to as NS, using a genetically inspired mechanism
        Mutate the parameters of NS
        Replace the worst solution in P with NS
    end for
end while
return The best solution attained within P

Fig. 1 Pseudocode of the introduced COAGO.

Simulation Setup

This works conducts simulations suing a real-world dataset collected in preceding works studying Parkinson's diagnosis. The complete datasets is publicly available[1]. This work focuses on the data collected by Galit Yogev et al., (Yogev et al., 2005) focusing on dual-task walking, as cognitive function and secondary activity are shown to have an effect on walking patterns of individuals. While the relation is not fully understood, researchers has observed a clear influence of neurodegenerative conditions on patient gait. Though the use of accelerometers inserted in a patient's shoes, these changes can be recorded and studied. A combined system of 16 accelerometer sensors is used, 8 per shoe to record changes in walking patterns. A total of 30 patients confirmed to be affected by Parkinson's are studied with a mean age of 71.8 years. A gender and age matched group of 28 healthy individuals are also studied to provide a control group. The data is sampled at 100 Hz while the patients are requested to walk in a well-lit, obstacle free, 25-meter long, 2-meter wide corridor while performing serial seven subtractions out loud, starting from 500 (Yogev et al., 2005).

---

[1] https://physionet.org/content/gaitpdb/1.0.0/

In this work, irregular patient's gait associated with Parkinson's is treated as a time-series classification problem. Gait data for three affected patients and three healthy patients is combined to formulate a continuous dataset with both control and Parkinson's patient data. Samples are combined under a uniform distribution and a balanced dataset is generated. This data is then formulated as a time series using the TensorFlow time-series generator. A total of 15 samples are used per time step. An initial 70% of data is used to train neural networks, while a latter 30% remained reserved for evaluations and model interpretation.

Several metaheursitic algorithms are tasked with optimizing LSTM neural networks for performance improvements. Both architecture and training parameters are considered during optimization. Algorithms included in the comparative analysis include all the optimization algorithms implemented under identical conditions to ensure a fair evaluation. The population size is limited to n individuals. Each algorithm is allowed Y iterations to improve population quality. Finally, to ensure statistical validity of the outcomes, experiments are carried out thorough 30 independent runs to account for the randomness associated with heuristic approaches.

The algorithms are evaluated on their ability to improve the diagnostic performance of LSTM networks using standard classification metrics including accuracy (Equation 17), recall (Equation 18), procession (Equation 19) and f1-score (Equation 20). In these equations, TP and FP represent true and false positives, while FN and TN denote false and true negatives.

Equation 17

$$\text{Accuracy} = \frac{TP + TN}{TP + TN + FP + FN}$$

Equation 18

$$\text{Recall} = \frac{TP}{TP + FN}$$

Equation 19

$$\text{Precision} = \frac{TP}{TP + FP}$$

Equation 20

$$F_1 \text{ Score} = 2 \times \frac{\text{Precision} \times \text{Recall}}{\text{Precision} + \text{Recall}}$$

Error rate is used as the objective function determined as given in Equation 21.

Equation 21

$$\text{Error} = 1 - \text{Accuracy}$$

Additionally, the indicator function used to track model performance is used. The function selected as the indicator is the Cohen's kappa determined according

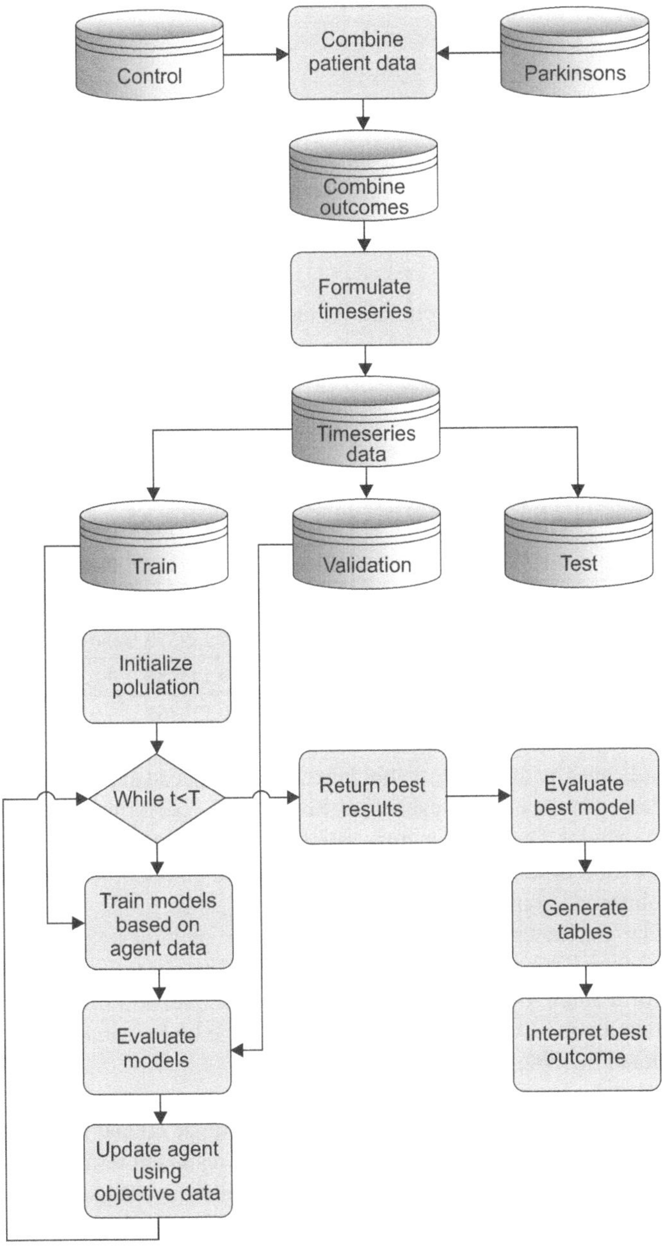

Fig. 2 Proposed experimental framework.

to Equation 22 due to the excellent performance of this metric when evaluating data with a dissemblance.

Equation 22

$$k = \frac{z_o \qquad z_e}{}$$

The described experimental framework is provided in Figure 2 in which z_o and z_e represent observed and expected classification values. Cohen's Kappa is used as the indicator function during optimizations.

Simulation Outcomes

Comparisons between each optimizer in terms of objective function are resented in Table 1.

Table 1   Overall Objective function scores attained by metaheuristic optimized model.

Method	Best	Worst	Mean	Median	Std	Var
LSTM-COAGO	**0.138829**	0.2147505	0.181205	0.185621	0.033134	0.001099
LSTM-COA	0.141618	**0.1710567**	**0.153651**	0.150966	0.012149	0.000148
LSTM-GA	0.177564	0.2021485	0.186138	0.182419	**0.009756**	**0.000095**
LSTM-PSO	0.169301	0.2309679	0.192800	0.185466	0.024928	0.000621
LSTM-ABC	0.167131	0.1997728	0.181515	0.179579	0.012785	0.000163
LSTM-HHO	0.140688	0.1983266	0.167344	**0.145181**	0.023823	0.000568
LSTM-RSA	0.161347	0.2002892	0.176790	0.172761	0.014887	0.000222

As indicated by the results in bold text, the introduced algorithms attained the best outcomes in the optimal execution; however, the performance of the original algorithm remains admirable demonstrating the optimal outcomes in the worst and mean execution cases. Furthermore, the GA shows an admirable stability that is further reinforced graphically in Figure 2.

Similar outcomes are demonstrated in terms of indicator function, in this study by Cohen Kappa. The introduced algorithm showcases the best performance in terms of a single best run; however, the best median and mean outcomes are those shown by the COA and HHO. Once again the highest stability, it is further demonstrated in Table 2.

As it can be observed in Figure 3, the introduced algorithm shows a low rate of stability. Suggesting that further modification may be crucial for improving the reliability of the algorithm. Nevertheless, the best results in terms of objective as well as indicator function are demonstrated by the introduced algorithm regardless. Further compassion of convergence rates for each algorithm are demonstrated in Figure 3. While algorithms certain algorithms focus on sub optional regions of the search space for extended periods of time, the introduced algorithm manages to

Table 2 Overall Cohen's Kappa scores attained by metaheuristic optimized model.

Method	Best	Worst	Mean	Median	Std	Var
LSTM-COAGO	**0.722484**	0.572899	0.638748	0.629804	0.065029	0.004229
LSTM-COA	0.717239	**0.659565**	**0.693414**	0.698427	0.023748	0.000564
LSTM-GA	0.646148	0.596342	0.628546	0.635846	**0.019568**	**0.000383**
LSTM-PSO	0.662465	0.539712	0.615360	0.629631	0.049544	0.002455
LSTM-ABC	0.666950	0.601798	0.638071	0.641768	0.025527	0.000652
LSTM-HHO	0.719036	0.604709	0.685821	**0.709769**	0.047130	0.002221
LSTM-RSA	0.678341	0.599839	0.647016	0.654945	0.029815	0.000889

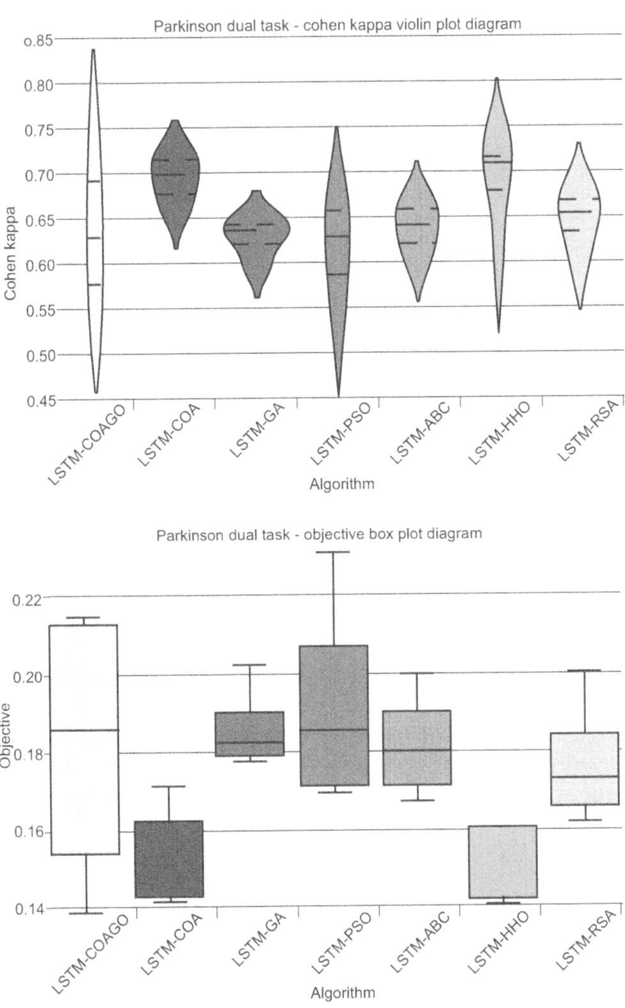

Fig. 3 Metaheuristic optimized model distribution in terms of objective and indicator functions.

overcome a relatively poor starting location and outperform all other algorithms in the best execution regardless. Detailed comparisons between the best preforming models generated by each optimization algorithm are provided in Table 3. Detailed metrics are considered including precision, recall, and f1 score. Furthermore, classifications are assessed for both cases of disease detection as well as control group classification.

Table 3 Best metaheursitic generated model performance detailed comparison.

Method	Metric	Control	Parkinson's	Accuracy	Macro avg	Weighted avg
LSTM-COAGO	precision	0.881485	**0.841668**	**0.861171**	**0.861576**	0.862076
	recall	**0.842402**	0.880907	**0.861171**	**0.861655**	**0.861171**
	f1-score	**0.861500**	0.860841	**0.861171**	**0.861171**	**0.861179**
LSTM-COA	precision	0.897498	0.824167	0.858382	0.860833	0.861753
	recall	0.817009	0.901886	0.858382	0.859448	0.858382
	f1-score	0.855364	0.861277	0.858382	0.858321	0.858247
LSTM-GA	precision	0.891194	0.770954	0.922436	0.831074	0.832583
	recall	0.744458	0.904429	0.822436	0.834443	0.822436
	f1-score	0.811244	0.832374	0.822436	0.821809	0.821544
LSTM-PSO	precision	0.892697	0.782569	0.830699	0.837633	0.839015
	recall	0.871185	0.903793	0.830699	0.832489	0.830699
	f1-score	0.821712	0.838824	0.830699	0.830268	0.830053
LSTM-ABC	precision	**0.904060**	0.779722	0.832869	0.841891	0.843452
	recall	0.753930	**0.915872**	0.832869	0.834901	0.832869
	f1-score	0.822198	0.842331	0.832869	0.832264	0.832012
LSTM-HHO	precision	0.894910	0.827640	0.859312	0.861275	**0.862119**
	recall	0.822048	0.989495	0.859312	0.860272	0.859312
	f1-score	0.856933	**0.861613**	0.859312	0.859273	0.859214
LSTM-RSA	precision	0.902844	0.789050	0.838653	0.845947	0.847375
	recall	0.767836	0.913117	0.838653	0.840476	0.838653
	f1-score	0.829885	0.846562	0.838653	0.838223	0.838014
	support	4962	4719			

As shown by the best results marked in bold text, the best performing model-generated by, the introduced COAGO algorithm demonstrated a clear superiority in terms of most metrics. Only slightly being outperformed by the ABC algorithm in terms of control group detection, precision, and Parkinson's detection in terms of recall. Finally, the HHO demonstrated the best f1 score for Parkinson's detection. An in-depth analysis of the best generated model is provided in terms of ROC and PR curves in Figure 4.

Classification comparisons and indicator objective function joint plots for the best performing model are also provided in Figure 5. Finally, to ensure experimental repeatability, the parameter selections for the respective best preforming models is provided in Table 4 for independent researchers seeking to replicated outcomes.

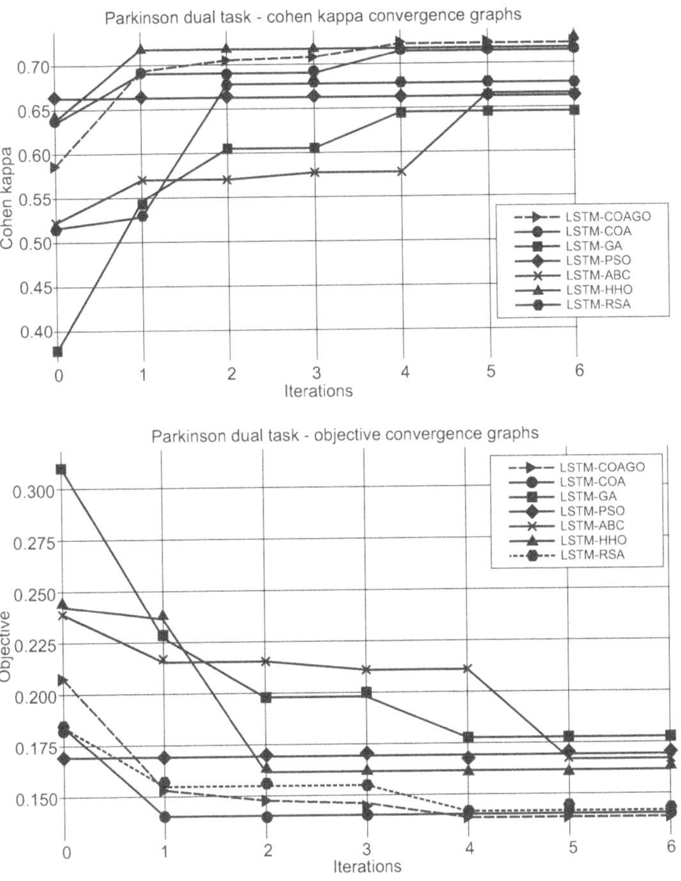

Fig. 4   Metaheuristic optimized model convergences in terms of objective and indicator functions.

The best performing model has also been subjected to SHAP analysis to determine the parameters that have the most significant contribution to proper detection (Lundberg et al., 2017).

SHAP (Shapley Additive exPlanations) stands out as a widely employed technique for elucidating the outcomes produced by ML models (Du et al., 2019). Inspired by Shapley values, a principle derived from cooperative game theory, SHAP values allocate the contribution of each feature to a specific prediction by examining all conceivable combinations of features (Molnar, 2020; Murdoch et al., 2019). In the realm of ML, features are the participants, and the payoff corresponds to the model's forecast for a particular instance. While making a prediction, SHAP values epitomize the average impact of each feature on the prediction across all potential feature combinations. The outcomes of the analysis are shown in Figure 6.

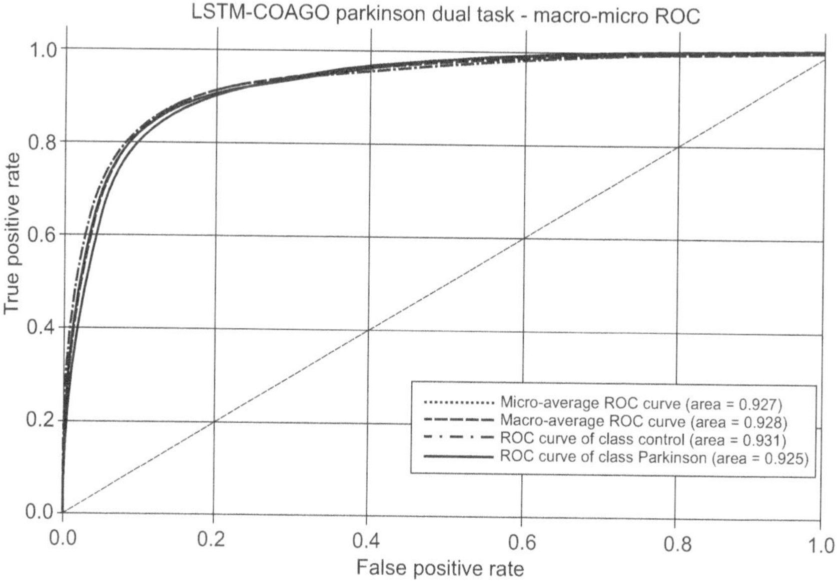

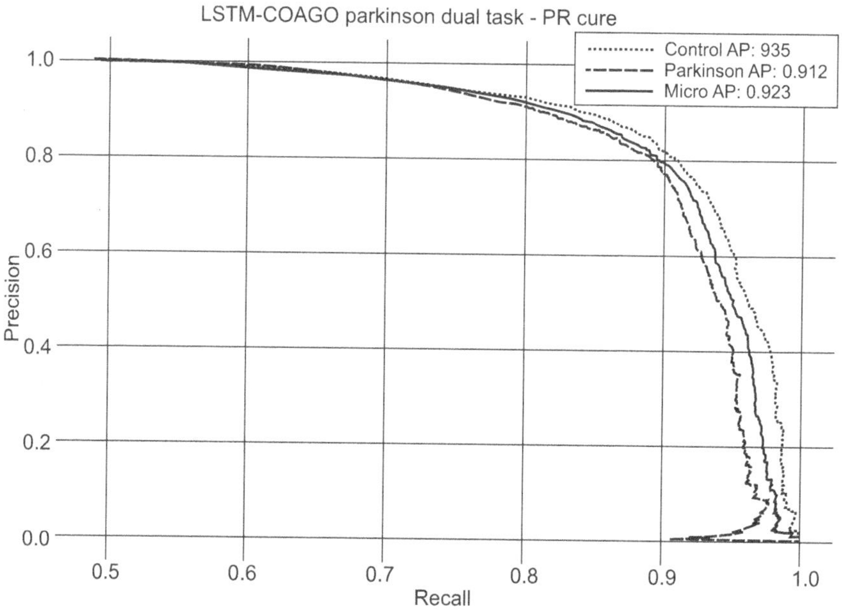

Fig. 5   "ROC AUC and PR AUC of best performing model".

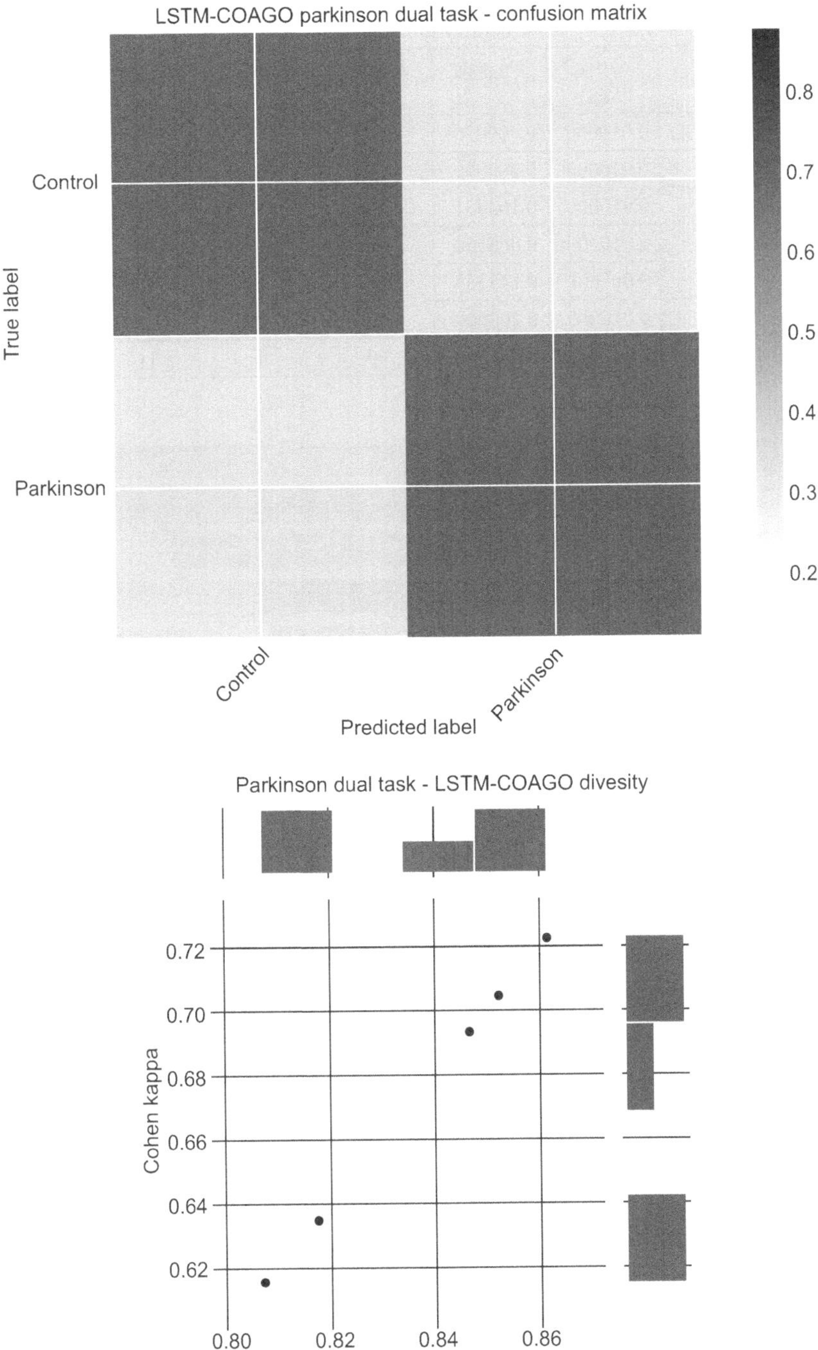

Fig. Best performing model confusion matrix and indicator convergence joint plot.

Table 4   Parameters selections for the best performing models generated by each optimization algorithm.

Method	Learning Rate	Dropout	Epochs	Layers	Neurons L1	Neurons L2
LSTM-COAGO	0.010000	0.200000	60	2	15	15
LSTM-COA	0.010000	0.200000	60	2	15	10
LSTM-GA	0.010000	0.103431	45	1	15	N/A
LSTM-PSO	0.010000	0.066550	60	2	13	15
LSTM-ABC	0.008608	0.143333	52	1	14	N/A
LSTM-HHO	0.010000	0.200000	60	2	15	11
LSTM-RSA	0.007443	0.084660	60	1	11	N/A

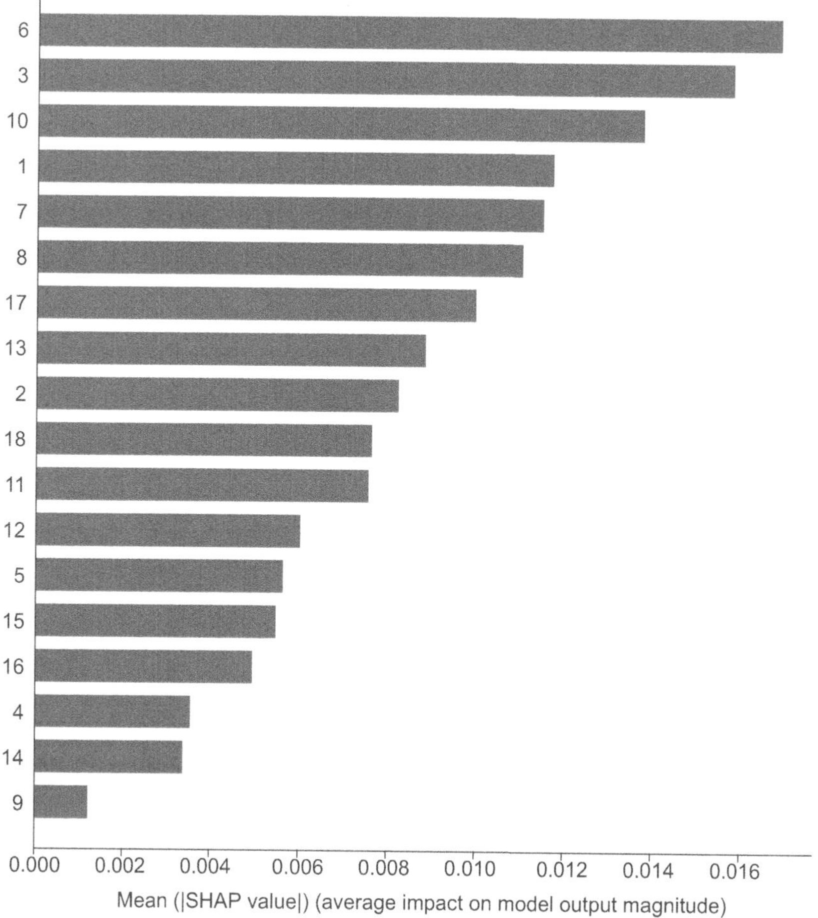

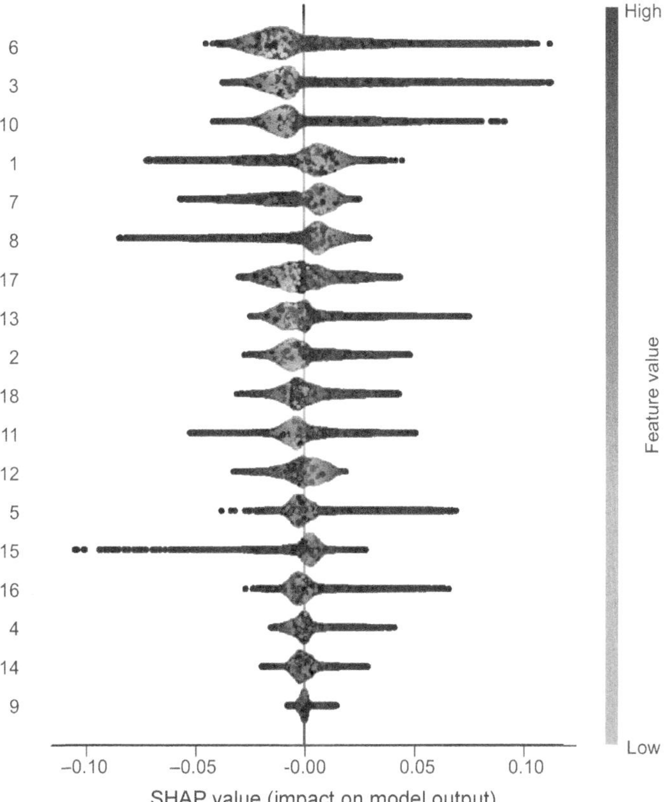

Fig. 6 SHAP Analysis outcomes.

Conclusion

This work explored the potential of the recently introduced COA for hyperparameter optimization. The algorithm is applied for the selection of LSTM parameters that yield optimal performance when detecting early symptoms of Parkinson's disease in patient gait during dual-task walking. Several state-of-the art metaheuristics are subjected to a comparative analysis implemented under identical conditions. A modified version of the original algorithm named the COAGO algorithm is proposed that combines genetic operators inspired by the GA. The introduced approach attained the best outcome; however, the COA still demonstrated admirable performance. A definite potential for improvement is observed. The best performing model demonstrated an accuracy of 86.157% showcasing potential for early detection of neurodegenerative disease using the introduced approach.

Certain limitations do exist within this work. Only a limited number of optimization algorithms have been compared and only the potential of LSTM is explored. Additionally, due to computational demands of optimization, smaller populations are observed and only a small number of iterations tested.

Future works will focus on further refining the introduced approach, evaluating other methods for sequential data classification. Additionally, the introduced optimizer will be applied to other pressing challenges in medicine and engineering.

References

Ahmed, U., Lin, J.C.W. and Srivastava, G. (2023). Multivariate time-series sensor vital sign forecasting of cardiovascular and chronic respiratory diseases. *Sustainable Computing: Informatics and Systems, 38*, 100868.

Al-Qaness, M.A., Ewees, A.A., Abualigah, L., Al-Rassas, A.M., Thanh, H.V. and Abd Elaziz, M. (2022). Evaluating the applications of dendritic neuron model with metaheuristic optimization algorithms for crude-oil-production forecasting. *Entropy, 24*(11), 1674.

Anikwe, C.V., Nweke, H.F., Ikegwu, A.C., Egwuonwu, C.A., Onu, F.U., Alo, U.R. and Teh, Y.W. (2022). Mobile and wearable sensors for data-driven health monitoring system: State-of-the-art and future prospect. *Expert Systems with Applications, 202*, 117362.

Ayaz, Z., Naz, S., Khan, N.H., Razzak, I. and Imran, M. (2023). Automated methods for diagnosis of Parkinson's disease and predicting severity level. *Neural Computing and Applications, 35*(20), 14499–14534.

Bacanin, N., Jovanovic, L., Zivkovic, M., Kandasamy, V., Antonijevic, M., Deveci, M. and Strumberger, I. (2023). Multivariate energy forecasting via metaheuristic tuned long-short term memory and gated recurrent unit neural networks. *Information Sciences, 642*, 119122.

Bacanin, N., Sarac, M., Budimirovic, N., Zivkovic, M., AlZubi, A.A. and Bashir, A.K. (2022). Smart wireless healthcare system using graph LSTM pollution prediction and dragonfly node localization. *Sustainable Computing: Informatics and Systems, 35*, 100711.

Bacanin, N., Stoean, C., Zivkovic, M., Rakic, M., Strulak-Wójcikiewicz, R. and Stoean, R. (2023). On the benefits of using metaheuristics in the hyperparameter tuning of deep learning models for energy load forecasting. *Energies, 16*(3), 1434.

Bacanin, N., Zivkovic, M., Bezdan, T., Venkatachalam, K. and Abouhawwash, M. (2022). Modified firefly algorithm for workflow scheduling in cloud-edge environment. *Neural Computing and Applications, 34*(11), 9043–68.

Baldacci, F., Lista, S., Vergallo, A., Palermo, G., Giorgi, F.S. and Hampel, H. (2019). A frontline defense against neurodegenerative diseases: The development of early disease detection methods. *Expert Review of Molecular Diagnostics, 19*(7), 559–63.

Barthel, H., Schroeter, M.L., Hoffmann, K.T. and Sabri, O. (2015, May). PET/MR in dementia and other neurodegenerative diseases. *In: Seminars in Nuclear Medicine, 45*(3), 224–33. WB Saunders.

Basile, L.J., Carbonara, N., Pellegrino, R. and Panniello, U. (2023). Business intelligence in the healthcare industry: The utilization of a data-driven approach to support clinical decision making. *Technovation, 120*, 102482.

Bezdan, T., Zivkovic, M., Bacanin, N., Chhabra, A. and Suresh, M. (2022). Feature selection by hybrid brain storm optimization algorithm for Covid-19 classification. *Journal of Computational Biology, 29*(6), 515–29.

Blasiak, A., Khong, J. and Kee, T. (2020). CURATE. AI: optimizing personalized medicine with artificial intelligence. *SLAS Technology: Translating Life Sciences Innovation, 25*(2), 95–105.

Boina, R. (2022). Assessing the Increasing Rate of Parkinson's Disease in the US and its Prevention Techniques. *International Journal of Biotechnology, 3*(1).

Budimirovic, N., Prabhu, E., Antonijevic, M., Zivkovic, M., Bacanin, N., Strumberger, I. and Venkatachalam, K. (2022). COVID-19 severity prediction using enhanced whale with SALP swarm feature classification. *Computers, Materials, & Continua, 72*(1).

Byrne, M.F., Parsa, N., Greenhill, A.T., Chahal, D., Ahmad, O. and Bagci, U. (Eds.). (2023). *AI in Clinical Medicine: A Practical Guide for Healthcare Professionals.* John Wiley & Sons.

Checkoway, H., Lundin, J.I. and Kelada, S.N. (2011). Neurodegenerative diseases. *IARC Scientific Publications, 163*, 407–19.

Chekani, F., Bali, V. and Aparasu, R.R. (2016). Quality of life of patients with Parkinson's disease and neurodegenerative dementia: A nationally representative study. *Research in Social and Administrative Pharmacy*, *12*(4), 604–13.

Chisci, L., Mavino, A., Perferi, G., Sciandrone, M., Anile, C., Colicchio, G. and Fuggetta, F. (2010). Real-time epileptic seizure prediction using AR models and support vector machines. *IEEE Transactions on Biomedical Engineering*, *57*(5), 1124–32.

Crowell, B., Cug, J. and Michalikova, K.F. (2022). Smart wearable internet of medical things technologies, artificial intelligence-based diagnostic algorithms, and real-time healthcare monitoring systems in COVID-19 detection and treatment. *American Journal of Medical Research*, *9*(1), 17–32.

Dai, Y., Tang, Z. and Wang, Y. (2019). Data driven intelligent diagnostics for Parkinson's disease. *IEEE Access*, *7*, 106941–106950.

Das, T., Kaur, H., Gour, P., Prasad, K., Lynn, A.M., Prakash, A. and Kumar, V. (2022). Intersection of network medicine and machine learning towards investigating the key biomarkers and pathways underlying amyotrophic lateral sclerosis: A systematic review. *Briefings in Bioinformatics*, *23*(6), bbac442.

Di Biase, L., Di Santo, A., Caminiti, M.L., De Liso, A., Shah, S.A., Ricci, L. and Di Lazzaro, V. (2020). Gait analysis in Parkinson's disease: An overview of the most accurate markers for diagnosis and symptoms monitoring. *Sensors*, *20*(12), 3529.

Dobrojevic, M., Zivkovic, M., Chhabra, A., Sani, N.S., Bacanin, N. and Amin, M.M. (2023). Addressing Internet of Things security by enhanced sine cosine metaheuristics tuned hybrid machine learning model and results interpretation based on SHAP approach. *PeerJ. Computer Science*, *9*, e1405.

Du, M., Liu, N. and Hu, X. (2019). Techniques for interpretable machine learning. *Communications of the ACM*, *63*(1), 68–77.

Dugger, B.N. and Dickson, D.W. (2017). Pathology of neurodegenerative diseases. *Cold Spring Harbor Perspectives in Biology*, *9*(7), a028035.

Gaba, D.M. (2018). Human Error in Dynamic Medical Domains 1. *In: Human Error in Medicine*, 97–224. CRC Press.

Gandin, I., Scagnetto, A., Romani, S. and Barbati, G. (2021). Interpretability of time-series deep learning models: A study in cardiovascular patients admitted to Intensive care unit. *Journal of Biomedical Informatics*, *121*, 103876.

Ghislieri, M., Agostini, V., Rizzi, L., Knaflitz, M. and Lanotte, M. (2021). Atypical gait cycles in Parkinson's disease. *Sensors*, *21*(15), 5079.

Godkin, F.E., Turner, E., Demnati, Y., Vert, A., Roberts, A., Swartz, R.H. and Van Ooteghem, K. (2022). Feasibility of a continuous, multi-sensor remote health monitoring approach in persons living with neurodegenerative disease. *Journal of Neurology*, 1–14.

Gupta, R. and Nalavade, J.E. (2023). Metaheuristic assisted hybrid classifier for bitcoin price prediction. *Cybernetics and Systems*, *54*(7), 1037–61.

Gurrola-Ramos, J., Hernàndez-Aguirre, A. and Dalmau-Cedeño, O. (2020, July). COLSHADE for real-world single-objective constrained optimization problems. *In: 2020 IEEE Congress on Evolutionary Computation (CEC)*, 1–8. IEEE.

Haleem, A., Vaishya, R., Javaid, M. and Khan, I.H. (2020). Artificial Intelligence (AI) applications in orthopaedics: An innovative technology to embrace. *Journal of Clinical Orthopaedics and Trauma*, *11*(Suppl 1), S80.

Heidari, A.A., Mirjalili, S., Faris, H., Aljarah, I., Mafarja, M. and Chen, H. (2019). Harris Hawks optimization: Algorithm and applications. *Future Generation Computer Systems*, *97*, 849–72.

Hess, S.P., Levin, M., Akram, F., Woo, K., Andersen, L., Trenkle, K., ... and Fleisher, J.E. (2022). The impact and feasibility of a brief, virtual, educational intervention for home healthcare professionals on Parkinson's Disease and Related Disorders: Pilot study of I SEE PD Home. *BMC Medical Education*, *22*(1), 506.

Hochreiter, S. (1998). The vanishing gradient problem during learning recurrent neural nets and problem solutions. *International Journal of Uncertainty, Fuzziness, and Knowledge-Based Systems*, *6*(02), 107–16.

Hochreiter, S. and Schmidhuber, J. (1997). Long short-term memory. *Neural Computation*, *9*(8), 1735–80.

Hunter, B., Hindocha, S. and Lee, R.W. (2022). The role of artificial intelligence in early cancer diagnosis. *Cancers*, *14*(6), 1524.

Jankovic, J. (2015). Gait disorders. *Neurologic Clinics, 33*(1), 249–68.

Jia, H., Rao, H., Wen, C. and Mirjalili, S. (2023). Crayfish optimization algorithm. *Artificial Intelligence Review, 56*(Suppl 2), 1919–79.

Johnson, M., Albizri, A. and Simsek, S. (2022). Artificial intelligence in healthcare operations to enhance treatment outcomes: A framework to predict lung cancer prognosis. *Annals of Operations Research*, 1–31.

Jovanovic, A., Dogandzic, T., Dobrojevic, M., Sarac, M., Bacanin, N. and Zivkovic, M. (2023, July). Gold prices forecasting using recurrent neural network with attention tuned by metaheuristics. *In: 2023 IEEE World Conference on Applied Intelligence and Computing (AIC)*, 345–50. IEEE.

Jovanovic, L., Bacanin, N., Jovancai, A., Jovanovic, D., Singh, D., Antonijevic, M., ... and Strumberger, I. (2022, November). Oil Price Prediction Approach Using Long Short-Term Memory Network Tuned by Improved Seagull Optimization Algorithm. *In: International Conference on Sustainable and Innovative Solutions for Current Challenges in Engineering & Technology*, 253–65. Singapore: Springer Nature Singapore.

Jovanovic, L., Bacanin, N., Zivkovic, M., Antonijevic, M., Jovanovic, B., Sretenovic, M.B. and Strumberger, I. (2023). Machine learning tuning by diversity oriented firefly metaheuristics for Industry 4.0. *Expert Systems*, e13293.

Jovanovic, L., Jovanovic, D., Antonijevic, M., Nikolic, B., Bacanin, N., Zivkovic, M. and Strumberger, I. (2023). Improving phishing website detection using a hybrid two-level framework for feature selection and xgboost tuning. *Journal of Web Engineering, 22*(3), 543–74.

Jovanovic, L., Jovanovic, D., Bacanin, N., Jovancai Stakic, A., Antonijevic, M., Magd, H., ... and Zivkovic, M. (2022). Multi-step crude oil price prediction based on LSTM approach tuned by Salp swarm algorithm with disputation operator. *Sustainability, 14*(21), 14616.

Jovanovic, L., Jovanovic, G., Perisic, M., Alimpic, F., Stanisic, S., Bacanin, N., ... and Stojic, A. (2023). The explainable potential of coupling metaheuristics-optimized-xgboost and SHAP in revealing vocs' environmental fate. *Atmosphere, 14*(1), 109.

Kamini and Rani, S. (2023). Artificial Intelligence and Machine Learning Models for Diagnosing Neurodegenerative Disorders. *In: Data Analysis for Neurodegenerative Disorders*, 15–48. Singapore: Springer Nature Singapore.

Kansara, S., Trivedi, A., Chen, S., Jankovic, J. and Le, W. (2013). Early diagnosis and therapy of Parkinson's disease: Can disease progression be curbed? *Journal of Neural Transmission, 120*, 197–210.

Karaboga, D. (2010). Artificial bee colony algorithm. *Scholarpedia, 5*(3), 6915.

Kennedy, J. and Eberhart, R. (1995, November). Particle swarm optimization. *In: Proceedings of ICNN'95-International Conference on Neural Networks, 4*, 1942–1948). IEEE.

Khaliq, F., Oberhauser, J., Wakhloo, D. and Mahajani, S. (2023). Decoding degeneration: The implementation of machine learning for clinical detection of neurodegenerative disorders. *Neural Regeneration Research, 18*(6), 1235.

Kiani, F., Seyyedabbasi, A., Nematzadeh, S., Candan, F., Çevik, T., Anka, F.A., ... and Muzirafuti, A. (2022). Adaptive metaheuristic-based methods for autonomous robot path planning: Sustainable agricultural applications. *Applied Sciences, 12*(3), 943.

Kim, J., Chae, M., Chang, H.J., Kim, Y.A. and Park, E. (2019). Predicting cardiac arrest and respiratory failure using feasible artificial intelligence with simple trajectories of patient data. *Journal of Clinical Medicine, 8*(9), 1336.

Kishor, A. and Chakraborty, C. (2022). Artificial intelligence and internet of things based healthcare 4.0 monitoring system. *Wireless Personal Communications, 127*(2), 1615–31.

Koikkalainen, J., Rhodius-Meester, H., Tolonen, A., Barkhof, F., Tijms, B., Lemstra, A.W., ... and Lötjönen, J. (2016). Differential diagnosis of neurodegenerative diseases using structural MRI data. *NeuroImage: Clinical, 11*, 435–49.

Kose, U. (2018). An ant-lion optimizer-trained artificial neural network system for chaotic electroencephalogram (EEG) prediction. *Applied Sciences, 8*(9), 1613.

Krishnamoorthy, S., Dua, A. and Gupta, S. (2023). Role of emerging technologies in future IoT-driven Healthcare 4.0 technologies: A survey, current challenges, and future directions. *Journal of Ambient Intelligence and Humanized Computing, 14*(1), 361–407.

Lundberg, S.M. and Lee, S.I. (2017). A unified approach to interpreting model predictions. *Advances in Neural Information Processing Systems, 30.*

Lundberg, S.M., Erion, G., Chen, H., DeGrave, A., Prutkin, J.M., Nair, B., ... and Lee, S. I. (2020). From local explanations to global understanding with explainable AI for trees. *Nature Machine Intelligence, 2*(1), 56–67.

Milicevic, M., Jovanovic, L., Bacanin, N., Zivkovic, M., Jovanovic, D., Antonijevic, M., ... and Strumberger, I. (2023). Optimizing Long Short-term Memory by Improved Teacher Learning-based Optimization for Ethereum Price Forecasting. *In*: *Mobile Computing and Sustainable Informatics: Proceedings of ICMCSI 2023*, 125–139. Singapore: Springer Nature Singapore.

Mirelman, A., Bonato, P., Camicioli, R., Ellis, T.D., Giladi, N., Hamilton, J.L., ... and Almeida, Q.J. (2019). Gait impairments in Parkinson's disease. *The Lancet Neurology, 18*(7), 697–708.

Mirjalili, S., and Mirjalili, S. (2019). Genetic algorithm. *Evolutionary Algorithms and Neural Networks: Theory and Applications*, 43–55.

Mirmohseni, S.M., Tang, C. and Javadpour, A. (2022). FPSO-GA: A fuzzy metaheuristic load balancing algorithm to reduce energy consumption in cloud networks. *Wireless Personal Communications, 127*(4), 2799–2821.

Molnar, C. (2020). Interpretable machine learning. Lulu. com.

Morris, M.E., Huxham, F., McGinley, J., Dodd, K. and Iansek, R. (2001). The biomechanics and motor control of gait in Parkinson disease. *Clinical Biomechanics, 16*(6), 459–70.

Mueller, S.G., Schuff, N. and Weiner, M.W. (2006). Evaluation of treatment effects in Alzheimer's and other neurodegenerative diseases by MRI and MRS. *NMR in Biomedicine: An International Journal Devoted to the Development and Application of Magnetic Resonance in vivo, 19*(6), 655–68.

Munavalli, J.R., Boersma, H.J., Rao, S.V. and Van Merode, G.G. (2021). Real-time capacity management and patient flow optimization in hospitals using AI methods. *Artificial Intelligence and Data Mining in Healthcare*, 55–69.

Murdoch, W.J., Singh, C., Kumbier, K., Abbasi-Asl, R. and Yu, B. (2019). Definitions, methods, and applications in interpretable machine learning. *Proceedings of the National Academy of Sciences, 116*(44), 22071–80.

Myszczynska, M.A., Ojamies, P.N., Lacoste, A.M., Neil, D., Saffari, A., Mead, R., ... and Ferraiuolo, L. (2020). Applications of machine learning to diagnosis and treatment of neurodegenerative diseases. *Nature Reviews Neurology, 16*(8), 440–56.

Nahavandi, D., Alizadehsani, R., Khosravi, A. and Acharya, U.R. (2022). Application of artificial intelligence in wearable devices: Opportunities and challenges. *Computer Methods and Programs in Biomedicine, 213*, 106541.

Nemat, H., Khadem, H., Eissa, M.R., Elliott, J. and Benaissa, M. (2022). Blood glucose level prediction: advanced deep-ensemble learning approach. *IEEE Journal of Biomedical and Health Informatics, 26*(6), 2758–69.

Pagan, F.L. (2012). Improving outcomes through early diagnosis of Parkinson's disease. *American Journal of Managed Care, 18*(7), S176.

Para, J., Del Ser, J. and Nebro, A.J. (2022). Energy-aware multi-objective job shop scheduling optimization with metaheuristics in manufacturing industries: A critical survey, results, and perspectives. *Applied Sciences, 12*(3), 1491.

Paul, S., Maindarkar, M., Saxena, S., Saba, L., Turk, M., Kalra, M., ... and Suri, J.S. (2022). Bias investigation in artificial intelligence systems for early detection of Parkinson's disease: A narrative review. *Diagnostics, 12*(1), 166.

Paulsen, J.S., Nance, M., Kim, J.I., Carlozzi, N.E., Panegyres, P.K., Erwin, C., ... and Williams, J.K. (2013). A review of quality of life after predictive testing for and earlier identification of neurodegenerative diseases. *Progress in Neurobiology, 110*, 2–28.

Perumal, S.V. and Sankar, R. (2016). Gait and tremor assessment for patients with Parkinson's disease using wearable sensors. *ICT Express, 2*(4), 168–74.

Petrovic, A., Jovanovic, L., Zivkovic, M., Bacanin, N., Budimirovic, N. and Marjanovic, M. (2023, January). Forecasting Bitcoin Price by Tuned Long Short-term Memory Model. *In*: *1st International Conference on Innovation in Information Technology and Business (ICIITB 2022)*, 187–202. Atlantis Press.

Pistacchi, M., Gioulis, M., Sanson, F., De Giovannini, E., Filippi, G., Rossetto, F. and Marsala, S.Z. (2017). Gait analysis and clinical correlations in early Parkinson's disease. *Functional Neurology*, *32*(1), 28.

Portela, F., Santos, M., Boas, M.V., Rua, F., Silva, Á. and Neves, J. (2010, October). Real-time Intelligent decision support in intensive medicine. *In: International Conference on Knowledge Management and Information Sharing*, *2*, 44–50. SCITEPRESS.

Rajpurkar, P., Chen, E., Banerjee, O. and Topol, E.J. (2022). AI in health and medicine. *Nature Medicine*, *28*(1), 31–38.

Rana, A., Dumka, A., Singh, R., Panda, M.K., Priyadarshi, N. and Twala, B. (2022). Imperative role of machine learning algorithm for detection of Parkinson's disease: Review, challenges, and recommendations. *Diagnostics*, *12*(8), 2003.

Rashid, J., Batool, S., Kim, J., Wasif Nisar, M., Hussain, A., Juneja, S. and Kushwaha, R. (2022). An augmented artificial intelligence approach for chronic diseases prediction. *Frontiers in Public Health*, *10*, 860396.

Recio-Garcia, J.A., Diaz-Agudo, B. and Acuaviva-Huertos, A. (2023). Becalm: Intelligent Monitoring of Respiratory Patients. *IEEE Journal of Biomedical and Health Informatics*, *27*(8), 3806–17.

Salb, M., Jovanovic, L., Bacanin, N., Kunjadic, G., Antonijevic, M., Zivkovic, M. and Devi, V.K. (2023, April). The Long Short-Term Memory Tuning for Multistep Ahead Wind Energy Forecasting Using Enhanced Sine Cosine Algorithm and Variation Mode Decomposition. *In: International Conference on Paradigms of Communication, Computing and Data Analytics*, 31–43. Singapore: Springer Nature Singapore.

Salehinejad, H., Sankar, S., Barfett, J., Colak, E. and Valaee, S. (2017). Recent advances in recurrent neural networks. *arXiv preprint. arXiv*:1801.01078.

Savanović, N., Toskovic, A., Petrovic, A., Zivkovic, M., Damaševičius, R., Jovanovic, L., ... and Nikolic, B. (2023). Intrusion detection in healthcare 4.0 internet of things systems via metaheuristics optimized machine learning. *Sustainability*, *15*(16), 12563.

Schnabel, R.B., Marinelli, E.A., Arbelo, E., Boriani, G., Boveda, S., Buckley, C.M., ... and Kirchhof, P. (2023). Early diagnosis and better rhythm management to improve outcomes in patients with atrial fibrillation: The 8th AFNET/EHRA Consensus Conference. *Europace*, *25*(1), 6–27.

Senturk, Z.K. (2020). Early diagnosis of Parkinson's disease using machine learning algorithms. *Medical Hypotheses*, *138*, 109603.

Shaik, T., Tao, X., Higgins, N., Li, L., Gururajan, R., Zhou, X. and Acharya, U.R. (2023). Remote patient monitoring using artificial intelligence: Current state, applications, and challenges. *Wiley Interdisciplinary Reviews: Data Mining and Knowledge Discovery*, *13*(2), e1485.

Shusharina, N., Yukhnenko, D., Botman, S., Sapunov, V., Savinov, V., Kamyshov, G., ... and Voznyuk, I. (2023). Modern Methods of Diagnostics and Treatment of Neurodegenerative Diseases and Depression. *Diagnostics*, *13*(3), 573.

Staffini, A., Svensson, T., Chung, U.I. and Svensson, A.K. (2021). Heart rate modeling and prediction using autoregressive models and deep learning. *Sensors*, *22*(1), 34.

Stankovic, M., Bacanin, N., Budimirovic, N., Zivkovic, M., Sarac, M. and Strumberger, I. (2023, April). Bi-Directional Long Short-Term Memory Optimization by Improved Teaching-Learning Based Algorithm for Univariate Gold Price Forecasting. *In: 2023 International Conference on Inventive Computation Technologies (ICICT)*, 1650–57. IEEE.

Stankovic, M., Gavrilovic, J., Jovanovic, D., Zivkovic, M., Antonijevic, M., Bacanin, N. and Stankovic, M. (2022). Tuning multi-layer perceptron by hybridized arithmetic optimization algorithm for healthcare 4.0. *Procedia Computer Science*, *215*, 51–60.

Stankovic, M., Jovanovic, L., Bacanin, N., Zivkovic, M., Antonijevic, M. and Bisevac, P. (2022, December). Tuned Long Short-Term Memory Model for Ethereum Price Forecasting through an Arithmetic Optimization Algorithm. *In: International Conference on Innovations in Bio-inspired Computing and Applications*, 327–37. Cham: Springer Nature Switzerland.

Staudemeyer, R.C. and Morris, E.R. (2019). Understanding LSTM: A tutorial into long short-term memory recurrent neural networks. *arXiv preprint. arXiv*:1909.09586.

Stoean, C., Zivkovic, M., Bozovic, A., Bacanin, N., Strulak-Wójcikiewicz, R., Antonijevic, M. and Stoean, R. (2023). Metaheuristic-Based Hyperparameter Tuning for Recurrent Deep Learning: Application to the Prediction of Solar Energy Generation. *Axioms*, *12*(3), 266.

Subasi, A. (2021). Disease Prediction Using Artificial Intelligence: A case study on epileptic seizure prediction. *Enhanced Telemedicine and e-Health: Advanced IoT Enabled Soft Computing Framework*, 289–14.

Subhan, F., Mirza, A., Su'ud, M.B.M., Alam, M.M., Nisar, S., Habib, U. and Iqbal, M.Z. (2023). AI-enabled wearable medical internet of things in healthcare system: A survey. *Applied Sciences*, *13*(3), 1394.

Tang, K.J.W., Ang, C.K.E., Constantinides, T., Rajinikanth, V., Acharya, U.R. and Cheong, K.H. (2021). Artificial intelligence and machine learning in emergency medicine. *Biocybernetics and Biomedical Engineering*, *41*(1), 156–72.

Thakur, A. and Goraya, M.S. (2022). RAFL: A hybrid metaheuristic based resource allocation framework for load balancing in cloud computing environment. *Simulation Modelling Practice and Theory*, *116*, 102485.

Von Coelln, R., Gruber-Baldini, A.L., Reich, S.G., Armstrong, M.J., Savitt, J.M. and Shulman, L.M. (2021). The inconsistency and instability of Parkinson's disease motor subtypes. *Parkinsonism & Related Disorders*, *88*, 13–18.

Wolpert, D.H. and Macready, W.G. (1997). No free lunch theorems for optimization. *IEEE Transactions on Evolutionary Computation*, *1*(1), 67–82.

Yang, W., Hamilton, J.L., Kopil, C., Beck, J.C., Tanner, C.M., Albin, R.L., ... and Thompson, T. (2020). Current and projected future economic burden of Parkinson's disease in the US. *NPJ Parkinson's Disease*, *6*(1), 15.

Yang, X.S. and Slowik, A. (2020). Firefly algorithm. *In: Swarm Intelligence Algorithms*, 163–74 CRC Press.

Yeasmin, S. (2019, May). Benefits of artificial intelligence in medicine. *In: 2019 2nd International Conference on Computer Applications & Information Security (ICCAIS)*, 1–6. IEEE.

Yogev, G., Giladi, N., Peretz, C., Springer, S., Simon, E.S. and Hausdorff, J.M. (2005). Dual tasking, gait rhythmicity, and Parkinson's disease: Which aspects of gait are attention demanding? *European Journal of Neuroscience*, *22*(5), 1248–56.

Zhang, Y., Hu, Y., Jiang, N. and Yetisen, A.K. (2023). Wearable artificial intelligence biosensor networks. *Biosensors and Bioelectronics*, *219*, 114825.

Zhao, J., Zhang, B., Guo, X., Qi, L. and Li, Z. (2022). Self-Adapting Spherical Search Algorithm with Differential Evolution for Global Optimization. *Mathematics*, *10*(23), 4519.

Zivkovic, M., Bacanin, N., Antonijevic, M., Nikolic, B., Kvascev, G., Marjanovic, M. and Savanovic, N. (2022). Hybrid CNN and XGBoost model tuned by modified arithmetic optimization algorithm for COVID-19 early diagnostics from X-ray images. *Electronics*, *11*(22), 3798.

Zivkovic, M., Bacanin, N., Venkatachalam, K., Nayyar, A., Djordjevic, A., Strumberger, I. and Al-Turjman, F. (2021). COVID-19 cases prediction by using hybrid machine learning and beetle antennae search approach. *Sustainable Cities and Society*, *66*, 102669.

Zivkovic, M., Bezdan, T., Strumberger, I., Bacanin, N. and Venkatachalam, K. (2021). Improved Harris Hawks optimization algorithm for workflow scheduling challenge in cloud–edge environment. *In: Computer Networks, Big Data and IoT: Proceedings of ICCBI 2020*, 87–102. Singapore: Springer Singapore.

Zivkovic, M., Petrovic, A., Venkatachalam, K., Strumberger, I., Jassim, H.S. and Bacanin, N. (2022). Novel chaotic best firefly algorithm: Covid-19 fake news detection application. *In: Advances in Swarm Intelligence: Variations and Adaptations for Optimization Problems*, 285–305. Cham: Springer International Publishing.

Zivkovic, M., Tair, M., Venkatachalam, K., Bacanin, N., Hubálovský, Š. and Trojovský, P. (2022). Novel hybrid firefly algorithm: An application to enhance XGBoost tuning for intrusion detection classification. *Peer J. Computer Science*, *8*, e956.

10

Emerald Evolution
Charting the Next Phase of Global Sustainability with Collective Intelligence

Uwe Seebacher[1*]

Introduction

In this chapter, we explore the concept of Green Deals, including the European Green Deal and the American Green New Deal, which represent ambitious efforts to tackle climate change and promote sustainability. These initiatives focus on transitioning to renewable energy, creating green jobs, and integrating environmental sustainability with socio-economic policies. However, despite their comprehensive agendas, these Green Deals face various challenges, such as economic feasibility, technological limitations, social and economic displacement, and the need for global cooperation. These shortcomings suggest a necessity for a novel approach, leading us to introduce the Emerald Deal. The Emerald Deal emerges as an alternative framework, aiming to address environmental challenges with a more globally integrated, technologically adaptive, and socially inclusive strategy. It emphasizes not just mitigation, but also adaptation to climate impacts, encompassing a broader range of sustainability aspects and prioritizing resilience and equity in its approach.

The Green Deals and Their Goals

The Green New Deal (Elliott et al., 2008) is a proposed package of United States legislation aimed at addressing climate change and economic inequality. Its core goals include:

[1]   University of Applied Sciences Munich, Germany.
*   Corresponding author: uwe.seebacher@hm.edu

1. **Transition to Renewable Energy:**   Aiming for a swift transition to renewable energy sources like solar and wind, reducing dependence on fossil fuels, and aiming for net-zero greenhouse gas emissions.
2. **Job Creation:**   Generating millions of new, high-paying jobs in clean energy industries, thus boosting the economy and reducing unemployment.
3. **Infrastructure and Innovation:**   Investing in energy-efficient infrastructure and innovative technologies to meet energy demands sustainably.
4. **Social and Economic Justice:**   Addressing systemic injustices by prioritizing the needs of underprivileged communities, workers, and indigenous peoples.
5. **Sustainable Agriculture:**   Supporting sustainable farming practices to reduce environmental impact and enhance food security.
6. **Conservation Efforts:**   Protecting and restoring natural ecosystems to maintain biodiversity and combat climate change impacts.

Essentially, the Green New Deal integrates environmental policy with social and economic reforms, striving for a sustainable future that is both environmentally sound and socially equitable. The European Green Deal is a set of policy initiatives by the European Union (EU) with the overarching aim of making Europe climate neutral by 2050. In conjunction with the American Green New Deal, it represents a significant global shift towards sustainable development. The EU Green Deal's key features include:

1. **Climate Neutrality by 2050:**   The central goal is to transform the EU into a modern, resource-efficient, and competitive economy, ensuring no net emissions of greenhouse gases by 2050 and decoupling economic growth from resource use.
2. **Clean Energy:**   Prioritizing renewable energy sources and increasing energy efficiency. The goal is to ensure a more secure, cleaner, and affordable energy supply for Europe.
3. **Sustainable Industry:**   Transforming EU industries for a clean and circular economy. It includes promoting recycling, reducing waste, and fostering innovation in clean technologies.
4. **Building and Renovating:**   Focusing on energy and resource efficiency in the building sector, which is key for reducing Europe's carbon footprint.
5. **Sustainable Mobility:**   Promoting more sustainable transport modes, making a significant switch to public transport, cycling, and walking, and reducing carbon emissions from vehicles.
6. **Biodiversity and Ecosystems:**   Aiming to protect and restore biodiversity and strengthen the natural systems upon which life depends, which in turn helps to mitigate climate change and prevent ecosystem collapse.
7. **From Farm to Fork:**   Transforming food systems for better health and environmental sustainability. This includes encouraging more sustainable agricultural practices, protecting marine resources, and ensuring food security.
8. **Just Transition:**   Recognizing that the transition must be inclusive and fair, providing support for regions and sectors most affected by the move towards the green economy.

The European Green Deal, like the American Green New Deal, integrates environmental sustainability with socio-economic policies, but it is distinct in its approach and specific targets, reflecting the unique challenges and opportunities within the European Union. Together, these initiatives represent a global trend towards prioritizing environmental sustainability in policymaking at the highest levels.

The Shortcomings of the Green Deals

While the European Green Deal and the American Green New Deal are ambitious initiatives aiming to address climate change and promote sustainability, they have faced criticisms (Péer et al., 2020) and perceived shortcomings, highlighting the need for a new, more holistic approach:

1. **Economic Feasibility and Cost:** Critics argue that the costs associated with these Green Deals are immense and could strain national budgets. The transition to renewable energy and sustainable practices requires significant investment, which can be challenging, especially in times of economic downturn.
2. **Technological Limitations:** Both deals heavily rely on current and future technologies for renewable energy and emissions reduction. However, some necessary technologies are still in development or not yet scalable or cost-effective, posing a significant challenge in meeting the set targets.
3. **Social and Economic Displacement:** The rapid shift away from fossil fuels and traditional industries could lead to job losses and economic displacement in certain sectors. While both deals propose job creation in new industries, the transition might not be smooth or fast enough to mitigate these effects.
4. **Political and Public Acceptance:** Implementing sweeping environmental policies often faces political resistance and requires significant public support. In democracies, changing administrations and public opinion can affect the continuity and effectiveness of long-term environmental policies.
5. **Global Impact and Cooperation:** Climate change is a global issue, yet the Green Deals are regional. Without similar commitments from other major economies and global cooperation, the overall impact on global emissions might be limited.
6. **Comprehensive Integration:** Both deals may not fully integrate or address all aspects of sustainability, such as sustainable consumption patterns, waste management, and the entire lifecycle impact of products and services.
7. **Adaptation Strategies:** While focusing heavily on mitigation, there may be insufficient emphasis on adaptation strategies for communities already facing the impacts of climate change.

When going through this concise listing of shortcomings, the reasoning for a need to adjust the focus of the Green Deals towards a new extended and adjusted becomes obvious—also and in especial when considering the many not successful projects and initiatives. Because both Green Deals mainly focus on content and technical issues and fall short on societal communicative, cooperative, and change management aspects such as listed in bullet points 3 to 7. Considering these challenges, a new approach must involve:

- **Enhanced Global Collaboration:** Strengthening international partnerships and agreements to ensure global action against climate change.
- **Balanced Transition Strategies:** Creating strategies that balance environmental goals with economic stability and social equity.
- **Inclusive Policymaking:** Involving a broader range of stakeholders, including affected communities, industries, and experts, in policy formulation.
- **Focus on Adaptation and Resilience:** Increasing emphasis on adaptation measures for communities already affected by climate change.
- **Continuous Innovation and Research:** Supporting ongoing research and development in sustainable technologies and practices.
- **Public Engagement, Interaction and Education:** Enhancing public awareness and education on sustainability issues to build stronger public support for green policies.

This new approach would aim to address the limitations of the current Green Deals, offering a more integrated, globally coordinated, and socially inclusive path towards sustainability. And this is why the author argues for the move from our current Green Deals towards an Emerald Deal closing the gap and offering a solution to the different shortcomings of the Green Deals.

Introducing the Emerald Deal

The Emerald Deal transcends the foundational principles of the Green Deals, incorporating the sophisticated capabilities of predictive intelligence (Seebacher, 2021; Harth et al., 2018) and the reengineering of corporate communication strategies (Seebacher, 2022). Predictive intelligence, an offshoot of advanced analytics (Bose, 2009), leverages data, statistical algorithms, and machine learning techniques (Mahesh, 2020) to identify the likelihood of future outcomes based on historical data (Sharda et al., 2021). In the context of environmental sustainability (Goodland 1995), this intelligence is crucial for forecasting environmental impacts, optimizing resource allocation, and developing proactive strategies for climate change mitigation (Fawzy et al., 2020).

In tandem with predictive intelligence, the reengineering of corporate communication emerges as a vital aspect of the Emerald Deal. Amidst the current social and political crisis, marked by a global pandemic, rising populism (McCarthy, 2019), and a proliferation of misinformation (Hills, 2019), the role of effective communication (Suter et al., 2009) has never been more critical. The surge in misinformation, especially around climate change and environmental policies, necessitates a strategic overhaul in how corporations and governments communicate their sustainability initiatives (Bennett & Livingston, 2018). This reengineering is not just about disseminating information but about fostering dialogue, enhancing transparency, and building public trust (Chanley, 2000).

In conclusion, the Emerald Deal is a call to action, a blueprint for a more aligned, accepting because of understanding, informed future as an authentic joined effort of all stakeholders acting on eye-level. What currently is an initiative of distrusted politicians and governments must become a collectively understood and accepted agenda—the Emerald Deal. It extends beyond the aspirations of the Green Deal, limited to defining the operational and content-driven roadmap and addressing the complexities of our current global challenges with a multidimensional approach. The Emerald Deal is a vision that combines the power of technology, the wisdom of collective intelligence, and the necessity of ethical, transparent communication to forge a path towards a sustainable, equitable, and resilient world.

It is the about the 'How' we can succeed in the society to engage, accept, and trust in the 'What' being defined by the Green Deal in the FIBS world. The Green Deal totally neglected the communicative and interactive component which in the context of such a huge societal change is a big failure. How big this failure is can be seen every day watching the news with fake news, populists, and conspiracists.

The Emerald Deal represents a transformative step beyond the European Union's Green Deal, introducing a holistic approach that integrates advanced technologies, collective intelligence, and societal engagement to foster a more profound and sophisticated practice of sustainability. This evolution is not merely an extension of existing environmental policies but a reimagining of them, adding crucial communicative, societal, and change management dimensions to address the challenges of misinformation (Aimeur et al., 2023), conspiracy theories (Nera et al., 2021), and populism (Flew, 2020), which are increasingly prevalent in today's society.

The Limitations of Green Deals

In the wake of the EU's and the America's ambitious Green New Deal (Rifkin, 2019; Hafner & Raimondim, 2020) a paradigm shift is becoming essential to emerge in the sense of proactively integrating the communicational and interactional dimension of this significant societal change program. This, as ethics and communication in science, must be placed highest on our agenda for the forthcoming years. Why? Because it becomes more than obvious every day in the context of fake news, filter bubbles, populists, and manipulation. Due to this, I am talking about the 'FIBS' environment or world being driven by the following phenomena:

- Fake News
- Isolating Filter Bubbles
- Burgeoning Populism (as a stand-in for "the renaissance of populism")
- Storms of Conspiracy Theories (in place of "conspiracy theory tsunamis")

'FIBS' not only captures the essence of each term but also carries the connotation of falsehood or deception, which is a common thread among these phenomena. Due to this the entire area of reengineering institutional as well as political communication must be placed on top of any agenda. The use of familiar and traditional methods of communication is no longer sufficient and up-to-date, as can easily be seen from the development towards a FIBS world and the massive increase in shitstorms.

Thus, we need to transit from the traditional content- and initiative-driven "Green New Deal" approach to what can be conceptualized as the "Emerald Deal" in which the ever more decisive for success and failure dimension of societal communication, information, and interaction must be stringently added to the game. If we remain on the pure content level of the Green New Deal, we will totally fail as can, already and anyway, be seen by the many pushbacks, delays, and development as the "Last Generation" and the "Climate Gluers". This transformation is not merely a change in color symbolism but represents a deeper, more comprehensive, and inclusive 360-degree approach to sustainability (Bonoli et al., 2021), integrating not only advanced technologies and collective intelligence (Malone & Bernstein, 2022) but in especial latest advancements in corporate and institutional communications and interactions (Seebacher, 2022).

The Green Deal, launched by the European Commission in 2019 and the New Green Deal in the USA (Galvin & Healy, 2020), set forth an integrated policy framework aiming to make the different regions climate-neutral within certain time frames (European Commission, 2019). While it marks a significant stride in environmental policy, the rapidly evolving global landscape from 1-2-1 communication towards All-2-All 24/7 interactions, the developments in the area of AI and the multiple global crisis causing the times of distrust (Gheorghe, 2023) necessitate a more advanced interaction-focused approach, embodied in the Emerald Deal.

The urgency to move from the Green Deal to the Emerald Deal is further underscored by current social and political events. The global pandemic has highlighted the interconnectedness of human (Ployhart, 2011), economic, and environmental health, revealing the fragility of our current systems (Hickel & Kallis, 2020). Additionally, the rise in populism and nationalist movements across the globe poses a significant challenge to multilateral environmental agreements and collective action against climate change (Norris & Inglehart, 2019). These events demonstrate the need for a more resilient, adaptive, and integrated approach to sustainability—one that leverages the collective intelligence of societies and technological advancements.

The Emerald Deal, therefore, represents a response to these multifaceted challenges. It calls for harnessing swarm intelligence systems (Eberhart et al., 2001) in environmental monitoring and management, enhancing predictive capabilities for more effective decision-making (Guo, 2020), and reengineering communication strategies to combat misinformation and foster a more informed public discourse to be able to engage and onboard the society, the people. Without them understanding and authentically accepting the enormous change urgency, we will not be able to master our next big societal and environmental challenge, the change from fossil to green. The transition to this new paradigm is not merely a technical or environmental undertaking but a comprehensive societal shift, embracing ethical considerations and collaborative efforts across disciplines and sectors.

Beyond the Green Deal: The Need for a Broader Perspective

The Green Deal, while groundbreaking in its ambition to make Europe climate—neutral by 2050, primarily focuses on environmental targets and industrial transformation (European Commission, 2019). However, the emergence of global challenges such as misinformation, societal polarization (Weber et al., 2021), and the rise of populism (Roth, 2017) highlights the need for a more inclusive approach. The Emerald Deal provides an agenda seeking to fill this gap by incorporating a more comprehensive understanding of societal dynamics (Nowak & Vallacher, 2019) into environmental policymaking (Feindt & Oels, 2005).

At the heart of the Emerald Deal is the concept of collective intelligence—the enhanced capacity achieved when groups of individuals, or even machines, collaborate to solve problems (Malone & Bernstein, 2022). This concept is expanded through the use of swarm systems, which mimic the behavior of natural systems like ant colonies or bird flocks, to create decentralized, self-organized, and adaptive problem-solving networks. In the context of sustainability, these systems can process vast amounts of environmental data, generate predictive models, and facilitate dynamic decision-making processes that are both scalable and resilient.

The inception of traditional environmental strategies can be traced back to the environmental movement of the 1960s and 1970s, which saw the emergence of policies aimed at addressing visible and acute environmental problems (McCormick, 1991). The primary objectives of these strategies were to reduce pollution, protect natural habitats, conserve resources, and ensure human health and safety. Policies such as the Clean Air Act (1970) and the Clean Water Act (1972) in the United States exemplify this approach, focusing on regulating and mitigating specific environmental harms (Hand et al., 2020).

Traditional environmental strategies often employed regulatory and policy-driven approaches. Governments typically enacted laws and regulations to control pollution, manage waste, and protect natural areas. These measures often involved setting standards and limits for pollutants, establishing protected areas, and promoting recycling and waste management practices. Another key component was the use of command-and-control mechanisms, where compliance with environmental standards was enforced through penalties and sanctions (Porter & van der Linde, 1995).

These traditional approaches have led to significant environmental improvements. For instance, air and water quality in many industrialized countries improved markedly in the latter half of the 20th century (Keene & Pullin, 2011). The establishment of protected areas has helped conserve biodiversity and safeguard critical habitats. Moreover, the promotion of recycling and waste management has contributed to more sustainable resource use and waste reduction.

However, traditional environmental strategies also exhibit notable limitations. Firstly, they often adopt a siloed approach, addressing specific issues in isolation rather than considering the interconnected nature of environmental systems (Folke et al., 2005). This can lead to unintended consequences, where solving one problem exacerbates another.

Additionally, these strategies frequently rely on reactive rather than proactive measures. They tend to address environmental issues after they have occurred rather than preventing them (Gunningham, 2009). This approach is less effective in dealing with complex, systemic challenges such as climate change, which require anticipatory and holistic strategies.

Moreover, traditional strategies often lack sufficient emphasis on behavioral and societal change. While regulatory and technical solutions are essential, long-term environmental sustainability necessitates changes in societal norms, values, and behaviors (Steg & Vlek, 2009).

Lastly, the global nature of many contemporary environmental challenges, such as climate change and biodiversity loss, requires international cooperation and coordination. Traditional strategies, often confined within national or regional boundaries, are ill-equipped to address these transboundary issues effectively (Victor, 2009).

In conclusion, while traditional environmental strategies have contributed to addressing some environmental challenges, their limitations highlight the need for more innovative, integrated, and holistic approaches. The transition to these approaches, which consider the complex interplay of ecological, social, and economic factors, is essential for effectively addressing the multifaceted environmental challenges of the 21st century.

Unprecedented Challenges Increasing the Urgency

The current environmental landscape is marked by unprecedented challenges, ranging from the escalating impacts of climate change to the complexities of global resource management. This evolving scenario underscores the urgent need for innovation in environmental sustainability. Innovative approaches are no longer a mere enhancement but a necessity to effectively address the multifaceted and dynamic environmental challenges of the 21st century.

One of the primary drivers for innovation is the escalating scale and complexity of environmental issues. Climate change, as evidenced by increasing global temperatures, rising sea levels, and extreme weather events, presents an existential threat that requires novel and effective solutions (IPCC, 2021). Traditional methods, while foundational in environmental protection, are often inadequate to address the systemic and interconnected nature of current environmental challenges. For instance, reducing carbon emissions necessitates not only regulatory measures but also technological advancements, such as the development of renewable energy sources and carbon capture technologies.

The rapid advancement of technology presents both a challenge and an opportunity in the realm of environmental sustainability. On one hand, technological progress, particularly in industries such as energy, manufacturing, and agriculture, has contributed to increased environmental pressures. On the other hand, it offers unprecedented opportunities for innovation. The integration of AI, big data analytics, and Internet of Things (IoT) technologies can significantly enhance environmental monitoring, predictive modeling, and resource management (Ojokoh et al., 2020). For example, AI can be utilized in optimizing energy use in buildings, predicting weather patterns for agriculture, and monitoring biodiversity.

The evolving socio-economic landscape also drives the need for innovative approaches in environmental sustainability. The global economic system, traditionally reliant on resource-intensive and linear production models, is increasingly being challenged to transition towards circular and sustainable practices. This shift necessitates innovative business models, production processes, and consumption patterns that prioritize sustainability (Geissdoerfer, et al., 2017). Furthermore, the growing global population and the rise of middle-class consumers in emerging economies add to the urgency for sustainable solutions that can meet the increasing demand for resources without degrading the environment.

Public awareness and concern about environmental issues have grown significantly, leading to increased demand for sustainable practices from both individuals and organizations. Consumers are more informed and concerned about the environmental impact of their choices, driving the market towards sustainable products and services (Nielsen, 2015). This shift in consumer behavior acts as a catalyst for companies to innovate in product design, supply chain management, and corporate sustainability initiatives.

The policy and regulatory environment is also evolving, with governments and international bodies implementing more stringent environmental regulations and setting ambitious sustainability goals. The Paris Agreement, for example, sets out a global framework to avoid dangerous climate change by limiting global warming to well below 2°C (UNFCCC, 2015). Meeting these targets requires innovative policy instruments, cross-sectoral collaboration, and the integration of sustainability into economic and social policies.

In conclusion, the emerging need for innovation in environmental sustainability is driven by a confluence of factors: escalating environmental challenges, technological advancements, shifting socio-economic paradigms, increasing public awareness, and a more stringent policy environment. Addressing these challenges effectively demands a shift from traditional approaches to more innovative, holistic, and integrated strategies. This shift is not only imperative for the health of the planet but also for the wellbeing and prosperity of current and future generations.

Analysis of Gaps in Traditional Strategies

Traditional environmental strategies (Ellis, 2005) have been instrumental in initiating the journey towards sustainability. Traditional environmental strategies are a set of practices and beliefs that have been developed by Indigenous peoples and local communities over generations to live sustainably with the natural world. These strategies are based on a deep understanding of the environment and a respect for all living things. Some examples of traditional environmental strategies include:

- **Sustainable agriculture:** Traditional agricultural practices such as crop rotation, intercropping, and organic pest control can help to improve soil health, conserve water, and reduce pollution.
- **Forest management:** Traditional Forest management practices such as selective logging, controlled burning, and sacred groves can help to protect forests and biodiversity.
- **Water management:** Traditional water management practices such as rainwater harvesting, irrigation systems, and water conservation measures can help to ensure a sustainable water supply.
- **Wildlife management:** Traditional wildlife management practices such as hunting and fishing regulations, totemic systems, and taboos can help to protect wildlife populations.

Traditional environmental strategies are often based on the concept of reciprocity, which means that humans have a responsibility to take care of the natural world in return for the benefits that it provides. This concept is often expressed in traditional stories, myths, and ceremonies.

Traditional environmental strategies can provide valuable insights for modern environmental management. For example, many of the principles of sustainable agriculture are based on traditional practices. Traditional forest management practices can help to protect forests from fire and disease. Traditional water management practices can help to conserve water during times of drought. And traditional wildlife management practices can help to protect endangered species.

In addition to their practical benefits, traditional environmental strategies can also have cultural and spiritual benefits. For Indigenous peoples and local communities, these strategies are often an integral part of their identity and way of life. They can provide a sense of connection to the natural world and a sense of responsibility for the planet.

However, as we grapple with increasingly complex and interconnected environmental challenges, the limitations of these strategies become more apparent. An in-depth analysis of these shortcomings reveals critical gaps, particularly in adaptability and addressing complex, systemic issues.

One of the principal shortcomings of traditional environmental strategies is their inability to effectively address the complexity and interconnectedness of modern environmental challenges and they also do not proactively consider societal interactional aspects being essential for acceptance, appreciation and understanding. Traditional approaches often focus on isolated issues, such as pollution reduction or species conservation, without considering the broader ecological and socio-economic contexts (Folke et al., 2005). This siloed approach fails to recognize the systemic nature of environmental problems, where factors such as climate change, biodiversity loss, and human social dynamics are deeply intertwined. For instance, efforts to reduce greenhouse gas emissions must consider economic impacts, societal behaviors, and technological capabilities.

Traditional environmental strategies often lack the adaptability and flexibility needed to respond to rapidly changing environmental conditions. Many of these strategies are based on static policies and regulations that are slow to evolve in response to new scientific findings or environmental shifts (Gunningham, 2009). This rigidity hinders the ability to implement timely and effective responses to emergent issues, such as the sudden onset of invasive species or the rapid advancement of environmentally harmful technologies.

Environmental issues are inherently nonlinear, characterized by complex feedback loops and unpredictable dynamics. Traditional environmental management often employs linear approaches, assuming direct cause-effect relationships (Meadows, 2008). Such methods are ineffective in dealing with nonlinear systems where changes can have disproportionate and unexpected effects. For instance, interventions in one part of an ecosystem can lead to unforeseen consequences in another, as seen in cases where the introduction of a species to control pests leads to new ecological imbalances.

Another significant gap in traditional strategies is their often short-term focus. Environmental policies and initiatives frequently prioritize immediate concerns and quick fixes over long-term sustainability goals (Sachs, 2015). This short-termism fails to address the underlying drivers of environmental degradation and neglects the need for sustainable, long-term solutions.

Traditional environmental strategies often lack meaningful stakeholder engagement and multi-sectoral collaboration. Environmental issues are complex and require the involvement of various stakeholders, including governments, businesses, communities, and individuals. Traditional approaches sometimes overlook the importance of engaging these stakeholders in decision-making processes, leading to solutions that are either impractical or lack public support (Reed, 2008).

In conclusion, while traditional environmental strategies have laid the groundwork for sustainability, their limitations in addressing the complexity, adaptability, non-linearity, and long-term aspects of environmental challenges are evident. The gaps in these strategies highlight the need for more dynamic, holistic, and inclusive approaches to environmental management. As we move forward, it is imperative to develop and implement strategies that can effectively navigate the intricacies of environmental issues, ensuring resilience and sustainability for future generations.

Impact of Global Trends on Traditional Models

The landscape of environmental sustainability is being reshaped by global trends such as rapid urbanization, technological advancements, and socio-economic changes. These trends are outpacing the capabilities of traditional environmental strategies, creating a pressing need for innovative approaches that can keep pace with the changing global context, not only from a content and technological perspective but especially from the interactional and change dimension, as this is put on the agenda by the Emerald Deal of the author.

Urbanization is one of the most significant global trends impacting traditional environmental strategies. The United Nations projects that, by 2050, nearly 68% of the world's population will reside in urban areas (United Nations, 2018). This rapid urbanization poses immense challenges for traditional environmental management, including increased pollution, higher energy demands, and greater strain on water and waste management systems. Traditional strategies often lack the scalability and adaptability required to manage these growing urban environmental pressures effectively through lack of the communicative and interactional dimensions. People as part of the societal dimension in the game feel continuously overruled and not or mis-informed. In the context of a world driven by community content creators (Seebacher, 2022) in which 24/7 All-2-All interact across arenas, this leads to protest, fake news, and populists being a wonderful nutrient medium for conspiracy theorists as we have seen raising during the COVID pandemic.

The unprecedented pace of technological advancement is another trend that traditional environmental models struggle to keep up with. Innovations in fields like AI, biotechnology, and material science have far-reaching implications for the environment. On one hand, technology offers potential solutions to environmental problems; on the other, it can lead to new forms of pollution and resource depletion. Traditional environmental strategies, which are often reactive rather than proactive, find it challenging to anticipate and manage the environmental impacts of these rapidly evolving technologies. And again, the society seems to be left behind when it comes to education, engagement, and information.

Socio-economic changes, including shifts in global economic power, changing consumer behaviors, and the rise of the middle class in developing countries, also significantly impact traditional environmental strategies. As global economic activity intensifies, so do the environmental pressures associated with it, such as increased resource consumption and waste production. Traditional environmental policies, which were designed in a different socio-economic context, often fall short in addressing the complex, globalized nature of today's economic activities. Moreover, the rise in consumerism and the global supply chain complicates the implementation of local or national environmental strategies.

Climate change, driven by global trends like fossil fuel use and deforestation, is exacerbating the limitations of traditional environmental strategies. Climate change brings about extreme weather events, sea-level rise, and shifting climate patterns, which require a dynamic and multifaceted approach to environmental management. Traditional methods, with their focus on specific, localized issues, are insufficient to tackle the global and systemic nature of climate change.

Global trends such as urbanization, technological advancements, and socio-economic changes are rapidly outpacing the capabilities of traditional environmental strategies. These trends not only bring new challenges but also exacerbate existing environmental issues. The limitations of traditional models in addressing these dynamic and interconnected challenges highlight the need for more flexible, comprehensive, and forward-thinking environmental strategies. As the world continues to evolve, so must our approaches to environmental sustainability, ensuring they are robust enough to withstand and adapt to the complexities of the 21st century.

The Emergence of Interconnected Solutions

In the face of escalating global environmental challenges, interconnected solutions in environmental sustainability has emerged as a vital approach. This paradigm shift, necessitated by the complexity and interdependence of contemporary environmental issues, offers a more holistic and effective strategy compared to traditional methods. These interconnected solutions emphasize the integration of ecological, social, and economic factors, providing a comprehensive framework to address the multifaceted nature of environmental challenges (Folke et al., 2005). This also means that these initiatives are generally more complex and thus need a greater shift towards the Emerald Deal putting the societal communication, information, and interaction at the core of the agenda. Only such a strategy can ensure that the society will not be falling by the wayside.

Traditional environmental strategies, emerging from the conservation and environmental movements of the 20th century, often focus on specific issues, such as pollution reduction or species conservation. These approaches typically operate within a linear cause-and-effect framework, relying heavily on regulatory and policy-driven solutions (Porter & van der Linde, 1995). However, the increasing complexity of environmental issues, such as climate change and biodiversity loss, has revealed the limitations of these traditional methods. They often fail to consider the dynamic interplay between different ecological processes, socio-economic factors, and global interdependencies.

In contrast, interconnected solutions advocate for an integrated approach, recognizing that environmental issues are part of a larger system that includes human, economic, and social dimensions. This approach calls for collaborative, multi-sectoral strategies that transcend traditional environmental management boundaries (MEA, 2005).

Interconnected environmental solutions are based on a holistic approach that considers the relationships and interactions within and between natural and human systems. These solutions are characterized by their emphasis on comprehensive problem-solving, stakeholder engagement, cross-sector collaboration, and adaptability. They aim to balance and optimize environmental, social, and economic outcomes, rather than viewing these objectives as separate or conflicting (Berkes & Folke, 1998).

Central to interconnected solutions is systems thinking, which involves understanding environmental issues as part of broader, interrelated systems. This approach shifts the focus from isolated problems to the connections and interactions within ecosystems, including human, political, and economic elements. Systems thinking facilitates the identification of effective intervention points and emphasizes resilience and long-term sustainability in environmental management (Meadows, 2008).

Real-world applications of interconnected solutions provide valuable insights into their effectiveness. The concept of Integrated Water Resources Management (IWRM) exemplifies this approach. IWRM considers the entire water cycle and integrates land and water management, emphasizing stakeholder participation and cross-sectoral collaboration. The implementation of IWRM in the Rhine River Basin has led to significant improvements in water quality and ecosystem health (Global Water Partnership, 2000).

Another example is the urban sustainability model, which combines green infrastructure, sustainable transportation, energy efficiency, and community engagement. Cities like Copenhagen and Singapore have adopted this integrated approach, successfully transforming urban areas into sustainable living spaces (Newman et al., 2009).

In addressing climate change, the interconnected approach is embodied by the Paris Agreement. This global framework acknowledges the need for a unified response to climate change, integrating national efforts, international collaboration, and the participation of non-state actors (UNFCCC, 2015). Interconnected solutions in environmental sustainability represent a significant evolution in addressing the complex and intertwined challenges of the modern world. By embracing systems thinking and integrating various aspects of sustainability, these solutions offer a pathway to achieving environmental, social, and economic well-being in a changing global landscape. But as such approaches are driven by an immense complexity, the 'Emerald' aspect becomes of even greater significance as the society must stringently be enabled to understand these new approaches as otherwise the complexity is considered as a "black box" leading to distrust causing fake news and conspiracy theorists being called on the plan.

Role of Technology in Enabling Interconnectedness

Technological advancements have emerged as powerful catalysts in the development of interconnected environmental solutions. Modern technologies, particularly in the domains of AI, IoT, and big data analytics, are providing unprecedented opportunities to address environmental challenges in more integrated and effective ways. These technologies enable the collection, analysis, and application of vast amounts of environmental data, leading to more informed and proactive decision-making (Kitchin, 2014).

AI, in particular, plays a crucial role in fostering interconnected environmental solutions. AI's ability to process and analyze large datasets has significant implications for environmental management and also facilitates activities in the area of interaction intelligence as part of the Emerald Deal. For instance, machine learning algorithms can predict environmental changes, identify patterns in ecological and societal data, and optimize resource management (Doshi-Velez et al.,

2017) as well as communication and information management (Seebacher, 2022). In climate change research, AI models have been employed to enhance climate modeling and forecasting, offering more accurate predictions of climate patterns and their impacts (Rolnick et al., 2019).

One specific application of AI in environmental management is in the area of biodiversity conservation. AI-powered image recognition systems are being used to monitor wildlife populations and track changes in biodiversity. This technology has proven invaluable in remote or inaccessible areas, providing data that are crucial for conservation efforts (Norouzzadeh et al., 2018).

The IoT and big data analytics are also revolutionizing environmental monitoring and management. IoT devices, such as sensors and cameras, can be deployed in various environments to collect real-time data on factors like air and water quality, temperature, and species movement (Ray, 2016). This data is then analyzed using big data analytics, providing insights that can inform environmental policies and strategies. With this information also valuable knowledge can be gained for extending purely content-driven activities as part of the Green Deal towards the Emerald Deal which is information and interactive driven to ensure broad understanding and acceptance in the society.

Big data analytics allow for the processing of vast and complex datasets, enabling a more nuanced understanding of environmental systems. For instance, big data tools can analyze patterns of energy consumption, identify areas for efficiency improvements, and contribute to the development of smart, sustainable cities (Bibri, 2018).

In water resource management, IoT and big data have been utilized to create smart water networks. These networks use sensors and data analytics to monitor water systems in real-time, helping to detect leaks, manage water quality, and optimize water distribution (Ghaffarian Hoseini et al., 2018).

In conclusion, technological innovations, especially in AI, IoT, and big data analytics, are playing a pivotal role in enabling interconnected environmental solutions. These technologies provide the tools needed to gather, analyze, and apply data in ways that enhance our understanding and management of environmental systems. As a result, they are facilitating more integrated, efficient, and effective approaches to addressing the complex environmental challenges of the 21st century.

Integration of Technology with Environmental Strategies

The integration of technology with environmental strategies is creating a synergy that is essential for sustainable development in the 21st century. This synergy is evident in how technological innovations are being utilized to enhance environmental conservation and management practices. The use of renewable energy technologies, such as solar and wind power, has become a cornerstone in the fight against climate change, reducing reliance on fossil fuels and lowering greenhouse gas emissions (Jacobson et al., 2017). Similarly, advancements in agricultural technologies, including precision farming and hydroponics, are optimizing resource use, and reducing environmental impacts (Hancke et al., 2012). Such advancements have also been realized in industries such as astrophysics, healthcare, manufacturing, supply chain management, and general industrial and managerial disciplines that can be seen on the different articles in this publication. But all these so-called essential activities and advancements can be deployed successfully only if the societal engagement is assured and secured by applying the principles of the Emerald Deal about stringent, continuous, aligned, authentic, empathetic and logic communication, information, and interaction. The principles of the Emerald Deal are based on the works of Frei and Morriss (2020) in the context of trust building communication and their triangle of trust.

The synergy between technology and sustainability is not just limited to these direct applications. It also extends to the way technology is transforming environmental monitoring and data analysis but also communication and information, which is crucial for informed decision-making in environmental management (Kitchin, 2014).

Integrating technology with environmental and interactional strategies, however, presents both challenges and opportunities. One significant challenge is the digital divide, which can limit access to technological solutions in less developed regions, thereby exacerbating existing inequalities in environmental and communicative management (Van Dijk, 2006). Additionally, the deployment of new technologies often requires substantial initial investment, which can be a barrier for many organizations and governments.

Despite these challenges, the opportunities presented by technological integration are vast. Technology offers tools for greater efficiency, accuracy, and scalability in environmental management supported by adequate interaction management in the context of societal engagement and acceptance. For instance, the use of AI and big data analytics in environmental monitoring can lead to more precise and timely interventions (Rolnick et al., 2022). The global reach of satellite technology also presents opportunities for monitoring environmental changes in remote and inaccessible areas, providing critical data for global conservation efforts (Pettorelli et al., 2014). If all such measures are accompanied by communicative activities in alignment with the principles of the Emerald Deal, then the efficiency and effectiveness of these initiatives can be leveraged to a totally new level as the societal positive impact is communicated and staged. It is not without reason that ethics and communication in science will be on the agenda in the coming years as the author stated at many different events during his key notes, because the neglect of the communicative aspect in all matters is more than omnipresent in our FIBS world.

Comprehensive Environmental Solutions through Technology

Technology enables a more holistic approach to environmental management. By facilitating the collection and analysis of data across various environmental parameters, technology provides a comprehensive view of ecological systems.

This holistic perspective is crucial for understanding the interdependencies within ecosystems and for developing integrated management strategies that address multiple environmental objectives simultaneously (MEA, 2005).

Predictive modeling, powered by AI and machine learning, is playing a pivotal role in environmental planning and communication. These models can predict the impacts of various environmental policies and climate change scenarios, allowing policymakers to evaluate different strategies and make informed decisions (Doshi-Velez et al., 2017). Furthermore, the use of adaptive management strategies, which rely on continuous monitoring and data analysis, enables environmental managers to respond dynamically to changing conditions and emerging challenges (Williams, 2011). When moving from the Green Deal towards the Emerald Deal, then *responding* turns into *proactive* and *predictive* action and *interaction* with relevant groups of information clients and stakeholders. And that is what we need for the new normality of a FIBS world.

The integration of technology with environmental strategies is creating synergies that are vital for addressing the complex environmental challenges of our time. And these challenges not only need to be addressed from a content point of view but also and, even more so, from a communicational and interactional perspective. While there are challenges to be overcome, particularly in terms of accessibility and cost, the opportunities for more effective, efficient, and holistic environmental management are significant. By leveraging technological innovations, we can develop comprehensive environmental solutions that are both sustainable and adaptable to changing conditions. But this will only work if the principles of the Emerald Deal of predictive trust building interaction are stringently and continuously applied and deployed.

The Emerald Deal—A New Paradigm

In this section, we delve into the concept of the Emerald Deal, a pioneering approach that seeks to address the shortcomings of traditional Green Deals. While Green Deals, such as those in the USA and Europe, have made significant strides towards addressing climate change and promoting renewable energy, they have often been criticized for their narrow focus and lack of comprehensive social and ecological considerations. The Emerald Deal emerges as a response to these limitations, offering a more holistic and inclusive framework.

The Emerald Deal is grounded in the principles of the Green Deals being added by ecosystem restoration, biodiversity conservation, social justice and the required societal interaction.

Unlike its predecessors, this new paradigm places a strong emphasis on restoring and preserving the natural environment as a fundamental component of sustainable development and proactively and predictively integrating societal communication, engagement, and information in their agenda settings. It recognizes both the 'hard' technological content dimension in the sense of the intrinsic value of biodiversity and the crucial role ecosystems play in maintaining the balance of our planet but also the 'soft' communicative engaging interactional component being at the core for a societal join sustainable advancement.

This model understands that effective and long-lasting environmental solutions cannot be achieved without considering the socio-economic contexts, societal communicative needs, and the well-being of all communities, especially those most affected by environmental degradation and climate change.

The potential benefits of the Emerald Deal are vast and transformative. By focusing on enhancing ecosystem resilience, it aims to create a more sustainable relationship between humans and nature. This approach promises not only environmental benefits but also improved human well-being and increased economic prosperity, achieved through sustainable practices and equitable resource distribution.

In this section, we will explore these core principles in detail, examining how the Emerald Deal addresses the limitations of earlier Green Deals and envisioning the impact it could have on our world. Through a comprehensive analysis, we will illuminate the ways in which the Emerald Deal could redefine our approach to environmental policy, economic development, and social equity.

One of the key components of the Emerald Deal is its emphasis on reengineering communication strategies (Seebacher, 2022) to foster a more informed and engaged society. Misinformation and conspiracy theories, particularly those undermining climate change science or opposing environmental initiatives, pose a significant threat to societal consensus (Lierse, 2022) on sustainability. The Emerald Deal proposes the use of advanced communication concept, automation tools and techniques, leveraging AI and machine learning for social listening (Stewart & Arnold, 2018), sentiment analysis (Medhat et al., 2014), and the identification and counteraction of false information. Seebacher's (2022: 84, 99) Communication Maturity Model (Figure 1) for corporates and organizations is the result of a research work with more than 1,000 organizations during 2021 and 2022 and provides a comprehensive, evidence-based framework for the realization of the Emerald Deal.

The model provides detailed measures for organization to effectively develop from one-dimensional reactive communication towards multidimensional-interagil interaction. This proactive approach in communication is essential for building public trust and consensus, which are crucial for the successful implementation of environmental policies. This model, in combination with the triangle of trust concept by Frei and Morriss (2020), provides the indispensable interactional components for sustainably and successfully master the Green Deal. Without these elements we will not be able to master the societal change of the forthcoming years for future generations.

The rise of populism and political polarization presents another challenge that the Green Deal alone does not fully address. The Emerald Deal, with its focus on inclusive and participatory decision-making (Bell & Reed, 2022), seeks to bridge the gap between diverse societal groups. By involving communities in the decision-making process and ensuring that their voices are heard and valued, this

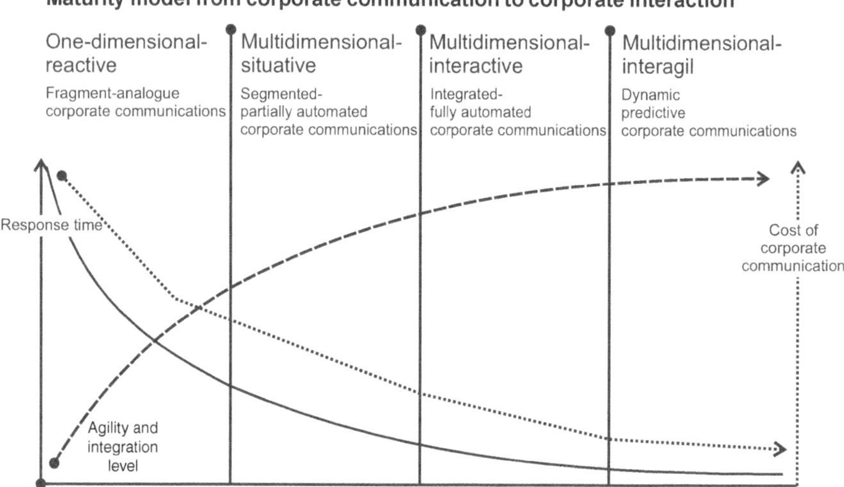

Fig. 1   Maturity Model for Institutional Communication
(*Source*: Seebacher 2022).

approach aims to build a broader societal consensus. Furthermore, it recognizes the importance of addressing the socio-economic concerns of all stakeholders, ensuring that the transition to sustainability is just and equitable. In this context, Seebacher defines the change from 1-2-1 communication on channels towards All-2-All (A2A) 24/7 interaction in arenas. This requires a total shift how, when, and with what we interact. Thus, data-driven interaction has to be established in order to predictively master collective communication and interaction.

Conventional communicative concepts, methods, and tools are no longer sufficient to combat the spread of conspiracy theories, misinformation, and populism. The rise of AI, collective intelligence, and digital twins provides a new opportunity to combat these threats in the context of our FIBS world.

The Role of Collective Intelligence in the Emerald Deal

Collective intelligence is the ability of a group of people to outperform any individual in a given task. In the context of combatting misinformation, collective intelligence can be used to crowdsource fact-checking, identify patterns in misinformation, and develop new strategies for communication.

Collective intelligence can be a powerful tool for reframing and optimizing community engagement and interaction in the context of reengineering corporate communication as part of the shift from the Green Deal to the new Emerald Deal. Here are some specific ways that collective intelligence can be used to achieve these goals:

1. **Gather and analyze insights from a broad range of stakeholders:** Collective intelligence can help to gather and analyze insights from a broad range of stakeholders, including employees, customers, partners, and community members. This can be done through surveys, social media listening, and other methods. By understanding the needs, wants, and concerns of these stakeholders, companies can better tailor their communication strategies to resonate with them.

2. **Identify and prioritize key issues:** Collective intelligence can also be used to identify and prioritize key issues that are important to stakeholders. This can be done by analyzing sentiment data, conducting focus groups, and other methods. By focusing on the most important issues, companies can ensure that their communication is focused and effective.

3. **Develop and test new communication strategies:** Collective intelligence can help to develop and test new communication strategies. This can be done by crowdsourcing ideas, conducting experiments, and gathering feedback. By using a data-driven approach, companies can ensure that their communication is fresh, relevant, and engaging.

4. **Measure and track the effectiveness of communication campaigns:** Collective intelligence can also help to measure and track the effectiveness of communication campaigns. This can be done by collecting data on engagement metrics, such as website traffic, social media shares, and customer sentiment. By analyzing this data, companies can identify what is working well and what needs to be improved.

5. **Foster a sense of community:** Collective intelligence can also be used to foster a sense of community among stakeholders. This can be done by creating online forums, hosting events, and encouraging collaboration. By creating a sense of community, companies can build stronger relationships with their stakeholders and gain their trust.

In addition to these specific ways, collective intelligence can also help to reframe and optimize community engagement and interaction in general by:

- **Democratizing the communication process:** Collective intelligence can help to democratize the communication process by giving stakeholders a voice in how they are communicated with. This can lead to more inclusive and representative communication strategies.

- **Enhancing transparency:** Collective intelligence can enhance transparency by making it easier for stakeholders to see how their feedback is being used. This can build trust and credibility with stakeholders.

- **Accelerating innovation:** Collective intelligence can accelerate innovation by bringing together a wide range of perspectives and expertise. This can lead to the development of more creative and effective communication strategies.

By leveraging collective intelligence, companies can reframe and optimize their community engagement and interaction strategies to better align with the shift from the Green Deal to the new Emerald Deal. This can help them to build stronger relationships with stakeholders, foster innovation, and achieve their sustainability goals.

How Digital Twins Facilitate the Emerald Deal

Digital twins are virtual representations of physical objects or systems. In the context of misinformation, digital twins can be used to track the spread of misinformation, identify its sources, and develop targeted interventions.

Digital twins can play a crucial role in reframing and optimizing community engagement and interaction in the context of reengineering corporate communication as part of the shift from the Green Deal to the new Emerald Deal. Here are some specific ways that digital twins can be used to achieve these goals:

1. **Creating personalized and immersive experiences:**　Digital twins can be used to create personalized and immersive experiences for stakeholders, such as virtual tours of facilities, interactive simulations, and customized messaging. This can help to connect stakeholders on a deeper level and make them feel more informed and engaged.
2. **Facilitating collaboration and crowdsourcing:**　Digital twins can provide a platform for stakeholders to collaborate and share ideas. This can be done through online forums, chat rooms, and other interactive tools. By enabling collaboration, companies can gather valuable feedback and identify new opportunities for innovation.
3. **Tracking and analyzing stakeholder engagement:**　Digital twins can be used to track and analyze stakeholder engagement in real-time. This data can be used to identify trends, measure the effectiveness of communication campaigns, and make data-driven decisions.
4. **Providing a centralized hub for information:**　Digital twins can serve as a centralized hub for information about the company, its products, and its sustainability commitments. This can make it easier for stakeholders to find the information they need and stay up-to-date on the latest developments.
5. **Promoting transparency and accountability:**　Digital twins can help to promote transparency and accountability by giving stakeholders access to real-time data and information. This can help build trust and credibility with stakeholders.
6. **Fostering a sense of community:**　Digital twins can help to foster a sense of community among stakeholders by providing a shared virtual space for interaction and collaboration. This can help to build stronger relationships with stakeholders and encourage a sense of shared purpose.
7. **Measuring and tracking the impact:**　Digital twins can help to measure and track the impact of corporate communication campaigns on stakeholder engagement and satisfaction. This data can be used to identify what is working well and what needs to be improved.

8. **Benchmarking against industry standards:**   Digital twins can be used to benchmark a company's community engagement and interaction strategies against industry standards. This can help to identify areas for improvement and ensure that the company is using best practices.

9. **Continuously refining and optimizing:**   Digital twins can be used to continuously refine and optimize community engagement and interaction strategies. This can be done by analyzing data, identifying trends, and making adjustments as needed.

10. **Empowering stakeholders to shape the future:**   Digital twins can empower stakeholders to shape the future of the company by providing them with opportunities to provide feedback and suggestions. This can help to ensure that the company's communication strategies are aligned with the needs and interests of its stakeholders.

How Artificial Intelligence Positively Impacts the Emerald Deal?

AI can be used to automate the detection and analysis of misinformation. This can help to free up human experts to focus on more complex tasks, such as developing new strategies for communication and engaging with audiences who may be susceptible to misinformation.

AI has the potential to revolutionize the way companies communicate with their stakeholders, helping to reframe and optimize their community engagement and interaction. Here are some specific ways that AI can be used to achieve these goals:

1. **Personalize communication:**   AI can be used to personalize communication with stakeholders, tailoring messages to their individual needs, interests, and preferences. This can be done by using natural language processing (NLP) to analyze data from social media, surveys, and other sources. By personalizing communication, companies can make their messaging more relevant and engaging.

2. **Predict stakeholder sentiment:**   AI can be used to predict stakeholder sentiment, helping companies to anticipate potential issues and address them proactively. This can be done by analyzing social media data, sentiment analysis, and other sources. By predicting stakeholder sentiment, companies can avoid surprises and maintain positive relationships with their stakeholders.

3. **Anticipate stakeholder needs:**   AI can be used to anticipate stakeholder needs, helping companies to identify new opportunities for engagement and collaboration. This can be done by analyzing data from customer support tickets, social media interactions, and other sources. By anticipating stakeholder needs, companies can demonstrate their commitment to stakeholder satisfaction and build stronger relationships.

4. **Detect and resolve conflicts:**   AI can be used to detect and resolve conflicts between stakeholders, helping companies to maintain a harmonious and productive environment. This can be done by analyzing online conversations,

identifying patterns of conflict, and suggesting mediation or other resolution strategies. By detecting and resolving conflicts, companies can avoid negative publicity and maintain a positive reputation.

5. **Facilitate collaboration and crowdsourcing:**   AI can be used to facilitate collaboration and crowdsourcing among stakeholders, helping companies to gather valuable feedback and ideas. This can be done by using chatbots, virtual assistants, and other AI-powered tools. By facilitating collaboration, companies can tap into the collective intelligence of their stakeholders and identify new opportunities for innovation.

6. **Enhance transparency and accountability:**   AI can be used to enhance transparency and accountability, making it easier for companies to track and demonstrate their performance against sustainability goals. This can be done by using AI-powered dashboards and analytics tools. By enhancing transparency, companies can build trust and credibility with stakeholders.

7. **Automate manual tasks:**   AI can be used to automate manual tasks, freeing up employees to focus on more strategic and creative activities. This can be done by using AI-powered chatbots, virtual assistants, and other tools. By automating manual tasks, companies can improve efficiency and productivity, allowing them to invest more in community engagement and interaction.

8. **Measure and track the impact:**   AI can be used to measure and track the impact of corporate communication campaigns on stakeholder engagement and satisfaction. This data can be used to identify what is working well and what needs to be improved. By measuring and tracking impact, companies can demonstrate the value of their communication efforts and justify their investment in AI.

By leveraging AI, companies can reframe and optimize their community engagement and interaction strategies to better align with the shift from the Green Deal to the new Emerald Deal. This can help them to build stronger relationships with stakeholders, foster innovation, and achieve their sustainability goals.

What Roles Does Communication Automation Play?

Communication automation can be used to disseminate accurate information and engage with audiences in real time. This can help to counter the speed at which misinformation can spread online.

Communication automation is a critical aspect of reengineering corporate communication as part of the shift from the Green Deal to the new Emerald Deal. By automating manual tasks and streamlining processes, companies can improve efficiency, productivity, and responsiveness, enabling them to engage with stakeholders more effectively.

Here are some specific ways that communication automation can help to reframe and optimize community engagement and interaction:

1. **Automate repetitive tasks:**   Communication automation can automate repetitive tasks such as responding to customer support inquiries, scheduling meetings, and generating reports. This frees up employees to focus on more strategic and creative activities, such as developing engaging content and building relationships with stakeholders.
2. **Streamline processes:**   Communication automation can streamline processes, such as the approval of content and the distribution of marketing materials. This can help to reduce errors and ensure that information is delivered to stakeholders in a timely and consistent manner.
3. **Improve efficiency:**   Communication automation can improve efficiency by automating tasks that would otherwise be done manually. This can save companies time and money, which can be reinvested in community engagement and interaction efforts.
4. **Enhance responsiveness:**   Communication automation can enhance responsiveness by enabling companies to respond to stakeholder inquiries and requests quickly and efficiently. This can help to build trust and credibility with stakeholders.
5. **Personalize communication:**   Communication automation can personalize communication by using data to understand individual stakeholder preferences and behaviors. This can help companies to tailor their messaging and interactions to the specific needs of each stakeholder.
6. **Gather and analyze feedback:**   Communication automation can gather and analyze feedback from stakeholders through surveys, chatbots, and other tools. This data can be used to identify areas for improvement and make data-driven decisions about communication strategies.
7. **Measure and track results:**   Communication automation can measure and track the results of communication campaigns, such as engagement metrics, sentiment analysis, and social media mentions. This data can be used to identify what is working well and what needs to be improved.
8. **Facilitate collaboration:**   Communication automation can facilitate collaboration among stakeholders by providing a platform for sharing information, exchanging ideas, and working together on projects. This can help to build stronger relationships and foster innovation.
9. **Promote transparency:**   Communication automation can promote transparency by making it easier for stakeholders to access information about the company, its products, and its sustainability commitments. This can help to build trust and credibility with stakeholders.
10. **Empower employees:**   Communication automation can empower employees by automating tasks that would otherwise be done manually, freeing them up to focus on more strategic and creative activities. This can help to create a more engaged and productive workforce.

By leveraging communication automation, companies can reframe and optimize their community engagement and interaction strategies to better align with the shift from the Green Deal to the new Emerald Deal. This can help them to build stronger relationships with stakeholders, foster innovation, and achieve their sustainability goals.

Here are some specific examples of how companies are using communication automation to reframe and optimize community engagement and interaction:

- Coca-Cola is using chatbots to answer customer questions and provide support.
- Nestlé is using social media listening tools to identify and respond to customer feedback.
- P&G is using predictive analytics to identify potential issues and address them proactively.
- IBM is using virtual assistants to provide personalized customer service.
- Amazon is using AI to personalize product recommendations and customer service.

As communication automation technology continues to develop, we can expect to see even more innovative ways for companies to engage with their stakeholders. This will help to create a more connected and responsive corporate communication landscape, which will be essential for companies to thrive in the years to come.

By combining these technologies, we can create a new approach to combating misinformation that is more effective and scalable than traditional methods. This new approach is necessary to address the growing threat of misinformation and disinformation in the digital age. Combining collective intelligence, digital twins, AI, and communication automation can have a profound impact on community engagement and interaction in the context of realizing the Emerald Deal. By leveraging these technologies and concepts, companies can proactively address stakeholder concerns, foster trust and credibility, and drive positive change. Here are some specific effects of combining these technologies:

1. **Enhanced understanding of stakeholder needs and expectations:** Collective intelligence tools can aggregate data from a variety of sources, including social media, surveys, and customer interactions, to provide companies with a comprehensive understanding of stakeholder needs and expectations. This information can be used to tailor communication strategies and initiatives to better resonate with stakeholders.

2. **Personalized and targeted engagement:** Digital twins and AI can be used to create personalized and targeted communication experiences for stakeholders. For example, chatbots can provide real-time support and answer questions based on individual preferences. Virtual assistants can schedule appointments, provide reminders, and facilitate collaboration.

3. **Proactive identification and resolution of issues:** AI can be used to analyze data from social media, customer support interactions, and other sources to identify potential issues before they escalate. By proactively addressing these issues, companies can build trust and credibility with stakeholders.

4. **Enhanced transparency and accountability:** Digital twins can provide a centralized hub for information about the company, its products, and its

sustainability commitments. This information can be accessed by stakeholders in real-time, promoting transparency and accountability.

5. **Streamlined and efficient collaboration:**   Communication automation tools can streamline and automate tasks, such as scheduling meetings, generating reports, and responding to inquiries. This frees up employees to focus on more strategic and creative activities, such as building relationships with stakeholders.

6. **Data-driven decision-making:**   AI can be used to analyze data from various sources, such as social media, customer support interactions, and website traffic, to identify trends and patterns. This information can be used to inform communication strategies and make data-driven decisions.

7. **Increased engagement and participation:**   By combining these technologies, companies can create more engaging and participatory experiences for stakeholders. This can lead to increased engagement, participation, and satisfaction.

8. **Fostered sense of community:**   Digital twins and AI can be used to create virtual communities where stakeholders can connect, collaborate, and share ideas. This can foster a sense of community and encourage stakeholder engagement.

9. **Enhanced reputation and brand image:**   By demonstrating its commitment to stakeholder engagement and sustainability, a company can enhance its reputation and brand image. This can lead to increased customer acquisition, loyalty, and advocacy.

10. **Achieved sustainability goals:**   By leveraging these technologies to foster stakeholder engagement and collaboration, companies can make progress towards achieving their sustainability goals. This can lead to a more sustainable future for the planet and its inhabitants.

In conclusion, combining collective intelligence, digital twins, AI, and communication automation can transform community engagement and interaction, enabling companies to better understand, connect with, and drive positive change among stakeholders. By leveraging these technologies, companies can play a leading role in realizing the Emerald Deal and creating a more sustainable future.

Fostering Societal Acceptance

Effective change management (Clegg & Walsh, 2004) is pivotal in the transition from the Green Deal to the Emerald Deal. This involves not only the implementation of new technologies and policies but also a cultural shift in how societies perceive and interact with their environment. The Emerald Deal advocates for an integrative approach to change management, one that encompasses education, community involvement (Batson et al., 2002), and the continuous adaptation of strategies in response to evolving challenges and opportunities.

In FIBS world where fake news and conspiracy theories can quickly spread and gain traction, the Emerald Deal approach to societal communication and information becomes ever more critical. By employing sophisticated data analysis (Pavlo et al., 2009) and AI-driven fact-checking systems (Walker et al., 2023), alongside transparent and responsible communication practices (Kitchen et al., 2001), this strategy aims to counteract misinformation effectively. Moreover, by engaging with the public through various channels and fostering an environment of open dialogue, the Emerald Deal seeks to dismantle conspiracy theories and build a foundation of factual, science-based understanding.

Achieving societal acceptance (Upham et al., 2015) is essential for the success of any large-scale environmental initiative. The Emerald Deal recognizes this and places a strong emphasis on education, public awareness campaigns, and community building (McNeely, 1999) and engagement. By involving citizens in sustainability projects, providing clear and accessible information, and demonstrating the tangible benefits of environmental actions, this approach aims to foster a deeper understanding and acceptance of sustainability measures.

The Emerald Deal represents a comprehensive and adaptive approach to sustainability, one that extends beyond environmental policy to encompass societal dynamics (Kleber et al., 2013), communication strategies, and change management. By integrating collective intelligence and swarm systems, it offers innovative solutions to complex environmental challenges, while also addressing the critical issues of misinformation, societal polarization (Levin et al., 2021), and the need for inclusive decision-making. As we navigate the challenges of the 21st century, the Emerald Deal provides a blueprint for a more sustainable, informed, and cohesive society (Perlik, 2018), firmly grounded in the principles of ethical and responsible stewardship of our planet.

Traditional environmental strategies have played a pivotal role in shaping our understanding and response to environmental challenges. Historically, these strategies have been characterized by a focus on specific issues such as pollution reduction, conservation of natural resources, and compliance with environmental regulations. While these approaches have yielded some success, they also exhibit limitations that necessitate a reevaluation in the face of contemporary environmental challenges.

From Challenges to Transition

Transitioning from traditional to more innovative environmental approaches encompass several challenges. These include resource allocation, adapting to policy shifts, and overcoming stakeholder resistance especially in the area of societal engagement and interaction. Allocating sufficient resources, both financial and human, is crucial for implementing new technologies and strategies. This often requires significant investment, which can be a barrier, particularly for less developed regions or smaller organizations (Sachs, 2015).

Policy shifts are another critical challenge. Moving towards more integrated environmental strategies combined with societal interaction measures necessitates changes in existing policies and regulatory frameworks. This can be a slow process, hindered by bureaucratic inertia and the need for consensus among diverse stakeholders (Meadows, 2008) but also the way how politicians are communicative and informing. Furthermore, there can be resistance from stakeholders who are accustomed to traditional approaches or who might be adversely affected by the new strategies, such as industries reliant on fossil fuels (Oreskes & Conway, 2010).

To overcome these challenges, a multifaceted strategy is needed. Financial incentives, such as grants and subsidies, can encourage the adoption of new technologies. Education and awareness campaigns can help shift public opinion and reduce resistance. Engaging stakeholders in the decision-making process ensures their concerns are addressed, which can facilitate smoother policy transitions (Reed, 2008).

Looking forward, the future of environmental strategies seems to be firmly rooted in integrated and interconnected solutions. The growing recognition of the interdependence of ecological, social, and economic systems is likely to drive further innovation in this area (Folke et al., 2005). Emerging technologies such as AI, IoT, and big data analytics will continue to play a critical role in advancing these integrated approaches.

There is a clear call to action for stakeholders, marketers and communicators, policymakers, and communities to actively participate in this transition. Embracing innovative and interconnected solutions is not just a matter of environmental responsibility but also a pathway to sustainable development that can benefit all sectors of society (United Nations, 2015).

In conclusion, transitioning to more innovative and interconnected environmental strategies is essential for addressing the complex and multifaceted challenges of our time. While there are significant challenges in this transition, including resource allocation, policy shifts, and stakeholder resistance, strategies exist to overcome these barriers. The future of environmental sustainability lies in integrated approaches that leverage the power of technology and embrace the interconnectedness of our world. The urgency and necessity of this shift cannot be overstated, as it is critical for achieving a sustainable and resilient future for all.

The Role of Collective Intelligence

In the context of the above thoughts and aspects, the transition from the principles of the Green Deal to the more expansive Emerald Deal represents a significant shift in the approach to environmental sustainability. Central to this transition is the evolution of collective intelligence, a concept that has gained substantial traction in addressing complex environmental challenges. As part of the transition from the Green Deal towards the Emerald Deal, the advancements in the area of collective intelligence can substantial leverage the entire societal communication and interaction dimension.

The Green Deal laid a foundation focusing on specific environmental issues like pollution control and conservation, primarily driven by top-down regulatory policies. These strategies, though effective in their time, often operated in silos, failing to address the interconnected nature of ecological systems. The Green approach was more about compliance and less about collaborative problem-solving, which is crucial in today's environmental context.

The Emerald Deal now broadens this perspective by integrating the principles of collective intelligence in order to enable the enrichment of the purely content and technology driven Green Deal ideology towards a more societal and change management one. Collective intelligence refers to the enhanced capacity achieved when diverse groups work together to solve problems, leveraging shared knowledge and diverse perspectives (Malone & Bernstein, 2015). In the realm of the Emerald Deal, this means that using collective intelligence an A2A 24/7 societal interaction can be designed, managed, and monitored which is of utmost significance for the sustainable efficient and effective deployment of the so urgently needed transition from fossil to green.

Modern environmental challenges are characterized by their complexity and interconnectedness. This must be explained thoroughly and properly to ensure the adequate societal acceptance and understanding. Issues like climate change, biodiversity loss, and sustainable resource management are not only global in scale but also involve multifaceted interactions between natural and human systems. Collective intelligence offers a framework for understanding and addressing these complexities in the right context, at the right time, and via the right channel or arena to the different information clients. By harnessing diverse knowledge and perspectives, more holistic and innovative solutions can be developed and communicatively managed that are beyond the reach of individual entities or traditional approaches.

The use of collective intelligence in environmental decision-making is exemplified in global initiatives like the Intergovernmental Panel on Climate Change (IPCC). The IPCC gathers experts worldwide to assess and synthesize scientific knowledge on climate change, providing comprehensive insights that inform global policy and action (IPCC, 2021). Similarly, the participatory approach in managing the Great Barrier Reef in Australia, involving scientists, local communities, and policymakers, has been crucial in developing effective conservation strategies (Great Barrier Reef Marine Park Authority, 2020). What all these initiatives fail to do, however, is to apply the use of collective intelligence to social interaction and acceptance management.

Integrating Technology with Collective Intelligence

The role of technology in enhancing collective intelligence for societal interaction management cannot be overstated. Advances in data analytics, AI, and communication sciences and platforms have significantly expanded the capacity for collective intelligence. These technologies facilitate the gathering and processing of vast environmental data, enable predictive modeling, and foster real-time collaboration among diverse groups but also 24/7 A2A predictive interaction management and intelligence (Seebacher 2022). For instance, platforms like Global

Forest Watch utilize satellite technology and crowdsourced data to monitor and manage forest resources globally, demonstrating the power of technology-enhanced collective intelligence in environmental conservation (Global Forest Watch, 2021). But the real added value as part of strategic initiatives in an FIBS world is their much deeper and stringent integration in societal communication and interaction management. Only then the adequate level of transparency will develop again building the basis for an authentic, empathetic, and logic societal interaction management leveraging the outcome and success of these important initiatives.

Despite its potential, implementing collective intelligence in environmental strategies enabling the transition towards the Emerald Deal is not without challenges. It requires an overarching communicative framework that supports collaboration while respecting diverse perspectives and interests. There's also the need to balance technological inputs with human insights to avoid over-reliance on data-driven approaches that may overlook local knowledge and socio-cultural factors.

Looking forward, the integration of collective intelligence into environmental strategies in the context of the Emerald Deal offers a promising path to addressing the daunting challenges of our FIBS time which advocates for more integrated, adaptive, and participatory environmental interaction and management practices. As we advance, the focus should be on fostering platforms and policies that enhance collaborative intelligence, integrating technological as well as societal interactional advancements with human insights, and ensuring that diverse voices are heard and valued in the quest for sustainable solutions.

Thus, the evolution from Green to Emerald in environmental strategies represents a critical shift towards leveraging collective intelligence for leveraging any kind of green initiative by its communicative and interactional dimension. This shift acknowledges the complexity and interconnectedness of modern environmental challenges and the necessity of collaborative, interdisciplinary approaches to address them. By embracing the principles of collective intelligence, augmented by technological advancements, we can develop more effective, inclusive, and sustainable solutions to the environmental crises of the 21st century as we can assure the required and ever more essential adequate societal communication, engagement, and information.

Integrating Advanced Technologies

AI and ML are at the forefront of this technological integration, offering powerful tools for environmental monitoring, analysis, and decision-making but also societal predictive interaction management. AI algorithms can process vast amounts of environmental and societal data, identify patterns, and make predictions that are invaluable for conservation and management practices (Rolnick et al., 2019). For example, AI has been instrumental in climate modeling, providing more accurate forecasts of climate patterns and their potential impacts (Abeshu & Chilamkurti, 2018). In the context of the works of Seebacher (2022) and Brunner (2023), the parallel and sequential combination of AI for predictive intelligence has opened up completely new opportunities in the area of predictive interaction management using interaction intelligence for automated alerts generation in case of emerging communicative crisis.

ML, a subset of AI, further enhances these capabilities. It can learn from data, adapt to new information, and improve over time, making it ideal for dynamic and complex environmental systems. A practical application of ML is seen in wildlife conservation, where algorithms analyze camera trap images to monitor animal populations and biodiversity (Norouzzadeh et al., 2018). ML in the context of communication intelligence was leveraged by the Austrian scientist Wolfgang Ecker-Lala as part of this works on the European Thinkers.ai application (Figure 2). With this tool 24/7, automatically any topic and use case can be monitored. This information is then forwarded to any kind of interactional intelligence engine then combining and evaluating this information with parameters from other sources for then deciding on how to predictively take measures to avoid a communicative crisis or leave the communicative flow fading out uninfluenced.

Swarm systems, inspired by the collective behavior of natural organisms like birds and insects, represent a new frontier in societal interaction efforts. These systems involve numerous agents working together to perform tasks, exhibiting a collective intelligence that can be applied to environmental interaction management. For instance, swarm robotics can be used as part of communication intelligence systems for 24/7 collecting communicational behaviors. This information can then be evaluated and tracked against pre-defined alert-thresholds for defining whether an intervention is required or not. Also, such swarm robotics can be used to monitor the development of keyword and context quantities in specially defined communities in order to then define the right wording and tone for specific messaging as part of environmental initiatives. Social Listening (Ballestar, 2020) and Community Building (Stewart & Arnold, 2018) are the terms and techniques also of relevance against this background.

Swarm intelligence also extends to decentralized energy systems, where multiple renewable energy sources work together, adjusting to changes in demand and supply. This approach enhances the resilience and efficiency of renewable energy systems, contributing to more sustainable energy management (Kammen & Sunter, 2016). The knowledge gains from the research on swarm intelligence in decentralized energy systems can be used for the societal communicative aspect in the context of collective intelligence in a number of ways.

One way is to use this knowledge to develop new communication protocols and algorithms that are more efficient and resilient than traditional ones. For example, swarm intelligence-inspired algorithms could be used to route traffic in communication networks, manage energy consumption in data centers, or distribute computing resources in cloud computing systems. This could lead to significant improvements in the performance, efficiency, and reliability of these systems.

Another way to use this knowledge is to develop new tools and platforms that enable people to collaborate and share information more effectively. For example, swarm intelligence-inspired algorithms could be used to develop new

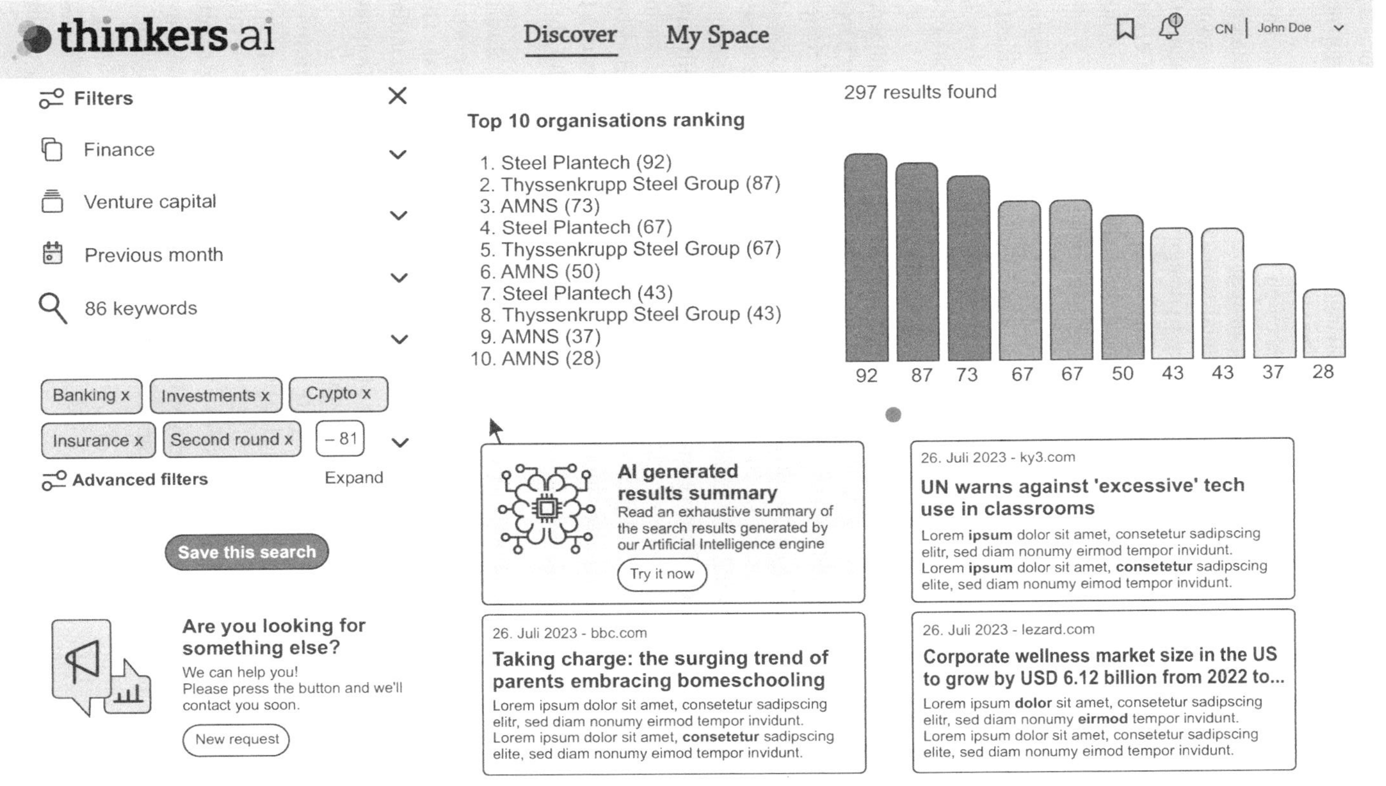

Fig. 2   Machine Learning based Communication Intelligence
(*Source:* thinkers.ai).

social media platforms that are better at filtering out misinformation and promoting civil discourse. Or, swarm intelligence-inspired tools could be used to develop new educational platforms that are more personalized and effective.

Finally, the knowledge gained from research on swarm intelligence can also be used to develop new ways to incentivize people to participate in collective intelligence initiatives as part of environmental projects. For example, swarm intelligence-inspired algorithms could be used to design new reward systems for crowdsourcing platforms or to develop new ways to distribute ownership of data and resources in decentralized systems. Here are some specific examples of how swarm intelligence could be used to improve societal communication:

- **Developing more resilient communication networks:**   Swarm intelligence could be used to develop communication networks that are more resilient to disruptions, such as natural disasters or cyberattacks. This could be done by using swarm intelligence algorithms to route traffic around disruptions or to dynamically allocate resources to critical nodes in the network.
- **Improving the efficiency of content delivery networks:**   Swarm intelligence could be used to improve the efficiency of content delivery networks (CDNs), which are used to deliver streaming video, music, and other types of content to users around the world. Swarm intelligence algorithms could be used to optimize the placement of CDN servers and to route traffic to the closest servers, which could reduce latency and improve performance for users.
- **Developing more effective social media platforms:**   Swarm intelligence could be used to develop social media platforms that are better at filtering out misinformation and promoting civil discourse. This could be done by using swarm intelligence algorithms to identify and flag harmful content, to promote high-quality content, and to connect users with people who share their interests.
- **Creating more personalized and effective educational platforms:**   Swarm intelligence could be used to create educational platforms that are more personalized and effective for society as part of the Emerald Deal. This could be done by using swarm intelligence algorithms to track understanding and acceptance progress and to recommend personalized information and communication materials and activities.
- **Incentivizing people to participate in collective intelligence initiatives:** Swarm intelligence could be used to develop new ways to incentivize people to participate in collective intelligence initiatives, such as crowdsourcing platforms or decentralized systems. This could be done by using swarm intelligence algorithms to design new reward systems or to develop new ways to distribute ownership of data and resources.

These are just a few examples of how swarm intelligence could be used to improve societal communication. As research on swarm intelligence continues, we can expect to see even more innovative and effective applications of this technology in the future.

Overcoming Challenges with Technology

Despite the potential of these technologies, deploying them for transitioning towards the Emerald Deal is not without challenges. There is a need for robust data infrastructure, technical expertise, and policies that support technology integration while addressing ethical and privacy concerns. Additionally, the risk of technological dependency and the potential for overlooking traditional knowledge and local solutions must be managed (Sachs, 2015).

Advanced technologies like AI, ML, and swarm systems play a critical role in enhancing societal communication efforts. They offer innovative solutions to complex communicative challenges and have the potential to transform traditional approaches to environmental management into much stronger and better accepted community-driven efforts. As we move towards a more sustainable future, it is imperative to harness these technologies responsibly and effectively, integrating them with traditional knowledge and ensuring their benefits are widely accessible in both dimensions of the Emerald Deal, the technology-content and also the societal interactional one.

The concept of swarm systems, rooted in the principles of swarm intelligence, is revolutionizing the approach to environmental stewardship. By leveraging the coordinated efforts of multiple agents, swarm systems are enhancing the efficiency and effectiveness of sustainability efforts.

1. **Swarm Intelligence: A New Paradigm in Environmental Monitoring:** Swarm intelligence refers to the collective behavior of decentralized, self-organized systems, typically natural or artificial, that work collaboratively to achieve specific goals. In environmental stewardship, swarm systems manifest as groups of sensors, robots, or drones that operate in coordination to monitor and manage ecological systems. These systems are particularly effective in large-scale and complex tasks where individual efforts are insufficient or impractical. For instance, in forest monitoring, swarm drones can be deployed to cover vast areas, collecting data on tree health, detecting illegal logging activities, or monitoring wildlife populations. These drones can communicate and collaborate, ensuring comprehensive and efficient coverage of the monitored area (Bayindir, 2016). Similarly, in oceanography, swarm robotics is used for tasks like seabed mapping, water quality monitoring, and studying marine life, which are otherwise challenging due to the vastness and depth of the oceans (Berlinger et al., 2018).

2. **Swarm Systems in Pollution Control and Cleanup Operations:** Swarm systems have shown remarkable potential in pollution control and cleanup operations. For example, in water bodies affected by oil spills or chemical contaminants, swarm robots can be employed to detect pollution sources, assess the extent of contamination, and even participate in the cleanup process. These robots can work tirelessly, covering larger areas more effectively than human-led efforts. Furthermore, they can operate in hazardous environments, reducing risks to human health (Sutantyo et al., 2010).

3. **Innovative Approaches to Renewable Energy Management:**   In the realm of renewable energy, swarm intelligence is playing a transformative role. One of the challenges with renewable energy sources, like wind and solar power, is their variability and dependence on environmental conditions. Swarm systems can help manage this variability. For instance, decentralized energy systems, where multiple renewable energy sources work together as a swarm, can adjust to fluctuations in energy demand and supply, enhancing the resilience and efficiency of the power grid (Kammen & Sunter, 2016).

4. **Agricultural Efficiency through Swarm Systems:**   Agriculture is another area where swarm systems are making significant contributions. Precision agriculture, which involves the use of advanced technologies to optimize field management and crop production, increasingly relies on swarm technologies. Swarm drones can perform tasks such as soil analysis, crop monitoring, and targeted pesticide application. By providing precise data and performing specific tasks, these systems reduce resource waste and increase agricultural efficiency.

Despite their potential, the implementation of swarm systems in environmental stewardship is not without challenges. Technical complexities, the need for robust communication networks, and the integration of these systems into existing environmental management frameworks are some of the hurdles. Additionally, there are concerns regarding privacy, ethical considerations, and the potential impact on employment in sectors traditionally reliant on human labor.

Successful Applications of Swarm Intelligence can be seen by different recent projects such as:

- **Forest Monitoring Using Drones:**   In tropical rainforests, where monitoring is crucial for conservation efforts, swarm drones have been used effectively for real-time monitoring and data collection, significantly enhancing forest management practices.

- **Ocean Cleanup Projects:**   Projects like The Ocean Cleanup have experimented with autonomous vehicles that work in swarms to collect plastic waste from oceans, demonstrating the potential of swarm systems in large-scale environmental restoration efforts.

Looking ahead, the role of swarm systems in environmental stewardship is set to grow. As these technologies become more advanced and accessible, their application in various environmental domains will likely expand. The future may see more integrated, intelligent swarm systems capable of autonomously executing complex environmental management tasks, from biodiversity conservation to climate change mitigation efforts.

In conclusion, the emergence of swarm systems represents a significant advancement in environmental stewardship. These systems offer new possibilities for environmental monitoring, management, and the implementation of innovative sustainability solutions. By harnessing the power of collective behavior and decentralized decision-making, swarm intelligence can address environmental challenges more effectively and efficiently. As we continue to explore and refine these technologies, the potential for their positive impact on the planet's ecological systems is immense.

The Emerald Deal Impact on Society

The Emerald Deal represents an advanced paradigm in environmental sustainability, transcending the foundational aspects of the Green Deal towards the new challenges of our FIBS world. The synergy between the Green Deal's environmental focus and the broader scope of the Emerald Deal creates a robust framework for tackling contemporary societal challenges. Central to this approach is the aspect of change management and trust-building communication. In an era where misinformation and public skepticism are rampant, establishing trust through transparent and effective communication is crucial. Predictive intelligence, another facet of the Emerald Deal, plays a key role in anticipating environmental trends, enabling proactive decision-making and policy formulation.

Reengineering corporate communication is also integral to this approach. By aligning corporate strategies with sustainable practices and transparent communication, businesses can lead by example, fostering a culture of sustainability and responsibility. This alignment is particularly relevant in combating misinformation and populism, where businesses and governments can act as sources of reliable and accurate information.

The impact of the Emerald Deal extends beyond environmental sustainability to influence broader societal dynamics. By promoting collaborative and interconnected approaches, this vision encourages a more participatory form of governance. Involving communities in decision-making processes and leveraging collective intelligence can lead to more inclusive and effective solutions to societal challenges.

Moreover, the use of advanced technologies in disseminating accurate information plays a pivotal role in countering fake news and conspiracy theories. For instance, AI and big data can be employed to track and debunk false information, particularly regarding climate change and environmental policies. This technology-driven approach, combined with effective communication strategies, can foster a more informed and rational public discourse.

Addressing global warming remains one of the most pressing challenges of our time, and the Emerald Deal offers innovative strategies in this regard. Swarm systems can enhance climate monitoring and data collection, providing accurate and comprehensive information crucial for climate action. Predictive models, powered by AI and ML, can forecast climate trends, informing adaptation and mitigation strategies.

In the fight against environmental misinformation, the principles of the Emerald Deal offer effective tools. By harnessing AI for social media monitoring and fact-checking, it is possible to identify and correct false narratives about climate change. Educational initiatives and public awareness campaigns, integral to this vision, further reinforce these efforts, promoting scientific literacy and critical thinking among the public.

The rise of disinformation, fake news, and populism poses significant challenges to environmental sustainability and societal stability. The Emerald Deal's emphasis on trust-building communication and reengineered corporate communication is crucial in this context. Corporations and governments can lead by setting standards for transparency and accountability in information dissemination. By providing accurate, science-based information and engaging with the public through various channels, these entities can build trust and counteract the influence of misinformation.

Combining collective intelligence, digital twins, AI, and communication automation can have a profound impact on community engagement and interaction in the context of realizing the Emerald Deal. By leveraging these technologies and concepts, companies can proactively address stakeholder concerns, foster trust and credibility, and drive positive change. The Emerald Deal, a revised and updated version of the Green Deals, places a strong emphasis on trust-building communication and interaction. This focus is crucial for achieving the societal impact of the Emerald Deal, which aims to foster a more sustainable and equitable future for all. Here are the cornerstones of the societal impact of the Emerald Deal, emphasizing trust-building communication and interaction:

1. **Fostering Transparency and Accountability:** The Emerald Deal calls for greater transparency and accountability from businesses and governments in their sustainability initiatives. This can be achieved through the use of AI, digital twins, and collective intelligence to track and measure progress, share data openly, and engage with stakeholders in a meaningful way.

2. **Empowering Stakeholder Engagement:** The Emerald Deal recognizes that stakeholder engagement is essential for achieving sustainability goals. By using AI and digital twins to better understand stakeholder needs and concerns, companies and governments can tailor their communication and initiatives to resonate with specific groups. Collective intelligence can be harnessed to gather and analyze feedback from stakeholders, ensuring that their voices are heard and their contributions valued.

3. **Prioritizing Social Justice and Equity:** The Emerald Deal places social justice and equity at the heart of its sustainability agenda. AI and digital twins can be used to identify and address social inequities, ensuring that the benefits of sustainability efforts are shared equitably across society. Collective intelligence can facilitate collaboration and knowledge-sharing among diverse stakeholders, promoting inclusive decision-making.

4. **Building Trust and Credibility:** Trust is a cornerstone of the Emerald Deal. By communicating openly and transparently, companies and governments can build trust with stakeholders and encourage their participation in sustainability efforts. AI, digital twins, and collective intelligence can be used to personalize communication, address concerns proactively, and demonstrate accountability.

5. **Fostering a Sense of Community:** The Emerald Deal seeks to cultivate a sense of community around sustainability. AI and digital twins can be used to create virtual communities where stakeholders can connect, collaborate, and share ideas. Collective intelligence can facilitate knowledge-sharing and peer-to-peer learning, strengthening community bonds.

6. **Empowering Individuals and Communities:** The Emerald Deal empowers individuals and communities to take action for sustainability. AI and digital twins can be used to provide personalized guidance and support, helping individuals make informed choices and take meaningful action. Collective intelligence can harness the collective knowledge and expertise of communities, fostering innovation and collaboration.

By embracing these cornerstones, the Emerald Deal can achieve its societal impact by fostering trust, collaboration, and action among stakeholders. This can lead to a more sustainable, equitable, and prosperous future for all.

The Role of Predictive Intelligence in Societal Decision-making

Predictive intelligence, a key component of the Emerald Deal, can significantly influence societal decision-making. By analyzing trends and forecasting future scenarios, policymakers can make informed decisions that anticipate and address potential challenges. This approach is particularly relevant in areas like public health, urban planning, and environmental policy, where foresight can lead to better outcomes.

Predictive intelligence plays a crucial role in achieving the societal impact of the Emerald Deal by enabling businesses and governments to anticipate potential communicative challenges, identify opportunities, and make informed decisions. This can help to accelerate the transition to a more sustainable and equitable future for all. Predictive Intelligence can also help to anticipate at a very early stage future developments of communital or collaborative behaviour. With this information communicative crisis and shitstorms can be avoided.

Here are some specific ways in which predictive intelligence can be used to enhance the societal impact of the Emerald Deal:

1. **Anticipating and mitigating risks:** Predictive intelligence can be used to analyze data and identify potential risks associated with sustainability initiatives. This can help businesses and governments to proactively address these risks, minimizing their impact and ensuring the success of their efforts. For example, predictive modeling can be used to forecast potential disruptions to supply chains, enabling businesses to plan accordingly and avoid disruptions.

2. **Identifying and seizing opportunities:** Predictive intelligence can also be used to identify emerging trends and opportunities that can be leveraged to further the goals of the Emerald Deal. For instance, AI can be used to analyze social media sentiment and market data to identify new consumer preferences

and market opportunities. This information can inform product development, marketing strategies, and investment decisions.

3. **Personalizing communication and engagement:** Predictive intelligence can be used to personalize communication and engagement with stakeholders, ensuring that they receive timely and relevant information tailored to their specific needs and interests. This can help to foster trust and engagement, and ultimately drive more positive outcomes for sustainability initiatives. For example, chatbots can be used to provide personalized advice and support to individuals based on their environmental footprint and lifestyle choices.

4. **Prioritizing resource allocation:** Predictive intelligence can be used to optimize resource allocation, ensuring that limited resources are directed to the most impactful efforts. This can help to maximize the efficiency of sustainability initiatives and maximize their impact. For example, AI can be used to identify the most vulnerable communities to climate change and allocate resources for adaptation and resilience measures.

5. **Enabling data-driven decision-making:** Predictive intelligence can provide businesses and governments with the insights necessary to make data-driven decisions that align with the goals of the Emerald Deal. This can help to ensure that resources are used wisely and that efforts are focused on the most pressing issues. For instance, ML can be used to analyze data from sensors and other sources to optimize energy consumption and resource utilization.

In conclusion, predictive intelligence plays a critical role in achieving the societal impact of the Emerald Deal. By enabling businesses and governments to anticipate risks, identify opportunities, personalize communication, optimize resource allocation, and make data-driven decisions, predictive intelligence can accelerate the transition to a more sustainable and equitable future for all.

Reengineering Corporate Communication for Societal Impact

Corporate communication strategies must be reengineered to align with the principles of sustainability and transparency and to enable politicians, institutions, and organizations to master the new FIBS world challenges and contingencies. They should not only focus on profit but also consider their environmental and societal impact. By adopting sustainable practices and communicating these efforts openly, corporations can play a pivotal role in driving societal change and combating misinformation.

Reengineering Corporate Communication (RCC) plays a vital role in achieving the societal impact of the Emerald Deal by transforming the way companies communicate with stakeholders, fostering trust, collaboration, and action. This can lead to a more sustainable, equitable, and prosperous future for all. Here are some specific ways in which RCC can be used to enhance the societal impact of the Emerald Deal:

1. **Reframing and optimizing communication strategy:**   RCC can be used to reframe and optimize communication strategies to better align with the goals of the Emerald Deal. This can involve shifting from traditional marketing-focused communication to a more stakeholder-centric approach that prioritizes transparency, accountability, and collaboration.

2. **Embracing technology for enhanced engagement:**   RCC can leverage emerging technologies like AI, digital twins, and collective intelligence to enhance stakeholder engagement. This can involve using chatbots, virtual assistants, and social media platforms to provide personalized and targeted communication experiences.

3. **Fostering trust and credibility:**   RCC can help to foster trust and credibility with stakeholders by demonstrating a genuine commitment to sustainability and open communication. This can involve sharing data, engaging in dialogue, and addressing concerns promptly and transparently.

4. **Promoting transparency and accountability:**   RCC can promote transparency and accountability by providing stakeholders with clear and accessible information about a company's sustainability practices. This can involve publishing sustainability reports, creating open-data platforms, and participating in stakeholder forums.

5. **Facilitating collaboration and innovation:**   RCC can facilitate collaboration and innovation among stakeholders by creating spaces for dialogue, knowledge-sharing, and co-creation. This can involve using online communities, hackathons, and other collaborative platforms.

6. **Measuring and tracking impact:**   RCC can measure and track the impact of communication efforts on stakeholder engagement, satisfaction, and behavior change. This data can be used to refine communication strategies and improve the overall impact of sustainability initiatives.

7. **Empowering stakeholders to take action:**   RCC can empower stakeholders to take action for sustainability by providing them with resources, guidance, and incentives. This can involve creating educational programs, offering volunteering opportunities, and aligning with stakeholders' interests and priorities.

8. **Driving behavioral change:**   RCC can drive behavioral change among stakeholders by promoting sustainable practices and inspiring individuals to make positive choices. This can involve using storytelling, gamification, and other persuasive communication techniques.

In conclusion, RCC plays a critical role in achieving the societal impact of the Emerald Deal. By reframing communication strategies, embracing technology, fostering trust, promoting transparency, facilitating collaboration, measuring impact, empowering stakeholders, and driving behavioral change, RCC can transform the way companies engage with society and contribute to a sustainable future.

Successful Pioneer Transition Projects

The transition from traditional environmental strategies based on the principles of the Green Deal to innovative and interconnected initiatives with proactive communicational measures for positive societal acceptance, engagement, and facilitation, akin to the principles of the Emerald Deal, has unconsciously already been demonstrated in various contexts globally. The following case studies provide valuable insights and evidence based on which the authors have elaborated and further specified the Emerald Deal into how embracing a holistic, integrated approach can yield better results compared to projects that solely focus on the technology and content aspects of the Green Deal —and this even though these Emerald Deal pioneering showcases did not dispose of modern concepts and the linked knowledge gains in the area of collective and predictive intelligence (Seebacher, 2021) or the results of the reengineering societal communication (Seebacher, 2022).

Copenhagen's Transition to a Green and Smart City

Copenhagen, the capital city of Denmark, has become a global leader in urban sustainability and is a prime example of successful implementation of Emerald Deal principles. The city's journey from conventional urban development to becoming one of the world's greenest and smartest cities is remarkable.

Strategy Implementation

- Copenhagen integrated sustainable urban planning with technological innovations, focusing on reducing carbon emissions, enhancing green spaces, and promoting sustainable transportation (City of Copenhagen, 2021).
- The city implemented a district heating system, utilizing waste heat from electricity production, and established one of the world's most efficient waste-to-energy plants.
- Emphasis on cycling infrastructure and electric public transport reduced reliance on fossil fuels.
- The city supported this initiative with proactive and intense communicational measures aiming at "know how" transfer, ensuring transparency and assuring the compliance with trust building communication.

Results

- Copenhagen aims to become the world's first carbon-neutral capital by 2025.
- The city has seen a reduction in carbon emissions by 42% since 2005, despite an increase in population and economic growth (Copenhagen 2025 Climate Plan).

Learning Points

- Integrating technology with sustainable urban planning can significantly reduce a city's carbon footprint.

- Community involvement and a strong political commitment are crucial for the success of such initiatives which can only be realized and ensured by taking measures in the context of the Emerald Deal principles.

Costa Rica's Pioneering Efforts in Renewable Energy

Costa Rica presents a successful case of national-level transition towards renewable energy, aligning with the Emerald Deal ethos. The country utilized its natural resources and innovative policies to shift from traditional energy sources to renewables.

Strategy Implementation

- Costa Rica invested heavily in hydropower, wind, solar, and geothermal energy, supported by national policies favoring renewable energy (IRENA, 2018).
- The government provided incentives for renewable energy projects and set ambitious targets for carbon neutrality.
- The entire state initiative was driven by a broad range of communication and information activities in order to ensure the societal acceptance and engagement.

Results

- By 2021, Costa Rica was generating more than 99% of its electricity from renewable sources (IRENA, 2021).
- The country is on track to achieve its goal of becoming carbon-neutral.

Learning Points

- Diverse and robust policy mechanisms can effectively drive a nation's transition to renewable energy.
- Utilizing a country's unique geographical and natural resources is key in shaping its sustainable energy landscape.
- Aside the technological and content-oriented strategies, a solid communication strategy is essential for the entire initiative ensuring the societal engagement and acceptance based on information allowing the community to be informed on eye-level and creating the awareness of being a substantial and integral part of the overall project success.

Singapore's Water Sustainability Revolution

Singapore's approach to water sustainability is an exemplary case of adopting interconnected solutions being facilitated with a sound and comprehensive communication strategy. Faced with limited natural water resources, Singapore innovated its water management strategies very successfully due to the intense and properly defined community communication and engagement.

Strategy Implementation

- The country developed the "Four National Taps" strategy, which includes water from local catchments, imported water, reclaimed water (NEWater), and desalinated water (PUB Singapore).
- Advanced technology, such as high-grade water recycling and desalination, has been pivotal in Singapore's water sustainability efforts.
- As part of the strategy implementation a sophisticated communication plan was deployed not only focusing on awareness creation, importance communication, but also results dissemination.

Results

- Singapore has become a global leader in water management, ensuring a sustainable and secure water supply for its population.
- NEWater meets up to 40% of the nation's water needs and is expected to fulfill up to 55% by 2060 (PUB Singapore, 2020).

Learning Points

- Diversification and technological innovation in water resources management can effectively address water scarcity.
- Long-term planning and investment in technology are crucial for water sustainability.
- Even an almost self-explanatory water initiative and its sustainable success can be significantly leveraged by a professional communicational strategy.

Rwanda's Ban on Plastic Bags

Rwanda's ban on plastic bags is an excellent example of a simple yet effective environmental strategy that aligns with Emerald Deal principles. Implemented in 2008, this policy was part of a broader initiative to improve environmental health and cleanliness. In 2008, the responsible managers already acknowledged the importance of societal engagement and communication and thus launched a broad ban campaigning focusing on the benefits for the people.

Strategy Implementation

- The government of Rwanda enforced a complete ban on the use of plastic bags, combined with public awareness campaigns (UNEP, 2018).
- Alternatives to plastic bags, such as paper and cloth bags, were promoted.
- All measures were supported with orchestrated communication and information measures being driven by a team responsible for community engagement and acceptance.

Results

- Rwanda is now one of the cleanest countries in Africa, with significantly reduced plastic waste.
- The ban has fostered a culture of environmental consciousness among the populace.

Learning Points

- Regulatory measures, when effectively enforced and combined with public engagement, can lead to significant environmental improvements.
- Fostering a culture of environmental responsibility is as important as implementing technological solutions.

The deployment of the principles of the Emerald Deal does not require today's technology and opportunities of modern communication automation and intelligence, but also works with very simple tools and technique if the responsible managers are aware of the importance of societal acceptance, engagement, and facilitation as the result of trust-building communication.

These case studies demonstrate that transitioning to innovative, interconnected environmental solutions, being substantially supported and facilitated by adequate societal communication and interaction, can lead to significant improvements over traditional strategies in the context of and based on the principles of the Green Deal. The key takeaways emphasize the importance of integrating technology with sustainable practices and societal interaction management, the role of policy and community engagement, and the need for a holistic approach to environmental challenges. As the FIBS world strives to achieve sustainable development goals, the lessons from these case studies can inform future strategies, ensuring that environmental initiatives are both effective and resilient when supported with adequate societal interaction management.

Furthermore, these non-European projects demonstrate the importance of integrating technology with sustainable practices, policy and community engagement, and the need for a holistic approach to environmental challenges, which are key aspects of both the Green Deal and Emerald Deal. But the real advancements with these case studies must be considered in the fact that they provide valuable insights into how a comprehensive approach, including societal interaction management, can enhance the effectiveness of environmental initiatives, resonating with the EU's objectives in its environmental policies.

The Roadmap to the Emerald Deal for Global Leaders

The transition from traditional green initiatives to the more advanced Emerald strategies marks a significant evolution in our approach to environmental sustainability. This shift, while promising, is fraught with challenges ranging from technological and ethical considerations to broader societal implications. Understanding these hurdles and exploring future prospects through interdisciplinary collaboration is essential for the successful implementation of the Emerald Deal.

The EU's Green Deal is a landmark initiative aimed at transforming it into a climate-neutral economy by 2050. The Emerald Deal, a revised and updated version of the Green Deal, places a strong emphasis on trust-building communication and interaction, using AI, digital twins, collective intelligence, and the learnings from Reengineering Corporate Communication to achieve its societal impact.

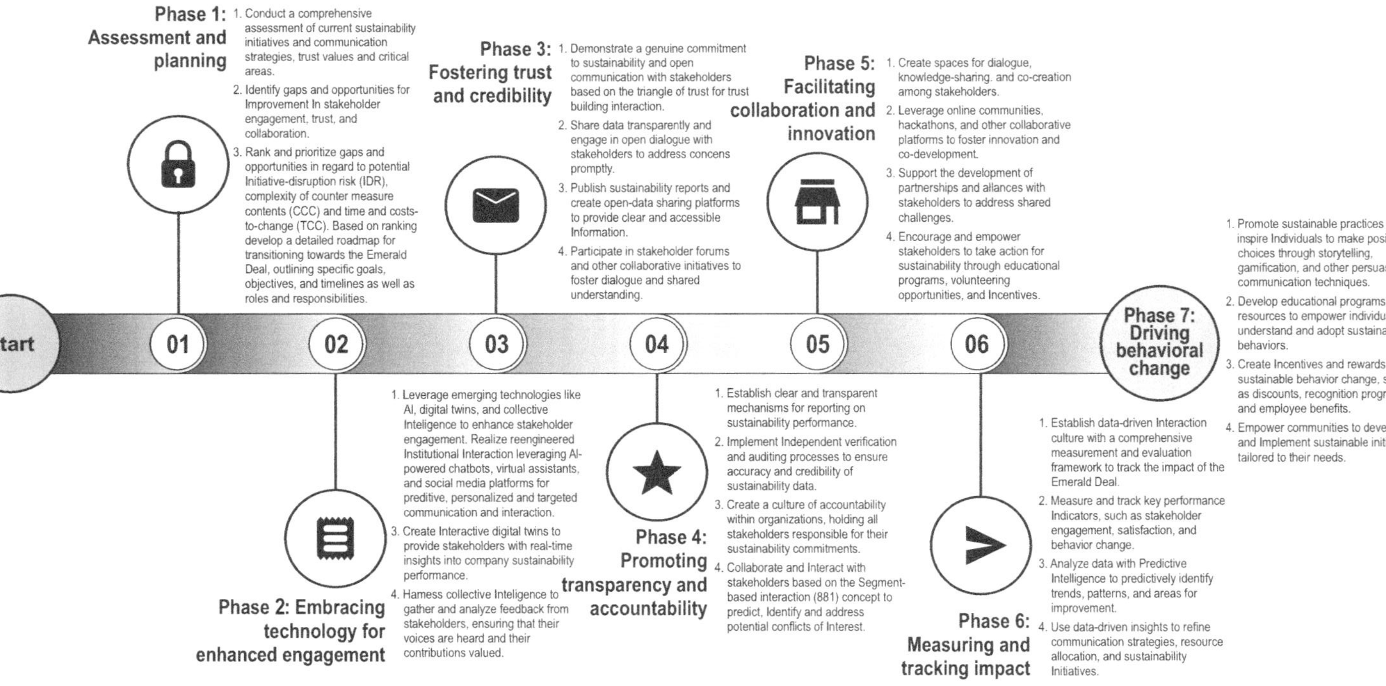

Fig. 3 The Roadmap towards the Emerald Deal
(*Source:* Seebacher 2024).

This roadmap (Figure 3) outlines a step-by-step approach for the global leaders to successfully transition from the Green Deal to the Emerald Deal.

Phase 1: Assessment and Planning

1. Conduct a comprehensive assessment of current sustainability initiatives and communication strategies, trust values, and critical areas.
2. Identify gaps and opportunities for improvement in stakeholder engagement, trust, and collaboration.
3. Rank and prioritize gaps and opportunities in regard to potential initiative-disruption risk (IDR), complexity of counter measure contents (CCC), and time and costs-to-change (TCC).
4. Based on ranking, develop a detailed roadmap for transitioning towards the Emerald Deal, outlining specific goals, objectives, and timelines as well as roles and responsibilities.
5. Establish clear governance structures and accountability mechanisms for the Emerald Deal.

Phase 2: Embracing Technology for Enhanced Engagement

1. Leverage emerging technologies like AI, digital twins, and collective intelligence to enhance stakeholder engagement.
2. Realize reengineered institutional interaction leveraging AI-powered chatbots, virtual assistants, and social media platforms for predictive, personalized, and targeted communication and interaction.
3. Create interactive digital twins to provide stakeholders with real-time insights into company sustainability performance.
4. Harness collective intelligence to gather and analyze feedback from stakeholders, ensuring that their voices are heard and their contributions valued.

Phase 3: Fostering Trust and Credibility

1. Demonstrate a genuine commitment to sustainability and open communication with stakeholders based on the triangle of trust for trust-building interaction.
2. Share data transparently and engage in open dialogue with stakeholders to address concerns promptly.
3. Publish sustainability reports and create open-data sharing platforms to provide clear and accessible information.
4. Participate in stakeholder forums and other collaborative initiatives to foster dialogue and shared understanding.

Phase 4: Promoting Transparency and Accountability

1. Establish clear and transparent mechanisms for reporting on sustainability performance.
2. Implement independent verification and auditing processes to ensure accuracy and credibility of sustainability data.

3. Create a culture of accountability within organizations, holding all stakeholders responsible for their sustainability commitments.
4. Collaborate and interact with stakeholders based on the Segment-based Interaction (SBI) concept to predict, identify, and address potential conflicts of interest.

Phase 5: Facilitating Collaboration and Innovation

1. Create spaces for dialogue, knowledge-sharing, and co-creation among stakeholders.
2. Leverage online communities, hackathons, and other collaborative platforms to foster innovation and co-development.
3. Support the development of partnerships and alliances with stakeholders to address shared challenges.
4. Encourage and empower stakeholders to take action for sustainability through educational programs, volunteering opportunities, and incentives.

Phase 6: Measuring and Tracking Impact

1. Establish data-driven interaction culture with a comprehensive measurement and evaluation framework to track the impact of the Emerald Deal.
2. Measure and track key performance indicators, such as stakeholder engagement, satisfaction, and behavior change.
3. Analyze data with Predictive Intelligence to predictively identify trends, patterns, and areas for improvement.
4. Use data-driven insights to refine communication strategies, resource allocation, and sustainability initiatives.

Phase 7: Driving Behavioral Change

1. Promote sustainable practices and inspire individuals to make positive choices through storytelling, gamification, and other persuasive communication techniques.
2. Develop educational programs and resources to empower individuals to understand and adopt sustainable behaviors.
3. Create incentives and rewards for sustainable behavior change, such as discounts, recognition programs, and employee benefits.
4. Empower communities to develop and implement sustainable initiatives tailored to their needs.

By implementing these steps, the EU and governments can effectively transition from the Green Deal to the Emerald Deal, driving a transformation towards a more sustainable, equitable, and prosperous future for all by acknowledging that 80% of such a significant societal change process has to be dedicated to communication, information, and interaction. In the following sections a deep dive into the different phases is realized to enable an easier deployment of the required activities. I also provide for each phase a set of key performance indicators and elaborate on 'how' and 'which' mechanisms are being used to ensure successful and effective transition.

Phase 1: Assessment and Planning for the Societal Impact of the Emerald Deal

Phase 1 of the roadmap for transitioning from the Green Deal to the Emerald Deal focuses on assessment and planning. This phase is crucial for laying a solid foundation for the Emerald Deal's success by identifying gaps and opportunities, prioritizing actions, and establishing clear governance structures. Collective intelligence can play a significant role in each of these aspects.

A comprehensive assessment of current sustainability initiatives and communication strategies is essential for understanding the current state of affairs and identifying areas for improvement. Collective intelligence platforms can be used to gather and analyze data from various sources, such as government reports, corporate sustainability reports, social media sentiment analysis, and stakeholder surveys. This data can provide insights into the effectiveness of current initiatives, stakeholder perceptions, and areas where improvements can be made.

By analyzing data gathered from the assessment, organizations can identify gaps and opportunities for improvement in stakeholder engagement, trust, and collaboration. Collective intelligence tools can help to identify trends, patterns, and correlations in data, providing insights into which stakeholders are most engaged, which initiatives are most effective, and where there is a need for further collaboration.

Once gaps and opportunities have been identified, they need to be ranked and prioritized based on their potential impact on the Emerald Deal. Collective intelligence tools can be used to model the impact of various scenarios, helping organizations to make informed decisions about which gaps to address first.

Based on the ranking and prioritization, a detailed roadmap can be developed that outlines specific goals, objectives, and timelines for transitioning towards the Emerald Deal. Collective intelligence tools can be used to track progress against the roadmap, providing real-time insights into the effectiveness of initiatives and identifying areas that require adjustments.

Clear governance structures and accountability mechanisms are essential for ensuring that the Emerald Deal is implemented effectively and that stakeholders are held accountable for their commitments. Collective intelligence tools can help to track stakeholder engagement, identify potential conflicts of interest, and ensure that all parties are working towards the common goal of achieving the Emerald Deal's objectives.

The effectiveness of Phase 1 can be measured by tracking the following metrics:

- Completion of the comprehensive assessment
- Identifying of gaps and opportunities
- Ranking and prioritization of gaps
- Development of a detailed roadmap
- Establishment of clear governance structures.

By measuring these metrics, organizations can track their progress towards achieving the goals of Phase 1 and make adjustments as needed.

Phase 2: Embracing Technology for Enhanced Engagement for the Societal Impact

Phase 2 of the roadmap for transitioning from the Green Deal to the Emerald Deal focuses on embracing technology for enhanced engagement. This phase highlights the role of emerging technologies like AI, digital twins, and collective intelligence in transforming stakeholder engagement and driving collaboration for sustainability goals.

Leveraging emerging technologies like AI, digital twins, and collective intelligence will revolutionize stakeholder engagement, enabling organizations to connect with stakeholders in more personalized, meaningful, and impactful ways. AI-powered chatbots and virtual assistants will provide personalized and targeted communication experiences for stakeholders, offering 24/7 support and assistance. Social media platforms will serve as hubs for real-time engagement, enabling organizations to respond to concerns and address issues promptly. AI can analyze social media sentiment and identify trends, providing valuable insights into stakeholder perceptions and preferences.

Digital twins provide a virtual representation of a physical asset, process, or system, allowing stakeholders to visualize and explore sustainability data in real-time. This can enhance transparency and accountability, empowering stakeholders to make informed decisions and contribute to sustainability initiatives. Collective intelligence platforms will gather and analyze feedback from a diverse range of stakeholders, providing a comprehensive understanding of their perspectives and concerns. This data can be used to refine sustainability initiatives, predictively address emerging issues, and foster a sense of shared ownership and responsibility.

The effectiveness of Phase 2 can be measured by tracking the following metrics:

- Increased stakeholder engagement
- Improved stakeholder satisfaction
- Increased trust and credibility
- Enhanced collaboration and knowledge sharing
- Greater transparency and accountability
- Effectiveness of sustainability initiatives
- Level of prediction precision

By measuring these metrics, organizations can track their progress towards achieving the goals of Phase 2 and make adjustments as needed. Collective Intelligence must be considered as a catalyst for enhanced engagement. Collective intelligence plays a crucial role in Phase 2 by enabling organizations to:

- Personalize and target communication to individual stakeholders
- Gather and analyze feedback from a diverse range of stakeholders
- Identify trends and patterns in stakeholder feedback
- Develop insights that inform sustainability initiatives
- Foster a sense of community and shared responsibility.

By harnessing collective intelligence, organizations can create a more inclusive and participatory approach to stakeholder engagement, driving collaboration and innovation towards a more sustainable future.

Phase 3: Fostering Trust and Credibility for the Societal Impact

Phase 3 of the roadmap for transitioning from the Green Deal to the Emerald Deal focuses on fostering trust and credibility. This phase emphasizes the importance of demonstrating a genuine commitment to sustainability, open communication, and transparency in order to build strong relationships with stakeholders and achieve shared sustainability goals.

To foster trust and credibility, organizations must demonstrate a genuine commitment to sustainability. This can be achieved by:

- Setting ambitious sustainability goals and taking concrete steps to achieve them.
- Making public statements and taking actions that align with the principles of the Emerald Deal.
- Investing in research and development to develop innovative solutions to sustainability challenges.
- Supporting initiatives that promote sustainability education and awareness.
- Engaging with stakeholders to gather feedback and incorporate their perspectives into sustainability strategies.

Open communication and transparency are essential for building trust and credibility. Organizations must:

- Communicate sustainability goals and progress openly and honestly to stakeholders.
- Respond promptly to stakeholder concerns and inquiries about sustainability practices.
- Publish comprehensive sustainability reports that provide detailed information about the organization's sustainability performance.
- Foster open dialogue with stakeholders through regular stakeholder engagement events and forums.
- Embrace digital technologies to enhance communication and transparency, such as using social media platforms to share sustainability updates and using chatbots to answer stakeholder questions.

Data transparency and open data sharing are crucial for building trust and credibility. Organizations must:

- Make sustainability data readily accessible to stakeholders through open-data platforms.
- Use data visualization tools to make sustainability data more understandable and engaging.
- Share data on a regular basis to provide stakeholders with up-to-date information on the organization's sustainability performance.

- Invest in data analytics to identify trends and patterns in sustainability data and use these insights to make informed decisions.
- Embrace collective intelligence platforms to gather and analyze data from a diverse range of stakeholders, providing a comprehensive understanding of stakeholder perspectives and concerns.

Active participation in stakeholder forums and collaborative initiatives fosters trust and credibility. Organizations must:

- Engage with stakeholders in ongoing dialogues and discussions about sustainability issues.
- Participate in collaborative initiatives that bring together stakeholders from different sectors to work together on sustainability challenges.
- Sponsor or support events and initiatives that promote sustainability education and awareness.
- Recognize and appreciate the contributions of stakeholders to the organization's sustainability efforts.
- Use collective intelligence tools to facilitate collaboration and knowledge sharing among stakeholders, leading to the development of innovative solutions to sustainability challenges.

The effectiveness of Phase 3 can be measured by tracking the following metrics:

- Increased stakeholder trust and credibility
- Reduced stakeholder skepticism and concerns
- Enhanced stakeholder engagement
- Increased willingness of stakeholders to support sustainability initiatives
- Improved public perception of the organization's sustainability commitment.

By tracking these metrics, organizations can assess their progress in fostering trust and credibility and make adjustments to their strategies as needed. Collective Intelligence in phase 3 acts as a catalyst for trust building by enabling organizations to:

- Gather and analyze stakeholder feedback to identify emerging trends and concerns
- Develop targeted communication strategies that address specific stakeholder needs
- Foster collaboration and knowledge sharing among stakeholders
- Track stakeholder sentiment and identify opportunities to improve trust and credibility
- Measure the impact of trust-building initiatives and make data-driven decisions.

Phase 3 of the roadmap for transitioning from the Green Deal to the Emerald Deal focuses on fostering trust and credibility, laying a foundation for sustainable success. By demonstrating a genuine commitment to sustainability, engaging in open communication and transparency, and sharing data openly, organizations can build strong relationships with stakeholders and achieve shared goals for a more sustainable future. As the Emerald Deal progresses, collective intelligence will continue to play a critical role in fostering trust, building communities, and unlocking innovative solutions to achieve the ambitious goals of the Emerald Deal.

Phase 4: Promoting Transparency and Accountability for the Societal Impact

Phase 4 of the roadmap for transitioning from the Green Deal to the Emerald Deal focuses on promoting transparency and accountability. This phase emphasizes the importance of establishing clear reporting mechanisms, implementing independent verification processes, fostering a culture of accountability, and leveraging collective intelligence to identify and address potential conflicts of interest.

Organizations must establish clear and transparent mechanisms for reporting on their sustainability performance. This includes:

- Developing comprehensive sustainability reporting frameworks that align with international standards, such as the Global Reporting Initiative (GRI).
- Regularly publishing sustainability reports that provide stakeholders with detailed information about the organization's sustainability performance.
- Making sustainability reports easily accessible to stakeholders through online platforms and other channels.
- Communicating sustainability performance metrics in a clear and understandable manner, using data visualization tools and other effective communication strategies.

Independent verification and auditing processes are essential to ensure the accuracy and credibility of sustainability data. This includes:

- Engaging independent auditors to verify the organization's sustainability data and practices.
- Making the results of independent audits publicly available to stakeholders.
- Establishing clear protocols for handling and addressing any discrepancies or concerns raised during audits.
- Investing in data analytics tools to identify potential data anomalies and ensure the integrity of sustainability data.

A culture of accountability within organizations is crucial for upholding sustainability commitments. This includes:

- Establishing clear accountability frameworks that assign responsibility for sustainability performance to specific individuals and teams within the organization.
- Measuring and tracking sustainability performance against established targets and metrics.
- Providing regular performance feedback to stakeholders and holding individuals and teams accountable for their contributions to sustainability goals.

- Cultivating a culture of transparency and open communication within the organization to foster trust and credibility.

Interest Segment-based interaction (SBI) is a data-driven approach to stakeholder engagement that can be used to identify and address potential conflicts of interest. This includes:

- Collecting and analyzing stakeholder data to identify key stakeholder segments based on their interests, motivations, and potential influence.
- Developing targeted communication strategies for each stakeholder segment to ensure that messages are tailored to their specific needs and concerns.
- Fostering open dialogue and collaboration with stakeholder segments to identify and address potential conflicts of interest proactively.
- Using collective intelligence tools to gather and analyze real-time information from stakeholders, allowing for rapid identification and resolution of conflicts.

The effectiveness of Phase 4 can be measured by tracking the following metrics:

- Increased stakeholder confidence in the organization's sustainability reporting and performance.
- Reduced stakeholder concerns about the accuracy and credibility of sustainability data.
- Greater alignment between sustainability commitments and actual performance.
- Increased accountability for sustainability outcomes among organizational leaders and employees.
- Reduced instances of conflicts of interest related to sustainability initiatives.

Collective intelligence plays a crucial role in Phase 4 by enabling organizations to:

- Gather and analyze large volumes of data from multiple sources to identify potential conflicts of interest.
- Develop predictive models to forecast and anticipate potential conflicts before they arise.
- Foster collaboration and knowledge sharing among stakeholders to identify and address potential conflicts proactively.
- Use data-driven insights to inform decision-making and optimize sustainability strategies.

Phase 5: Facilitating Collaboration and Innovation for the Societal Impact

Phase 5 of the roadmap for transitioning from the Green Deal to the Emerald Deal focuses on facilitating collaboration and innovation. This phase emphasizes the importance of creating enabling environments for collaboration, leveraging collective intelligence platforms, fostering partnerships, and empowering stakeholders to take action for sustainability.

Establishing spaces for dialogue, knowledge-sharing, and co-creation among stakeholders is crucial for fostering innovation and driving sustainable solutions. This includes:

- Creating online and offline forums and platforms where stakeholders can connect, share ideas, and collaborate.
- Organizing workshops, seminars, and conferences that bring together stakeholders from diverse backgrounds to exchange knowledge and expertise.
- Fostering a culture of open innovation and experimentation that encourages stakeholders to share their ideas and contribute to the development of innovative solutions.

Online communities, hackathons, and other collaborative platforms can be powerful tools for facilitating innovation and co-development among stakeholders. This includes:

- Creating online communities where stakeholders can share their expertise, collaborate on projects, and learn from each other.
- Hosting hackathons and other events where stakeholders can come together to tackle specific sustainability challenges and develop innovative solutions.
- Leveraging collective intelligence platforms to gather and analyze ideas, insights, and expertise from a wide range of stakeholders.

Partnerships and alliances can create synergies and leverage the strengths of different stakeholders to address shared sustainability challenges. This includes:

- Identifying and connecting stakeholders with complementary skills and expertise to form effective partnerships.
- Developing joint action plans and roadmaps to guide collaboration and ensure alignment of goals.
- Creating mechanisms for regular communication and coordination among partners to facilitate progress and address challenges.

Educating, empowering, and engaging stakeholders is essential for driving collective action towards sustainability. This includes:

- Developing educational programs and resources that raise awareness of sustainability challenges and solutions.
- Creating volunteering opportunities that allow stakeholders to contribute their time and skills to sustainability initiatives.
- Implementing incentive programs that reward stakeholders for their contributions to sustainability initiatives.

The effectiveness of Phase 5 can be measured by tracking the following metrics:

- Increased collaboration among stakeholders on sustainability initiatives.
- Greater innovation and development of new sustainability solutions.
- Strengthening of partnerships and alliances to address shared sustainability challenges.
- Increased engagement of stakeholders in taking action for sustainability.

- Improved outcomes of sustainability initiatives as a result of stakeholder contributions.

Collective intelligence plays a crucial role in Phase 5 by enabling organizations to:

- Facilitate the creation of knowledge-sharing hubs and collaborative platforms.
- Identify and connect stakeholders with complementary skills and expertise.
- Gather and analyze insights from a diverse range of stakeholders.
- Foster innovation and co-creation through hackathons and other events.
- Develop personalized and targeted communication strategies to engage stakeholders effectively.

Phase 6: Measuring and Tracking Impact for the Societal Impact of the Emerald Deal

Phase 6 of the roadmap for transitioning from the Green Deal to the Emerald Deal focuses on measuring and tracking impact. This phase emphasizes the importance of establishing a data-driven culture, developing a comprehensive measurement and evaluation framework, tracking key performance indicators, leveraging predictive intelligence, and using data-driven insights to refine strategies and initiatives.

A data-driven interaction culture is essential for effective measurement and tracking of the Emerald Deal's impact. This includes:

- Collecting and analyzing data from multiple sources, including sustainability reports, stakeholder feedback, and operational data.
- Using data visualization tools to present information in a clear and understandable manner.
- Sharing data insights with stakeholders to foster transparency and accountability.
- Using data to inform decision-making and refine sustainability strategies.

A comprehensive measurement and evaluation framework is crucial for tracking the impact of the Emerald Deal. This includes:

- Defining clear and measurable goals and objectives for the Emerald Deal.
- Identifying key performance indicators (KPIs) that align with the goals and objectives.
- Developing a data collection plan to gather data on the KPIs.
- Establishing a data analysis plan to analyze the collected data and identify trends and patterns.
- Developing a reporting plan to communicate the results of the analysis to stakeholders.

KPIs are used to track the progress of the Emerald Deal and assess its impact. Some key KPIs for the Emerald Deal include:

- *Stakeholder engagement:*   This refers to the level of involvement and participation of stakeholders in sustainability initiatives.

- *Stakeholder satisfaction:*   This refers to the level of satisfaction of stakeholders with the Emerald Deal and its impact.
- *Behavior change:*   This refers to the changes in behavior of individuals and organizations as a result of the Emerald Deal.

Predictive intelligence can be used to identify trends, patterns, and areas for improvement in the Emerald Deal. This includes:

- Using ML algorithms to forecast future trends and identify potential risks.
- Analyzing data to identify anomalies and patterns that may indicate problems or opportunities.
- Developing predictive models to forecast the impact of future actions and initiatives.

Data-driven insights can be used to refine communication strategies, resource allocation, and sustainability initiatives. This includes:

- Using data to identify the most effective communication channels and messages for different stakeholders.
- Allocating resources based on data-driven needs and priorities.
- Developing new or refining existing sustainability initiatives based on data insights.

The effectiveness of Phase 6 can be measured by tracking the following metrics:

- Increased utilization of data-driven decision-making across the Emerald Deal.
- Improved clarity and alignment of goals, objectives, and KPIs.
- Increased effectiveness of communication strategies and stakeholder engagement.
- Greater alignment of resource allocation with data-driven insights.
- Enhanced innovation and development of new sustainability solutions.

Collective Intelligence in Phase 6 acts as a catalyst for measurement and tracking by enabling organizations to:

- Gather and analyze data from multiple sources in real-time.
- Leverage ML and AI to identify patterns and trends.
- Develop predictive models to forecast future outcomes.
- Visualize data in a way that is easy to understand and actionable.
- Share insights with stakeholders to drive collective action.

Phase 7: Driving Behavioral Change for the Societal Impact of the Emerald Deal

Phase 7 of the roadmap for transitioning from the Green Deal to the Emerald Deal focuses on driving behavioral change and will be an ongoing one. This phase emphasizes the importance of using persuasive communication techniques, developing educational resources, creating incentives and rewards, and empowering communities to adopt sustainable practices.

This phase is about promoting sustainable practices and inspiring individuals to make positive choices through storytelling, gamification, and other persuasive interaction techniques. These techniques can be powerful tools for inspiring individuals to make positive choices for sustainability. This includes:

- Developing compelling narratives that highlight the personal and community benefits of sustainable behaviors.
- Creating gamified experiences that make sustainability fun and engaging.
- Tailoring communication strategies to different audiences and using culturally appropriate messaging.
- Using social media and other digital platforms to reach a wider audience and amplify the message.

Educational programs and resources can play a crucial role in equipping individuals with the knowledge and skills they need to adopt sustainable behaviors. This includes:

- Creating accessible and engaging educational materials that are tailored to different age groups and learning styles.
- Providing opportunities for individuals to learn from experts and peers through workshops, seminars, and online forums.
- Creating a culture of continuous learning and development around sustainability.
- Fostering partnerships with educational institutions to integrate sustainability education into the curriculum.

Incentives and rewards can be powerful motivators for individuals to adopt sustainable behaviors. This includes:

- Offering discounts or rebates for individuals who make sustainable choices, such as using public transportation or purchasing energy-efficient appliances.
- Developing recognition programs that celebrate individuals and organizations that are making a positive impact on sustainability.
- Creating employee benefit programs that reward sustainable behaviors, such as providing discounts on gym memberships or offering flexible work arrangements that encourage carpooling or cycling.

Communities can play a key role in driving sustainable behavior change. This includes:

- Empowering communities to identify and prioritize their own sustainability challenges and goals.
- Providing funding and technical support to communities to develop and implement sustainable initiatives.
- Fostering collaboration and knowledge sharing among communities to share best practices.
- Recognizing and celebrating the achievements of communities that are making a positive impact on sustainability.

The effectiveness of Phase 7 can be measured by tracking the following metrics:

- Increased awareness of sustainable practices among individuals.
- Adoption of sustainable behaviors among individuals and households.
- Reduced environmental impact as a result of behavioral change.
- Community engagement and participation in sustainable initiatives.
- Sustainability initiatives developed and implemented by communities tailored to their needs.

Collective intelligence plays a crucial role in Phase 7 by enabling organizations to:

- Gather and analyze data from individuals and communities to understand their needs and preferences.
- Develop targeted communication strategies that resonate with different audiences.
- Facilitate collaboration and knowledge sharing among individuals and communities.
- Identify and share best practices for sustainable behavior change.

By harnessing collective intelligence, organizations can create a more informed, engaged, and empowered society that is committed to sustainable practices. Phase 7 of the roadmap for transitioning from the Green Deal to the Emerald Deal is essential for driving the societal impact of the Emerald Deal. By promoting sustainable practices, developing educational programs, creating incentives and rewards, and empowering communities, we can inspire individuals to make positive choices and create a more sustainable future for all.

Conclusion and Call to Action

In this exploration of the transition from the Green Deal to the Emerald Deal, we have delved into the transformative potential of collective intelligence and swarm systems in addressing contemporary environmental and societal challenges. These advanced strategies offer a more holistic, integrated approach to sustainability, transcending traditional methods and embracing technological innovation, ethical considerations, and social inclusivity.

However, the path to realizing this vision is not without its challenges. From technological complexities and data privacy concerns to ethical dilemmas and the need for robust policy frameworks, the road ahead demands careful navigation. To overcome these obstacles, a collaborative effort is essential. This collaboration must span disciplines, sectors, and communities, combining diverse expertise and perspectives to unlock the full potential of these innovative approaches.

The 4 Essential Success Factors

The transition towards the Emerald Deal, a more ambitious version of the existing Green Deal, is a critical step towards achieving global climate goals. To ensure a successful transition, politicians and governments need to focus on four core success factors:

1. **Establish Clear and Ambitious Targets:**   The first step towards success is to set clear and ambitious targets for reducing greenhouse gas emissions and transitioning to a sustainable economy. These targets should be based on the latest scientific evidence and align with the Paris Agreement's goals of limiting global warming to well below 2 degrees Celsius above pre-industrial levels, and preferably to 1.5 degrees Celsius.

2. **Implement Strong and Effective Policies:**   Once targets have been set, governments need to implement strong and effective policies to achieve them. This includes policies that promote renewable energy, energy efficiency, sustainable transportation, sustainable agriculture, and sustainable forestry. Governments should also invest in research and development to support the development of new technologies that can help reduce emissions.

3. **Create a Level Playing Field:**   To ensure that the transition is fair and equitable, governments need to create a level playing field for all businesses and individuals. This includes policies that support the development of clean technologies, provide incentives for sustainable behavior, and ensure that the costs of pollution are reflected in the price of goods and services.

4. **Build Public Support:**   A successful transition towards the Emerald Deal will require broad public support. Governments need to communicate the benefits of sustainability to the public and engage with them in the transition process. This includes providing clear information about the impacts of climate change, the benefits of a sustainable economy, and the role that individuals can play in the transition.

In addition to these four core success factors, politicians and governments also need to ensure that the transition is equitable, inclusive, and sustainable across all sectors of society. This includes addressing the needs of vulnerable groups, such as low-income communities and indigenous peoples, and ensuring that the transition does not lead to job losses or social unrest.

The 4 Essential Pitfalls and Mistakes Politicians and Governments Must Avoid

The failure of the Green Deal and the recent crises have highlighted the need for politicians and governments to avoid certain pitfalls and mistakes when making the transition towards the Emerald Deal. These pitfalls include:

1. **Lack of Ambition:**   The Green Deal's goals were not ambitious enough to achieve the necessary reductions in greenhouse gas emissions. The Emerald Deal must set more ambitious targets and provide clear roadmaps for achieving them.

2. **Reliance on Market Forces:** The Green Deal placed too much faith in market forces to drive the transition to a sustainable economy. This approach has not been effective, and the Emerald Deal must rely on a more proactive role for government intervention.

3. **Failure to Address Social and Equity Issues:** The Green Deal did not adequately address the social and equity implications of the transition to a sustainable economy. The Emerald Deal must ensure that the transition is fair and equitable for all, including vulnerable groups such as low-income communities and indigenous peoples.

4. **Inadequate Communication and Public Engagement:** The Green Deal was not adequately communicated to the public, leading to a lack of awareness and support for the transition. The Emerald Deal must invest in effective communication and public engagement strategies to build broad support for the transition.

In addition to avoiding these pitfalls, politicians and governments need to learn from the crises of recent years. These crises have highlighted the need for:

1. **Resilience and Adaptability:** The Emerald Deal must build in resilience and adaptability to address the challenges of climate change and other crises. This includes anticipating and preparing for future shocks, and developing flexible and scalable solutions.

2. **Collaboration and Cooperation:** The Emerald Deal must foster collaboration and cooperation among governments, businesses, civil society, and individuals. This is essential to share knowledge, resources, and best practices, and to build a global movement for sustainability.

3. **Innovation and Technological Advancement:** The Emerald Deal must invest in innovation and technological advancement to develop new solutions for sustainability. This includes supporting research and development, and creating an environment that encourages entrepreneurship and innovation.

By learning from the mistakes of the past and the challenges of recent years, politicians and governments can make the transition towards the Emerald Deal a success. This transition is crucial to ensuring a sustainable future for all.

Closing Remarks

As we stand at the crossroads of environmental and societal change, the call to action is clear: Embrace the Emerald Deal with an open mind and a commitment to collaboration. Invest in research and development, advocate for supportive policies, engage with communities, and foster a culture of transparency and ethical responsibility. By doing so, we can harness the power of collective intelligence and swarm systems to create a sustainable, resilient, and informed society.

This journey is not just a responsibility but an opportunity—an opportunity to reshape our world for the better, leveraging the best of technology and human ingenuity. Let us embrace this challenge.

References

Abeshu, A. and Chilamkurti, N. (2018). Deep learning: The frontier for distributed attack detection in fog-to-things computing. *IEEE Communications Magazine, 56*(2), 169–75.

Aïmeur, E., Amri, S. and Brassard, G. (2023). Fake news, disinformation, and misinformation in social media: A review. *Social Network Analysis and Mining, 13*(1), 30.

Ballestar, M.T., Cuerdo-Mir, M. and Freire-Rubio, M.T. (2020). The concept of sustainability on social media: A social listening approach. *Sustainability, 12*(5), 2122.

Batson, C.D., Ahmad, N. and Tsang, J.A. (2002). Four motives for community involvement. *Journal of Social Issues, 58*(3), 429–45.

Bayındır, L. (2016). A review of swarm robotics tasks. *Neurocomputing, 172,* 292–21. Bell, K., and Reed, M. (2022). The tree of participation: A new model for inclusive decision-making. *Community Development Journal, 57*(4), 595–14.

Bennett, W.L. and Livingston, S. (2018). The disinformation order: Disruptive communication and the decline of democratic institutions. *European Journal of Communication, 33*(2), 122–39.

Berkes, F., Kislalioglu, M., Folke, C. and Gadgil, M. (1998). Minireviews: Exploring the basic ecological unit: Ecosystem-like concepts in traditional societies. *Ecosystems, 1,* 409–15.

Berlinger, F., Duduta, M., Gloria, H., Clarke, D., Nagpal, R. and Wood, R. (2018, May). A modular dielectric elastomer actuator to drive miniature autonomous underwater vehicles. *In: 2018 IEEE International Conference on Robotics and Automation* (*ICRA*), 3429–35. IEEE.

Bibri, S.E. (2018). The IoT for smart sustainable cities of the future: An analytical framework for sensor-based big data applications for environmental sustainability. *Sustainable Cities and Society, 38,* 230–53.

Bonoli, A., Zanni, S. and Serrano-Bernardo, F. (2021). Sustainability in building and construction within the framework of circular cities and European new green deal. The contribution of concrete recycling. *Sustainability, 13*(4), 2139.

Bose, R. (2009). Advanced analytics: Opportunities and challenges. *Industrial Management & Data Systems, 109*(2), 155–72.

Brown, M.T., Cohen, M.J. and Sweeney, S. (2009). Predicting national sustainability: The convergence of energetic, economic, and environmental realities. *Ecological Modelling, 220*(23), 3424–38.

Chanley, V.A., Rudolph, T.J. and Rahn, W.M. (2000). The origins and consequences of public trust in government: A time series analysis. *Public Opinion Quarterly, 64*(3), 239–56.

City of Copenhagen. (2021). *Copenhagen 2025 Climate Plan.*

Clegg, C. and Walsh, S. (2004). Change management: Time for a change! *European Journal of Work and Organizational Psychology, 13*(2), 217–39.

Doshi-Velez, F., Kortz, M., Budish, R., Bavitz, C., Gershman, S., O'Brien, D., ... and Wood, A. (2017). Accountability of AI under the law: The role of explanation. *arXiv preprint. arXiv:1711.01134.*

Eberhart, R.C., Shi, Y. and Kennedy, J. 2001. *Swarm Intelligence.* Elsevier.

Ellis, S.C. (2005). Meaningful consideration? A review of traditional knowledge in environmental decision-making. *Arctic,* 66–77.

Elliott, L., Colin Hines, Tony Juniper, Jeremy Leggett, Caroline Lucas, Richard Murphy, Ann Pettifor, Charles Secrett and Andrew Simms. (2008) *A Green New Deal: Joined-up Policies to Solve the Triple Crunch of the Credit Crisis, Climate Change, and High Oil Prices.* The first report of the Green New Deal Group. Hrsg.: Green New Deal Group. New Economics Foundation, London.,

European Commission. (2019). *The European Green Deal.*

Fawzy, S., Osman, A.I., Doran, J. and Rooney, D.W. (2020). Strategies for mitigation of climate change: A review. *Environmental Chemistry Letters, 18,* 2069–94.

Feindt, P.H. and Oels, A. (2005). Does discourse matter? Discourse analysis in environmental policy making. *Journal of Environmental Policy and Planning, 7*(3), 161–73.

Flew, T. (2020). Globalization, neo-globalization and post-globalization: The challenge of populism and the return of the national. *Global Media and Communication, 16*(1), 19-39.

Folke, C., Hahn, T., Olsson, P. and Norberg, J. (2005). Adaptive governance of social-ecological systems. *Annu. Rev. Environ. Resour., 30,* 441–73.

Frei, F.X. and Morriss, A. (2020). Begin with trust. *Harvard Business Review, 98*(3), 112–21.

Galvin, R. and Healy, N. (2020). The Green New Deal in the United States: What it is and how to pay for it. *Energy Research and Social Science, 67*, 101529.

Geissdoerfer, M., Savaget, P., Bocken, N.M. and Hultink, E.J. (2017). The Circular Economy: A new sustainability paradigm? *Journal of Cleaner Production, 143*, 757–68.

Ghaffarianhoseini, A., AlWaer, H., Ghaffarianhoseini, A., Clements-Croome, D., Berardi, U., Raahemifar, K. and Tookey, J. (2018). Intelligent or smart cities and buildings: A critical exposition and a way forward. *Intelligent Buildings International, 10*(2), 122–29.

Gheorghe, R. (2023). Why economic optimism collapses? The business environment: The only competent and ethical global institution. *Internal Auditing & Risk Management, 18*(2).

Global Forest Watch. (2021). *Global Forest Watch: Monitoring Forests in Near Real-Time.*

Global Water Partnership. (2000). *Integrated Water Resources Management.* TAC Background Papers.

Goodland, R. (1995). The concept of environmental sustainability. *Annual Review of Ecology and Systematics, 26*(1), 1–24.

Great Barrier Reef Marine Park Authority. (2020). *Reef 2050 Long-Term Sustainability Plan.*

Gunningham, N. (2009). Environment law, regulation, and governance: Shifting architectures. *Journal of Environmental Law, 21*(2), 179–12.

Guo, K.L. (2020). DECIDE: A decision-making model for more effective decision-making by health care managers. *The Health Care Manager, 39*(3), 133–41.

Hackfort, S. (2021). Patterns of inequalities in digital agriculture: A systematic literature review. *Sustainability, 13*(22), 12345.

Hafner, M. and Raimondi, P.P. (2020). Priorities and challenges of the EU energy transition: From the European Green Package to the new Green Deal. *Russian Journal of Economics, 6*(4), 374–89.

Hancke, G.P., de Carvalho e Silva, B. and Hancke Jr., G.P. (2012). The role of advanced sensing in smart cities. *Sensors, 13*(1), 393–25.

Hand, J.L., Prenni, A.J., Copeland, S., Schichtel, B.A. and Malm, W.C. (2020). Thirty years of the Clean Air Act Amendments: Impacts on haze in remote regions of the United States (1990–2018). *Atmospheric Environment, 243*, 117865.

Harth, N., Anagnostopoulos, C. and Pezaros, D. (2018). Predictive intelligence to the edge: Impact on edge analytics. *Evolving Systems, 9*, 95–118.

Hickel, J. and Kallis, G. (2020). Is green growth possible? *New Political Economy, 25*(4), 469–86.

Hills, T.T. (2019). The dark side of information proliferation. *Perspectives on Psychological Science, 14*(3), 323–30.

Intergovernmental Panel on Climate Change (IPCC). (2021). *Climate Change 2021: The Physical Science Basis.*

International Renewable Energy Agency (IRENA). (2018). *Renewable Energy Prospects: Costa Rica.*

International Renewable Energy Agency (IRENA). (2021). *Renewable Capacity Statistics 2021.*

IPCC. (2021). *Climate Change 2021: The Physical Science Basis.*

Jacobson, M.Z., Delucchi, M.A., Bauer, Z.A., Goodman, S.C., Chapman, W.E., Cameron, M.A., ... and Yachanin, A.S. (2017). 100% clean and renewable wind, water, and sunlight, all-sector energy roadmaps for 139 countries of the world. *Joule, 1*(1), 108–21.

Kammen, D.M. and Sunter, D.A. (2016). City-integrated renewable energy for urban sustainability. *Science, 352*(6288), 922–28.

Keene, M. and Pullin, A.S. (2011). Realizing an effectiveness revolution in environmental management. *Journal of Environmental Management, 92*(9), 2130–35.

Kitchen, P.J., Schultz, D.E. and Varey, R.J. (2001). Responsive and responsible communication practices: A pluralist perspective. *In: Raising the Corporate Umbrella: Corporate Communications in the 21st Century*, 62–81. Springer Link.

Kitchin, R. (2014). *The Data Revolution: Big Data, Open Data, Data Infrastructures, and Their Consequences*, 1–240. Sage Publications.

Kleber, R.J., Figley, C.R. and Gersons, B.P. (Eds.). (2013). *Beyond Trauma: Cultural and Societal Dynamics.* Springer Science & Business Media.

Levin, S.A., Milner, H.V. and Perrings, C. (2021). The dynamics of political polarization. *Proceedings of the National Academy of Sciences, 118*(50), e2116950118.

Lierse, H. (2022). Globalization and the societal consensus of wealth tax cuts. *Journal of European Public Policy, 29*(5), 748–66.

Mahesh, B. (2020). Machine learning algorithms: A review. *International Journal of Science and Research (IJSR). [Internet]*, *9*(1), 381–86.

Malone, T.W. and Bernstein, M.S. (Eds.). (2022). *Handbook of Collective intelligence*. MIT Press.

McCarthy, J. (2019). Authoritarianism, populism, and the environment: Comparative experiences, insights, and perspectives. *Annals of the American Association of Geographers*, *109*(2), 301–13.

McCormick, J. (1991). *Reclaiming Paradise: The Global Environmental Movement* (Vol. 660). Indiana University Press.

McNeely, J. (1999). Community building. *Journal of Community Psychology*, *27*(6), 741–50.

Millennium Ecosystem Assessment (MEA). (2005). *Ecosystems and Human Well-being: Synthesis*. Island Press.

Meadows, D. (2008). *Thinking in Systems: A Primer*. Chelsea Green Publishing.

Medhat, W., Hassan, A. and Korashy, H. (2014). Sentiment analysis algorithms and applications: A survey. *Ain Shams Engineering Journal*, *5*(4), 1093–13.

Nera, K., Wagner-Egger, P., Bertin, P., Douglas, K.M. and Klein, O. (2021). A power-challenging theory of society, or a conservative mindset? Upward and downward conspiracy theories as ideologically distinct beliefs. *European Journal of Social Psychology*, *51*(4–5), 740–57.

Newman, P., Beatley, T. and Boyer, H. (2009). Resilient cities. Responding to Peak Oil and Climate Change. *Journal of Urban Design*, *17*(2).

Nielsen. (2015). The Sustainability Imperative. *In: 2016 Global Sustainability Summit*, 10–12 August, New Orleans, LA.

Norouzzadeh, M.S., Nguyen, A., Kosmala, M., Swanson, A., Palmer, M.S., Packer, C. and Clune, J. (2018). Automatically identifying, counting, and describing wild animals in camera-trap images with deep learning. *Proceedings of the National Academy of Sciences*, *115*(25), E5716–E5725.

Norris, P. and Inglehart, R. (2019). *Cultural Backlash: Trump, Brexit, and Authoritarian Populism*. Cambridge University Press.

Nowak, A. and Vallacher, R.R. (2019). Nonlinear societal change: The perspective of dynamical systems. *British Journal of Social Psychology*, *58*(1), 105–28.

Ojokoh, B.A., Samuel, O.W., Omisore, O.M., Sarumi, O.A., Idowu, P.A., Chimusa, E.R., Darwish, A., Adekoya, A.F. and Katsriku, F.A. (2020). Big data, analytics, and artificial intelligence for sustainability. *Scientific African*, *9*, e00551.

Oreskes, N. (2015). The fact of uncertainty, the uncertainty of facts, and the cultural resonance of doubt. *Philosophical Transactions of the Royal Society A: Mathematical, Physical, and Engineering Sciences*, *373*(2055), 20140455.

Pavlo, A., Paulson, E., Rasin, A., Abadi, D.J., DeWitt, D.J., Madden, S. and Stonebraker, M. (2009, June). A comparison of approaches to large-scale data analysis. *In: Proceedings of the 2009 ACM SIGMOD International Conference on Management of Data*, 165–78.

Pe'er, G., Bonn, A., Bruelheide, H., Dieker, P., Eisenhauer, N., Feindt, P. H., ... and Lakner, S. (2020). Action needed for the EU Common Agricultural Policy to address sustainability challenges. *People and Nature*, *2*(2), 305–16.

Perlik, M. (2018). Less Regional Rhetoric, More Diversity. Urbanised Alps in the Interest of Cohesive Societies. *Journal of Alpine Research| Revue de géographie alpine*, 106–12.

Pettorelli, N., Laurance, W.F., O'Brien, T.G., Wegmann, M., Nagendra, H. and Turner, W. (2014). Satellite remote sensing for applied ecologists: Opportunities and challenges. *Journal of Applied Ecology*, *51*(4), 839–48.

Ployhart, R.E., Van Iddekinge, C.H. and MacKenzie Jr., W.I. (2011). Acquiring and developing human capital in service contexts: The interconnectedness of human capital resources. *Academy of Management Journal*, *54*(2), 353–68.

Porter, M.E. and Linde, C.V.D. (1995). Toward a new conception of the environment-competitiveness relationship. *Journal of Economic Perspectives*, *9*(4), 97–118.

PUB Singapore. (2020). *Singapore's National Water Agency*.

Ray, P.P. (2018). A survey on Internet of Things architectures. *Journal of King Saud University-Computer and Information Sciences*, *30*(3), 291–19.

Reed, M.S. (2008). Stakeholder participation for environmental management: A literature review. *Biological Conservation*, *141*(10), 2417–31.

Rifkin, J. (2019). *The Green New Deal: Why the Fossil Fuel Civilization Will Collapse by 2028, and the Bold Economic Plan to Save Life on Earth.* St. Martin's Press.

Rolnick, D., Donti, P.L., Kaack, L.H., Kochanski, K., Lacoste, A., Sankaran, K., ... and Bengio, Y. (2022). Tackling climate change with machine learning. *ACM Computing Surveys (CSUR), 55*(2), 1–96.

Roth, K. (2017). The dangerous rise of populism: Global attacks on human rights values. *Journal of International Affairs*, 79–84.

Sachs, J.D. (2015). *The Age of Sustainable Development.* Columbia University Press.

Seebacher, U. (2021). *Predictive Intelligence for Data-driven Managers.* Springer International Publishing.

Seebacher, U. (2022). How to Reengineer Corporate Communication. *In: Reengineering Corporate Communication: A Marketer's Perspective Offering New Concepts, Processes, Tools, and Templates*, 83–138. Cham: Springer International Publishing.

Sharda, R., Delen, D. and Turban, E. (2021). *Analytics, Data Science, & Artificial Intelligence: Systems for Decision Support.* Harlow: Pearson.

Stewart, M.C. and Arnold, C.L. (2018). Defining social listening: Recognizing an emerging dimension of listening. *International Journal of Listening, 32*(2), 85–100.

Sutantyo, D.K., Kernbach, S., Levi, P. and Nepomnyashchikh, V.A. (2010, July). Multi-robot searching algorithm using Lévy flight and artificial potential field. *In: 2010 IEEE Safety Security and Rescue Robotics*, 1–6. IEEE.

Steg, L. and Vlek, C. (2009). Encouraging pro-environmental behavior: An integrative review and research agenda. *Journal of Environmental Psychology, 29*(3), 309–17.

Stewart, M.C. and Arnold, C.L. (2018). Defining social listening: Recognizing an emerging dimension of listening. *International Journal of Listening, 32*(2), 85–100.

Suter, E., Arndt, J., Arthur, N., Parboosingh, J., Taylor, E. and Deutschlander, S. (2009). Role understanding and effective communication as core competencies for collaborative practice. *Journal of Interprofessional Care, 23*(1), 41–51.

United Nations Environment Programme (UNEP). (2018). *Rwanda's Journey to a Clean Environment.*

United Nations Framework Convention on Climate Change (UNFCCC). (2015). *The Paris Agreement.*

United Nations. (2018). *World Urbanization Prospects: The 2018 Revision.*

Upham, P., Oltra, C. and Boso, À. (2015). Towards a cross-paradigmatic framework of the social acceptance of energy systems. *Energy Research & Social Science, 8*, 100–12.

Van Dijk, J.A. (2006). Digital divide research, achievements, and shortcomings. *Poetics, 34*(4–5), 221–35.

Victor, D.G. (2009). *The Politics of Fossil-fuel Subsidies.* Available at SSRN 1520984.

Walker, J., Thuermer, G., Vicens, J. and Simperl, E. (2023, August.) AI Art and Misinformation: Approaches and Strategies for Media Literacy and Fact Checking. *In: Proceedings of the 2023 AAAI/ACM Conference on AI, Ethics, and Society*, 26–37.

Weber, T.J., Hydock, C., Ding, W., Gardner, M., Jacob, P., Mandel, N., Sprott, D.E. and Van Steenburg, E. (2021). Political polarization: Challenges, opportunities, and hope for consumer welfare, marketers, and public policy. *Journal of Public Policy & Marketing, 40*(2), 184–205.

Williams, B.K. (2011). Adaptive management of natural resources: Framework and issues. *Journal of Environmental Management.* Elsevier.

Closing Remarks and Predictions

As we stand on the precipice of 2040, it is with a profound sense of anticipation and responsibility that I reflect on the journey of Collective Intelligence. The findings and advancements from Predictores.ai have opened a myriad of doors into uncharted territories of human collaboration and machine partnership. The predictive models and analytical tools which have been developed, aided by the relentless curiosity and innovation of students and experts from universities such as the University of Applied Sciences Munich, the Technical University of Applied Sciences Augsburg, and the University of Applied Sciences Vienna, have propelled us into a new era of understanding and capability.

The foresight provided by Predictive Intelligence has enabled us to envision a 2040 where Collective Intelligence transcends today's limitations. It is a world where the seamless integration of Artificial Intelligence and human insight has revolutionized every aspect of our lives—from the way we cultivate our crops to the management of global supply chains, and even to the nuanced realms of healthcare. The collaborative efforts of the international community, including the Academy of Management, the Indian Academy of Management, and the Federal German Association for Industrial Communication, have been invaluable in this endeavor, ensuring that our advancements are both ethically grounded, societal responsible, and widely accessible.

Vijay Primalani and his team at CRC Press, along with the diligent members of Predictores.ai, deserve our heartfelt gratitude. Their support has not only brought this book into being but has also played a crucial role in catalyzing the practical applications of the research. Looking towards 2040, it is evident that the symbiosis of human and Artificial Intelligence will continue to flourish, driving sustainable knowledge gains and fostering an ever-more resilient and adaptive global community.

The future promises a world where decision-making is bolstered by the collective wisdom of diverse perspectives, where the challenges of sustainability, healthcare, and societal well-being are approached with unprecedented agility and precision. As we strive to expand the limits of Collective Intelligence, this book aims to guide readers towards a more connected, intelligent, and empathetic global society. The horizon of 2040 is bright with the promise of collective intelligence, a testament to the power of unity in thought and purpose.

By 2040, Collective Intelligence is likely to be an intrinsic part of societal and technological infrastructure, playing a pivotal role in how we solve complex problems, make decisions, and innovate. Here's a speculative outlook:

1. **Convergence of Artificial Intelligence and Internet of Things:** Collective Intelligence will play a crucial role for integrating Artificial intelligence in the Internet of Things, creating highly adaptive and responsive environments. This integration will allow for real-time data gathering, processing, and analysis, enabling communities and machines to make informed decisions quickly.

2. **Blurring of Virtual and Actual Reality:** With recent advancements in augmented/mixed reality and the development of digital metaverses, Collective Intelligence will play an important role from technical but also societal perspective. The creation of digital twins representing physical elements of the actual reality are forming a complementary Collective Intelligence in the digital space, blurring the lines between digital and physical reality and making (autonomous and assisted) decision-making pervasive.

3. **Decentralized Governance:** With advancements in distributed ledger technologies such as blockchain and similar technologies, Collective Intelligence could revolutionize governance and decision-making processes, potentially leading to decentralized and more democratic systems.

4. **Crisis Management:** In the face of global challenges such as climate change, pandemics, and humanitarian crises, Collective Intelligence could enable a more coordinated and effective response, leveraging the knowledge and resources of a global network.

5. **Personalized Education:** Education systems could become more personalized and adaptive, utilizing Collective Intelligence to tailor learning experiences to individual needs and learning styles.

6. **Healthcare:** In healthcare, collective intelligence will enable the global medical community to quickly share insights and data, leading to faster breakthroughs in treatments and better patient outcomes.

7. **Business and Management:** Businesses will employ predictive collective intelligence for market analysis, strategic planning, and innovation, relying on the aggregated insights of employees, customers, and AI systems.

8. **Sustainability and Conservation:** It will also play a significant role in environmental conservation, with multiple AI-enhanced ecosystems that are managed by collective intelligence ensuring sustainable practices.

9. **Space Exploration and Colonization:** As humanity pushes further into space, collective intelligence will be vital in managing the complexities of interplanetary travel and colonization efforts.

In each of these realms, the underlying theme will be the ability of Collective Intelligence to synthesize vast amounts of information from diverse sources, enabling humanity to understand and react to complex patterns and changes with greater wisdom and speed than ever before. In this context, the editors in the name of all esteemed authors look forward to all your valuable feedback and inputs.

Index

USA 246

V

Value-at-Risk (VaR) model 145, 149
Value chain 158-161, 164, 165, 167
Variance with Skewness (VwS) model 145
Variants of PSO 146
Virtual assistants 278
Visual odometry 129

W

Walking patterns 205, 206, 213
Water management 239
Wavelet Neural Network (WNN) 148
Wearable sensors 207, 224, 227
Weight coefficients 208, 209
Wildlife management 239